# Innovation in Life Cycle Engineering and Sustainable Development

*Edited by*

DANIEL BRISSAUD
*Université de Grenoble, France*

SERGE TICHKIEWITCH
*Université de Grenoble, France*

and

PEGGY ZWOLINSKI
*Université de Grenoble, France*

Springer

A C.I.P. Catalogue record for this book is available from the Library of Congress.

ISBN-10 1-4020-4601-4 (HB)
ISBN-13 978-1-4020-4601-8 (HB)
ISBN-10 1-4020-4617-0 ( e-book)
ISBN-13 978-1-4020-4617-9 (e-book)

---

Published by Springer,
P.O. Box 17, 3300 AA Dordrecht, The Netherlands.

*www.springer.com*

*Printed on acid-free paper*



Printed in the Netherlands.

# Contents



## Business Models



## End-of-Life Strategies

## Product Development for Sustainability

## Product Life Cycle Management

# Introduction to

# *Innovation in Life Cycle Engineering and Sustainable Development*

This book presents a selection of papers related to the 12th CIRP International conference on Life Cycle Engineering, held at the university of Grenoble, France, in April 2005. The CIRP Life Cycle Engineering conference is an annual event and contributes to a continuous updating of research in the field of environmental issues in mechanical engineering, design and manufacturing.

The scientific committee members of the conference have selected all the lectures from complete papers, which is the guarantee for the conference of quite an outstanding scientific level. After that, a new selection has been carried out to retain the best publications, which establish in a book, a state-of-the-art analysis as regards Innovation in Life Cycle Engineering and Sustainable Development. The 30 papers in the book, were selected from the 71 papers presented during the conference.

Sustainable development is more and more at the core of governments and industries policy. Industrial production and consumption culture are facing dramatic changes due to pollution and waste problems, exhaustion of available non-renewable resources and rapid growth in world population. So, the environmental focus has shifted from production processes to the product's entire life cycle. The potential of technology to create synergies between environmental protection and economic growth has been recognized. Life cycle engineering aims at providing engineering tools targeted towards cleaner product-oriented activities for improving the environment while contributing to competitiveness and growth.

It is within this context that the following themes were highlighted during the conference:

- **Business models,**

which aims at determining how efficient can be a life cycle strategy from different point of view (customer, policy, environment, economics,…).

- **End of life strategies,**

presenting recent approaches and technological solutions for end-of life treatments.

- **Product development for sustainability,**

which aims at showing how designers integrate environmental considerations to improve their solutions.

- **Product life cycle management,**

dealing with methods and tools to support life cycle considerations.

This book is divided into four sections reflecting the above themes and will be of interest to academics, students and practitioners, specializing in environmental issues in mechanical engineering. We hope that you will find it of the greatest interest to compare your various points of view within the field broached throughout the conference. We hope you all enjoy reading this book, which aims to be a reference textbook for all researchers in this particular field and for the teaching staff confronted with training methodologies in integrated design and environment. It will allow you to assess the scope of the development prospects in an extremely wide ranging field.

Finally, we would like to highlight the very significant input of the members of the organizing committee for the success of the conference and to express our sincere appreciation to all the authors and to the members of the international program committee.

Daniel Brissaud

Serge Tichkiewitch

Peggy Zwolinski

# Integrated Design Centre

## *3S Laboratory University of Grenoble, France*

The *Integrated Design Centre* is part of the 3S Research Laboratory sponsored by the technological university of Grenoble (INPG), the scientific university of Grenoble (UJF) and the National Centre for the Scientific Research (CNRS).

**Objectives**

For more than ten years, the activity of the *Integrated Design Centre* has focused on the improvement of engineering design and on the development of mechanical products/systems. New models, methods and computer tools dedicated to Integrated Design into a context of Concurrent Engineering have been developed. This aims at contributing to a better understanding of engineering design and to an improvement of the performance of design considered as a collective activity and a complex process. As seen by the research team, Design is a collaborative and distributed activity, that covers the whole product life cycle. It incorporates analysis as well as synthesis activities. Such a context involves 'skilled actors' having points of view, knowledge and tools, interactions with an organization, shared knowledge, decision-making processes, as well as technologies for cooperation activities, for the multi-representation and for shape generation of a product. The main topics are the product-process integration and the integration of downstream activities within the design process (manufacturing, assembly, recycling, etc.), CAD and geometric modelling, design methodologies and collaborative engineering involving new IT technologies. Research is carried out strongly connected to other disciplines (i.e. industrial sociology, cognitive ergonomics, computer sciences, applied mathematics, etc.).

**3 main research directions**

*Methodologies for Integrated Design, Collaborative Engineering, Innovation.* This research axis is based on observations conducted on real design processes. On-site studies in companies, experiments about the design activity to set up methodologies for analysing the design process are realised. They lead to methods for incorporating innovative solutions in a design process and they help for the management of the concept of value throughout the design process. Software tools and models for defining the product model and an integrated design environment for multi-actors: network-based, plug and play approaches, are proposed. Some experiments are conducted to characterized synchronous and asynchronous tasks among distant or co-located designers. Concepts for providing common work environments between designers having different skills are proposed. Experiments for the evaluation of software tools used in a collaborative context help to validate these concepts.

*Product-process integration in design.* This second research direction focuses on the formalization of knowledge and on methods related to process and production skills in mechanical design. The different explored fields are manufacturing processes (forging, assembly, machining, process planning, aluminium extrusion, composite materials), end of life processes (disassembly, recycling, re-use), tolerancing. Models for the dynamic behaviour of a machining system to improve the design process, to set up new technologies for the drilling process are examples of detailed contributions whereas concepts of product-process co-development, of product life cycle address a global level of the design process. Development of software demonstrators for product-process integration are realized to help the validation of the proposed approaches.

*Digital Mock-ups for Integrated Design.* This direction is necessary to provide methods for performing shape changes with a digital product models. Geometry simplification, adaptation for design and downstream processes, design data adaptation and idealization for mechanical simulations, free-form shape parameterisation and deformation, shape optimisation, surface mesh generation and adaptation are examples of researches to produce such mock-ups. Methods for knowledge, know-how and services management around digital mock-ups are also addressed through the capitalization, re-use of models concerning mechanical analyses of the products. Thus, new concepts for the use of digital mock-ups in design can be evaluated through the development of software demonstrators and libraries for digital mock-ups of products.

**The whole team in April 2005.** Prof. Daniel Brissaud, Prof. Jean-Claude Léon, Dr. Jean-Luc Marcellin, Dr. Philippe M Marin, Dr. Philippe R Marin, Mrs Fadila Messaoud, Dr. Peter Mitrouchev, Dr. Gabriel Moreau, Dr. Frédéric Noel, Dr. Henri Paris, Dr. Frank Pourroy, Dr. Guy Prudhomme, Dr. Guillaume Thomann, Prof. Serge Tichkiewitch, Prof. François Villeneuve, Dr. Peggy Zwolinski.

Sabeur Bettaieb, Cyrille Beylier, Vincent Capponi, Vincent Cheutet, Alexandre Ciclet, Lidia Dejeu, Guillaume Drieux, Rosalinda Ferrandes, Gilles Foucault, Alexis Gehin, Okba Hamri, Nizar Haoues, Denis Lovinfosse, El-Hadi Mechekour, Chaiwat Noomtong, Kusol Pimapunsri, Bruno Radulescu, Vincent Riboulet, German Ruiz, Aurélie Vacher, Frédéric Vignat, Omar Zirmi, Said Zirmi.

# Eco-development of products and sustainable manufacturing strategies

## The key interests of the *Integrated Design Centre*

The *Integrated Design Centre* objective is to:

- promote the change toward sustainable manufacturing paradigms
  - lifecycle orientation of the manufacturing industries
  - development of product oriented services
- increase engineering competence
  - methods and tools for design and engineering
- contribute to environmental goals
  - protection and conservation of natural resources: increase of material life time
  - clean processes: sustainable protection and consumption practices
  - eco-product development: optimization of products' life cycle

This paper aims at presenting the recent and on-going research studies in the field at the *Integrated Design Centre*.

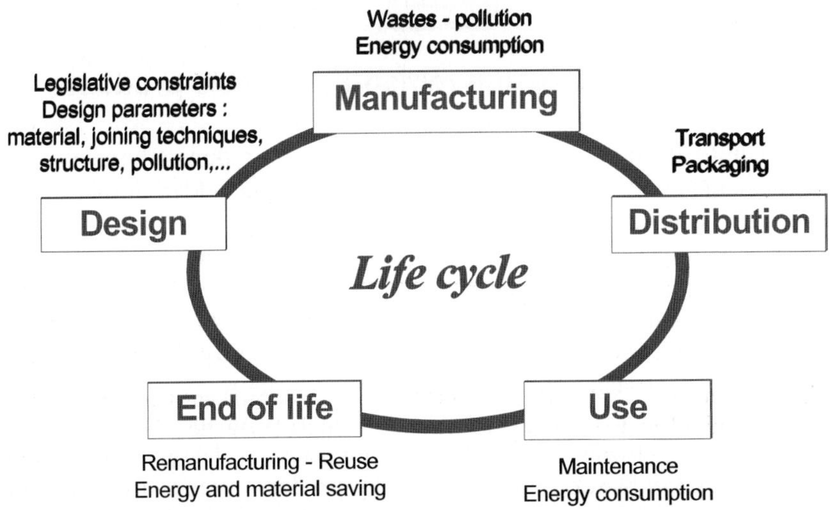

## Environment and lifecycle product design

The optimization of the product life cycle stages needs an integrated definition of the product throughout designing. New environmental objectives lead to support new design situations calling upon new competences, knowledge and tools. It opens the way to research on new design methodologies based on the entire lifecycle of the product, with benefit to the environmental impacts, a clean consumption and a sustainable production, and resulting in an increase of the innovative power of companies.

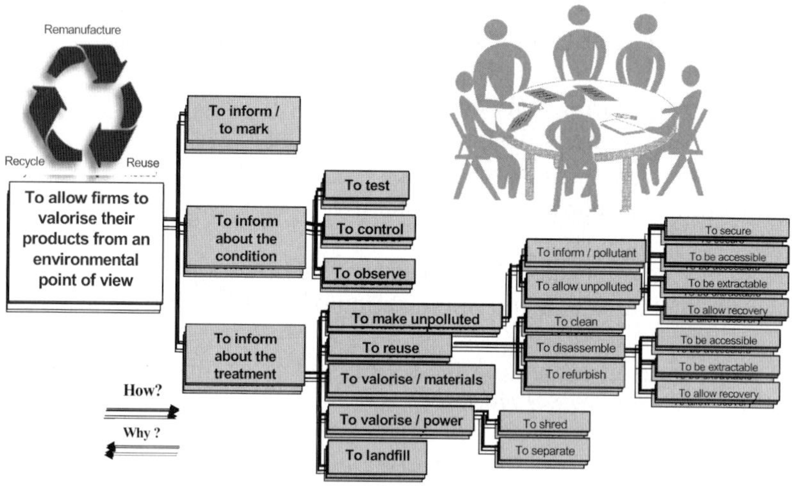


## Product Profiles to design remanufacturable products

The remanufacturing is an end of life strategy that reduces the use of raw materials and saves energy while preserving the value added during the design and manufacturing processes. But, in most of the cases, remanufacturing processes must be adapted to existing products because products have not been designed to be remanufacturable. However, the process adaptations increase costs and this can lead the overall benefits obtained with the remanufacturing process to be reconsidered. The aim of our research was to propose an approach for the designers to integrate remanufacturing constraints throughout the design process and mainly in the earliest phases. For the product profiles definition, 8 categories of design criteria were identified based on a survey of about thirty products successfully remanufactured. The methodology is supported by REPRO² (REmanufacturig with PROduct PROfiles), a tool developed for a real integrated design of remanufacturable products.

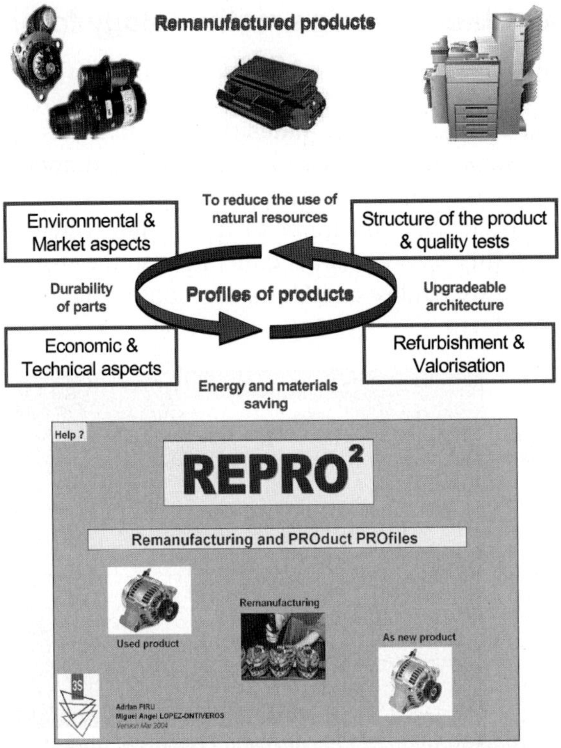


From *Integration of remanufacturing constraints during the design process of products,* PhD thesis, 2004, Miguel Angel LOPEZ-ONTIVEROS

## Change in manufacturing industry to support 'sustainable products'

A 'sustainable product' is a product which life is extended by partly or entirely re-using for a new life cycle. It contributes to a world where material is used up to its final properties and where energy consumption and wastes are minimized by keeping product added-value in use. It also impacts new ways of consumption. The project aims at examining the necessary condition to transform the industrial organisation to develop and manufacture sustainable products. It focuses on products remanufacturing and reuse strategies and covers the new economic and social paradigm to implement, the environmental impact to minimize and the engineering methodology and tools to develop.

From *Integrated design of sustainable products,* PhD thesis, in progress, Alexis GEHIN

## Environment-based design methodology for innovative products

A 2 years observation of the industrial practice has been realised at RENAULT. It was a mean to identify the tools and approaches used in a firm. The aim of the study was to define the industrial concept of recyclability to translate this constraint to the design. Integration resulted in adding Recyclability as constraints equivalent to the usual three industrial constraints - Quality, Delay and Costs –to control the development process of a new product.

From *Integration of the recycling constraint during the design process,* PhD thesis, 2000, Thierry GAUCHERON

## Disassembly for product recycling

Recent actions engaged by manufacturers are focused on the recycling rate required by the European Directives on EEE. Along with traditional recycling process as shredding, improving product end-of-life treatment needs the disassembly of the product or part of it. A "Noble" recycling must be managed easily from this. Product disassembly offers new solutions for valorisation: reuse of components, high quality material recycling with better economical indicators and decrease of landfilled wastes. The design work should be assisted by methodologies early in the conceptual design phase of the product and supported by CAD/CAM software identifying and evaluating assembly/disassembly sequences and issues.

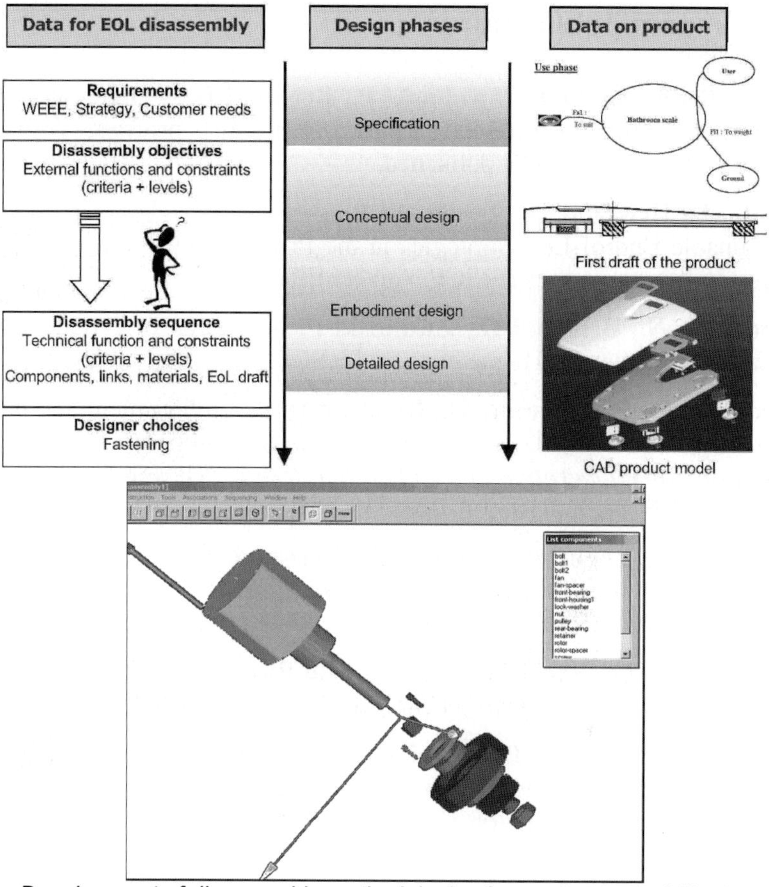


From *Development of disassembly methodologies for product end of life: Application to electric and electronic products,* PhD thesis in progress, Nizar HAOUES

## Other studies

- A new manufacturing paradigm: the extended products and their design, PhD thesis in progress, Nicolas Maussang
- Trajectories to implement eco-design support in a company practices, PhD thesis in progress, Tatiana Reyes
- The validation and improvement of a product remanufacturability assessment and design aid, Engineering final project, Jamie O'Hare

## Main recent publications

**Lopez-Ontiveros M.A., Zwolinski P., Brissaud D.** — Integrated design of remanufacturable products based on product profiles, *Journal of Cleaner Production,* to be published.

**Gehin A., Brissaud D., Zwolinski P.** — Imaging a Tool to Implement Sustainable End-of-Life Strategies in the Product Development Phase, in: Actes du 10ème congrès ERSCP, Anvers ERSCP 2005.

**Maussang N., Brissaud D., Zwolinski P.** — Design of Product-Service Systems, in: Actes du 10ème congrès ERSCP, Anvers ERSCP 2005.

**Zwolinski P., Prudhomme G., Brissaud D.** — Environment and design : Toward methods and tools for integration and co-operation, in *Methods and tools for cooperative and integrated design*, Kluwer Academic Publishers –ISBN 1-4020-1889-4, 223-232, 2004.

**Zwolinski P., Lopez-Ontiveros M.A., Brissaud D.** — Product end of life characterisation for integrated design, *International Journal of Production Engineering and Computers*, vol.6, No 7, 85-89, 2004.

**Brissaud D., Zwolinski P.** — End-of-Life-Based Negotiation Throughout the Design Process, *Annals of CIRP*, vol 53/1/2004, 155-158, 2004.

**Haoues N., Zwolinski P., Cornier A., Brissaud D.** — How to integrate end of life disassembly constraints in the early design phases? *International Journal of Production Engineering and Computers*, vol.6, No 7, 59-64, 2004.

**Haoues, N., Froelich, D., Zwolinski, P.** — Disassembly for Valorization in Conceptual Design. In proceedings of the SPIE : International Conference on Environmentally Conscious Manufacturing IV, Philadelphia, October 2004., Volume 5583, 31-42, 2004.

**Lopez M., Zwolinski P., Brissaud D.**— Profile of products for the creation of remanufacturable products during the conceptual design phase. Proceedings of CIRP seminar on life cycle engineering 2003, Copenhagen (Denmark), 11p, 2003.

**Prudhomme G., Zwolinski P., Brissaud D.** — Integrating into the design process the needs of those involved in the product life cycle, *Journal of Engineering Design*, vol. 14, No 3, 333-353, 2003.

Contact: Daniel.Brissaud@inpg.fr, Peggy.Zwolinski@hmg.inpg.fr

# Business Models

# Feasibility and scope of life cycle approaches to sustainable consumption

E.G. Hertwich and G.P. Peters
*Industrial Ecology Programme, Norwegian University of Science and Technology, NO-7491 Trondheim, Norway, edgar.hertwich@ntnu.no, www.indecol.ntnu.no*

Abstract: International policy documents call for the application of life-cycle assessment (LCA) in the development of sustainable patterns of consumption and production. We describe how LCA, in combination with input-output analysis and household consumption data can be used to determine the environmental impact of consumption patterns. We shortly review results from studies focussing on CO2 emissions as environmental indicator. This household environmental impact (HEI) assessment can be used to analyse differences among households according to socioeconomic variables, age, settlement characteristics, and lifestyles. The change of the total environmental impact over time can be attributed to underlying variables such as efficiency gains, changes in consumption levels, structural changes, and changes in the population. This provides an understanding of causal factors influencing the development of HEIs, which can be used in scenario modelling. In addition, we describe how HEI assessment can be used to inform policy implementation, both through ex-ante analysis of the possible effect of an intended measure and through the ex-post evaluation of implemented policy measures.

Keywords: Life-Cycle Assessment, Sustainable Consumption, Environmental Impacts

## 1 Introduction

At the World Summit for Sustainable Development (WSSD) in Johannesburg, world leaders recognized that it is necessary to 'chang[e] unsustainable patterns of consumption and production'. In the **Plan of Implementation,** the main document to emerge from the WSSD, world leaders call for "fundamental changes in the way societies produce and consume" [1, § 13]. They resolve to 'encourage and promote the development of a 10-year framework of programmes in support of regional and national initiatives to accelerate the shift towards sustainable



*D. Brissaud et al. (eds.), Innovation in Life Cycle Engineering and Sustainable Development,* 3–16.

consumption and production […]' [1, §14]. They also call for the adoption of tools, policies and assessment mechanisms based on **life-cycle analysis** to promote sustainable patterns of production and consumption and to increase the eco-efficiency of products and services [1, §13].

This paper focuses on the analytical support for sustainable consumption. A critical review indicates that traditional, process-based life-cycle assessment (LCA) is insufficient to capture household environmental impacts (HEI) [2]. Instead, a hybrid approach combining process analysis with economic input-output analysis and survey data on consumption patterns is required This approach originates from cumulative energy analysis, one of the antecedents to LCA developed in the 1970s. We describe key results and insights provided by such analyses, but also indicate limitations with the work done to date. Support for the implementation of sustainable consumption policies requires new types of analysis, new research designs, which can make use of LCA-based HEI assessments. It also requires the development of databases and an improvement and standardization of hybrid LCA databases. The interface between sociological analyses of consumption patterns and LCA-type assessments is currently poorly developed and requires improvement.

## 2 Assessment of household environmental impacts

HEI assessment started with the assessment of direct and indirect energy use of households [3, 4]. Since these early analyses, the assessment combines residential and transportation energy use in physical units with an input-output modelling of energy embodied in the goods and services purchased by households in monetary units. For energy, these two items are in fact equally important. Attempts that use only process analysis end up with incomplete inventories and partial assessments [5, 6]. Input-output analysis is not able to capture impacts occurring directly in the households. The development of hybrid LCA through integration of economic input-output analysis (IOA) in LCA supports HEI assessment. For linear systems, LCA can be formally written in matrix form as

Emission factors  Process model

$$I_{LC} = \mathbf{C}\mathbf{S}(\mathbf{I}-\mathbf{A})^{-1} y$$

Inventory analysis  Functional unit

Impact assessment  (1)

Where $I_{LC}$ is the vector of category indicators for the product investigated.

That means, the process description of production, distribution, use and disposal is all integrated into the normalized process flow matrix A. While in the LCA of an individual product, the demand vector y contains a single entry, the LCA of a household requires the assessment of many different goods. This means that the demand vector y needs to represent the entire basket of goods **h** consumed, and the process model needs to encompass all these products [2]. Commonly, the basket of goods is derived from a consumer expenditure survey, which national statistical bureaus conduct at regular intervals to update their inflation measurement. It can be interesting to group consumption items by activity or to compare different items individually to evaluate the relative importance. This is done by diagonalizing the household expenditure vector.

$$\mathbf{I}_h = \mathbf{CS}\left(\mathbf{I} - \mathbf{A}\right)^{-1} \hat{\mathbf{h}} \tag{2}$$

The grouping of different goods according to consumption activities can be instructive. Fig. 1 displays the contribution to household CO2 emissions of various countries. These assessments are based on the average household expenditure and commonly assume domestic production patterns also for imported goods. The studies differ in their treatment of capital investment, including roads and buildings, which are sometimes left out, and in the exact classification of activities. These variations are reflected in the size of the categories 'other' and 'other shelter.' In all countries, however, shelter (including heating/cooling), transportation, and food are the most important energy uses and causes of CO2 emissions. Emissions due to transport are growing fastest.

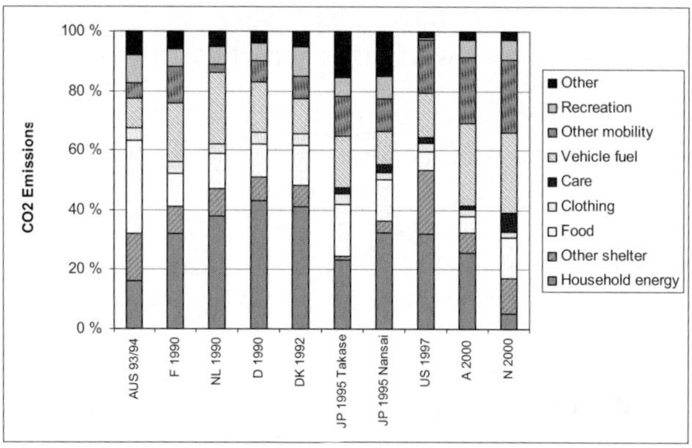


**Fig. 1.** CO2 emission due to different household consumption activities in different countries, assuming domestic technologies for imports

Consumer expenditure surveys provide different population subgroups, and the importance of these groups can be compared in various manners, e.g. looking at the contribution of different age groups or regions to the total national impact.

$$\mathbf{I}_{\mathbf{cons}} = \mathbf{CS}\left(\mathbf{I} - \mathbf{A}\right)^{-1} \mathbf{H}\hat{\mathbf{p}} \quad (3)$$

Here, the matrix H includes the expenditure vectors for the different household types, and the diagonalized vector p the size of the different subpopulations. Figure 2 provides an illustration from our own calculations. The figure shows the CO2 emissions by household size in Norway in 2000. It indicates that while total emissions increase with household size, they do not do so proportionally. The per capita emissions hence decrease with household size. For energy consumption, such a pattern has been observed in several countries [4]. Many studies show that energy use and CO2 emissions increase with income, but only indirect emissions are proportional to income; direct emissions increase less than proportionally. The increase of total household energy consumption with a doubling of income varies between 67% for India 1993-94 [7] and 90% for Denmark 1995 [8]. The ownership of a first and second car are important predictors of direct household energy use [9]. Urban populations have a lower energy consumption than rural or suburban populations, because they use less energy for transportation and have smaller flats in multifamily buildings and thus lower heating requirements. The strength of this urban advantage, however, varies. There are some indications that urban families have more leisure travel. With increasing wealth, air travel becomes important. Air travel has very high emissions per unit expenditure, so can easily dwarf other activities [10].

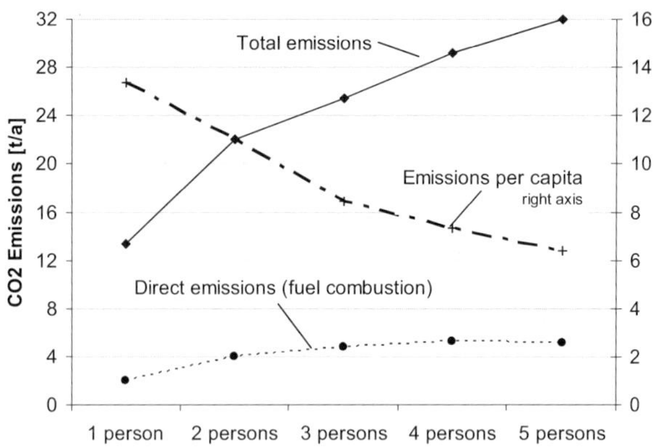


**Fig. 2.** Norwegian household energy consumption as a function of household size

It should be noted that even with income, location, car ownership and household size as explanatory variables, a significant portion of the overall per capita energy consumption is not explained. This indicates that there are differences in behaviour and household technology which are important.

# 3 Applications of HEI assessment

We review a number of applications of household impact assessment, focusing mostly on the assessment of the entire basket of goods. Other applications of LCA relate for example to environmental product declarations and ecolabels.

## 3.1 Consumer advice

An understanding of the items that shape household consumption can be used to inform consumers about the overall environmental impacts and options to reduce impacts. An assignment of overall environmental impact to different household activities as represented in Figure 1 can be useful because it allows consumers to search for environmentally friendly behaviour in the activities with the highest impact. In addition, it is useful to inform consumers about the impact intensity, i.e. the impact per unit expenditure, because shifting the expenditure from high-impact items to low-impact ones can result in effective impact reduction. Targeted information about specific, low-impact alternatives can be provided. In the United States, a consumer guide based on this type of analysis has been published [11]. There are several internet-based tools to estimate your ecological footprint or greenhouse gas emissions, but these often work with very crude data and contain many assumptions about individual lifestyles. They are not based on real life-cycle assessments. Students at the University of Tokyo have developed computerized personal consumption advisor called EcoLife, which has been customized to specific population groups tested among different users [12]. The tool, while not yet fully developed, presents customized information which helps interested consumers to evaluate their consumption patterns. It also provides specific advice on how to reduce environmental impacts through changes in consumption. If it can be adapted to European culture, such a tool also presents a promising possibility for informing consumers in Europe, not at least for experimenting with developing new, environmentally sustainable consumption patterns.

Advice to consumers is most useful when it is based on an understanding of the factors that influence the overall environmental impact, as outlined in the previous paragraph. Household environmental impacts are shaped by different factors: (1) the global production networks which produce a substantial portion of products and services, from manufacturing to entertainment; (2) the national economy, whose structure and regulation shapes e.g. the supply of electricity and a significant portion of the food; and (3) local conditions such as infrastructure, service availability, attractiveness of local recreation opportunities, which shape the daily life of the consumer. The influence of the national economy is well captured in today's assessments. The influence of local factors is not yet well understood.

There are first attempts to model global production networks. We have investigated the role of imports in Norwegian consumption, and out results indicate that imports account for half of the CO2 emissions caused by household consumption, including direct fuel combustion. Only 10% in value comes from developing countries, but these imports are responsible for 50% of the CO2 emissions embodied in import. Most studies use domestic emissions intensities for also for imports. This would underestimate the emissions embodied in imports to Norway by nearly a factor of 3 [13]. Our results also indicate that consumption of clothing is much more important than studies based on data from industrialized countries show. This is likely the result not of less efficient production conditions but of a much cheaper production in developing countries, which implies that we can buy many more pieces of clothing (or toys or electronic gadgets) for a given amount of money than if we buy a similar product produced in industrialized countries [14].

## 3.2 Understanding structural changes in production-consumption systems

Energy analysts have observed that, in industrialised countries on average, the growth in energy demand has been 1% per year lower than that of GDP. At the same time, the energy mix has shifted to lower-carbon fuels, reducing the CO2 emissions per unit energy by almost 0.5% per year. This change has commonly been attributed to technological improvements. There are other potential explanations, however: (1) structural change in the basket of goods consumed and a corresponding change in industry; (2) the export of energy-intensive production to developing countries. With development, we have seen a shift of economic activity from agriculture first to industry and now to services. In industrialised countries, services comprise approximately 70% of economic output. It is not that less food or industrial products are produced and consumed, but their production has become efficient and hence cheap. New consumption is in the service sector. In addition, existing services have grown comparatively more expensive. Since services often require less energy to produce, a predominant growth of output on the service sector will on the aggregate look like an efficiency gain, but it is not. The same is true for an export of energy-intensive production.

HEI studies show that household income and household energy use and respective $CO_2$ emissions are strongly related. It should be noted that luxury consumption may be less impacting than estimated by these studies, as luxury goods usually require more "design" rather than more materials. If growth means that the poor consume like the wealthy did before, we would not expect that GDP and aggregate energy consumption or $CO_2$ emissions are proportional.

Structural change happens also in production: today products contain more software than before. The role of technical improvements and the role of structural change in the production system and in the consumption basket can be distinguished from each other through a so-called structural decomposition analysis. Structural decomposition analysis consists of investigating the changes of the various elements of an identity equation like equation (3) over time. It measures the effect of the change in a single variable while keeping the other variables constant. Munksgaard et al. [15] present a structural decomposition analysis for household $CO_2$ emissions for the period 1966-1992. They show that most of the increase in HEI is due to the increased consumption of commodities. There is a 47% increase in consumption, and a 4.1% decrease due to changes in the aggregate commodity mix. At the same time, the energy intensity of companies has decreased by 31%, while the change in the energy mix of industry increases $CO_2$ emissions by 3%. Changes in industry structure do not have a large, unidirectional influence on $CO_2$ emissions. Overall, $CO_2$ emissions from indirect household consumption increased by 15%, which is much less than the 47% increase in consumption. Direct household emissions (due to fuel combustion) increased marginally, because a change in the energy mix and emission factors compensated for the 18.5% increase in energy consumption.

Additional information about direct emissions in both households and manufacturing can be gained from a decomposition as practiced in energy analysis [16]. Unander et al. [17] present such an analysis for residential energy use in Denmark. The time period covered is 1973-1999, which is different from the time period covered by Munksgaard, but it is nonetheless interesting to compare the results. Residential energy use fell by 24%, which is mostly due to a fall in heating energy by 40%. The change in housing structure (area per dwelling and persons per dwelling) indicate that the activity level or service level (floor space heated) grew substantially in the same time period. For the period of 1972-1990, the decomposition shows an overall efficiency increase of >3.5% p.a., compensating the 1.5% p.a. increase in consumption. There has been little change in either of the variables since 1990. The comparison between the two types of decomposition analyses shows that structural decomposition analysis is better in looking at aggregate household consumption and also takes into account upstream (industry) emissions, but the energy analysis decomposition, by being more specific to a specific demand, can better measure the level of service provided to the consumer and specific technology efficiencies. It would, in principle, also be possible to include household activities and levels of service provided in a structural decomposition. Measurement issues and data requirements, however, are substantial.

An understanding of the reasons for changes in the energy intensity and pollution intensity of economic activities and consumption patterns is important, because it indicates what changes we can expect in the future, and what changes are desirable or undesirable.

## 3.3 Developing scenarios

Scenarios can be developed both to explore what a continuation of trends implies for the environment, and to evaluate how sets of policies or social developments affect the attainment of policy goals. The more common way of using scenario analysis is to explore possible and likely future developments. It requires the development of internally consistent story lines about what happens in each scenario in terms of economic, demographic, social and technological changes. There is relatively little work on scenario analysis, although some of the past and ongoing work is quite promising [18, 19]. In their work on Germany, France and the Netherlands, Weber and Perrels [18] develop following scenarios: stagnation, business as usual, sustainable technology and sustainable consumption. Of these scenarios, stagnation is the worst, whereas sustainable consumption and sustainable technology offer the largest emissions reductions, with certain differences among countries. Scenario analysis can also be used to evaluate possible courses of action. Duchin [20] evaluates the impact of a radical change in diets towards more healthy, low-meat 'Mediterranean' diets, typical of Greece in the 1950s. Such diets would have a substantial health benefit and, depending on details, could also reduce pollution and land use. Duchin analyses the effects of a Mediterranean diet on agricultural production patterns in the US and identifies who benefits and looses from such a development. She shows that the production of unhealthy crops is systematically subsidised by the US government and that strong agricultural interests are connected to the current production system. Takase et al. [21] evaluate the increased use of trains instead of cars and the shift of eating at home instead of in restaurants. They find that the first action results in a clear improvement, while the second action is positive only because the cost of eating in restaurants leads to the reduction in purchases of other goods. Hertwich [22] discusses such ripple effects on a more general basis. The use of scenarios for evaluating options is very promising and maybe even more important than the development of 'business-as-usual' and other storyline scenarios. Scenarios for radical change can be used to evaluate the impacts on the environment and the production system of different technologies, infrastructures or behavioural changes. They quantify the potential contribution of policy options and thus indicate the desirability of different actions.

## 3.4 Guiding and evaluating implementation efforts

As policy makers have committed themselves in Johannesburg to implement sustainable development, tools for developing and selecting measures to implement are required. At the same time, implementation efforts should be monitored, evaluated and benchmarked. A number of studies in Germany indicate how this can be done [23, 24]. These studies focused on the conversion of military installations to settlements or the development of new settlements. In this case, LCA-type analysis was used to assess different options, for example energy systems, building materials, and transportation infrastructure. These analyses were used to guide the implementation of the project and inform the discussion with the future tenants on various choices. An evaluation of the projects indicates that the efforts have been successful in reducing emissions from what they would have been otherwise. To our knowledge, however, there has not been an independent evaluation of these projects. To compare different implementation efforts, it is important to develop consistent methods which use standardized indicators. This would allow a benchmarking of different policy measures and an evaluation of the relative effectiveness of the measures. In principle, it is possible to compare measures addressing consumption patterns with those addressing production methods, technology, or infrastructure. In practice, such comparisons are today not yet possible, because different studies use widely differing methods, systems boundaries, indicators, and nomenclatures.

**Table 1.** Different ground transportation patterns and their effect on the household CO2 emissions assuming a constant expenditure level and re-spending of the saved money on (A) the average good and (B) air transport to holiday destinations

| | Car only | Bus & car | Bus &coach | Car share light | Car share intensive |
|---|---|---|---|---|---|
| Car - distance (vkm) | 19 206 | 10 755 | 0 | 0 | 0 |
| Bus - distance (pkm) | 0 | 10 492 | 15 787 | 11 276 | 13 297 |
| Coach - distance (pkm) | 0 | 0 | 5 216 | 0 | 0 |
| Car share - distance (vkm) | 0 | 0 | 0 | 2 268 | 6 371 |
| (A) Total travel distance (pkm) | 27 769 | 26 041 | 21 003 | 16 493 | 24 445 |
| (A) Total travel cost (NOK) | 46 740 | 48 931 | 15 126 | 18 777 | 31 543 |
| (A) Non-transport GWP | 8 399 | 8 326 | 9 450 | 9 329 | 8 904 |
| (A) Transport GWP | 5 589 | 4 315 | 1 971 | 1 934 | 3 356 |
| (A) Total GWP | 13 988 | 12 641 | 11 421 | 11 263 | 12 261 |
| (B) Air - distance (pkm) | 28 980 | 24 434 | 94 549 | 86 976 | 60 498 |
| (B) Total travel distance (pkm) | 56 748 | 50 475 | 115 552 | 103 469 | 108 127 |
| (B) Total travel cost (NOK) | 60 712 | 60 712 | 60 712 | 60 712 | 60 712 |
| (B) Non-transport GWP | 7 934 | 7 934 | 7 934 | 7 934 | 7 934 |
| (B) Transport GWP | 14 370 | 11 719 | 30 619 | 28 288 | 24 307 |
| (B) Total GWP | 22 304 | 19 653 | 38 553 | 36 222 | 32 241 |

## 4 Example: Car Sharing

Car sharing is a much-cited example of a sustainable consumption measure. We have evaluated car sharing in Norway using life-cycle approaches [25]. In addition, we have studied a car-free housing project in Vienna, where tenants have access to car-sharing [26]. Table 1 presents the results of two different rebound scenarios regarding transportation patterns and car sharing. The investigation is based on information from the car-sharing cooperatives in Trondheim, Oslo and Bergen, as well as the national transportation survey. We develop characteristic travel patterns for persons who own a car, who participate in car sharing at different intensities, and who do not use cars at all. Car ownership leads to increased travel distances. Assuming a constant income, we provide two scenarios for the rebound effect, i.e. the emissions due to spending the money saved by not operating a car. In scenario A, all the money is spent to purchase more of the average basket of goods. In scenario B, all of the savings are spent on air travel. In scenario A, light car sharing and not using a car at all lead to per-capita greenhouse gas emissions of nearly 20% less than a car owner who does not use public transportation. In scenario B, however, public transportation users and car sharers have 70% and 60% higher greenhouse gas emissions than the car owner, respectively. Our analysis indicates that it is important to understand the rebound effect.

We have conducted an evaluation of a car-free housing project in Vienna. This building complex has 244 flats and is designed to encourage the use of bikes and public transportation. In addition, it has a number of other features such as common facilities, roof-top gardens, and solar hot-water collectors. We have compared a sample of 42 households in this complex with a sample of 46 households in a nearby, similar building complex. The data collected includes socioeconomic variables, income, transport behaviour, utility bills, and information on food consumption. We assumed that the consumption areas for which we did not collect data for was similar to those of the rest of Austria, as contained in the Austrian consumer expenditure survey.

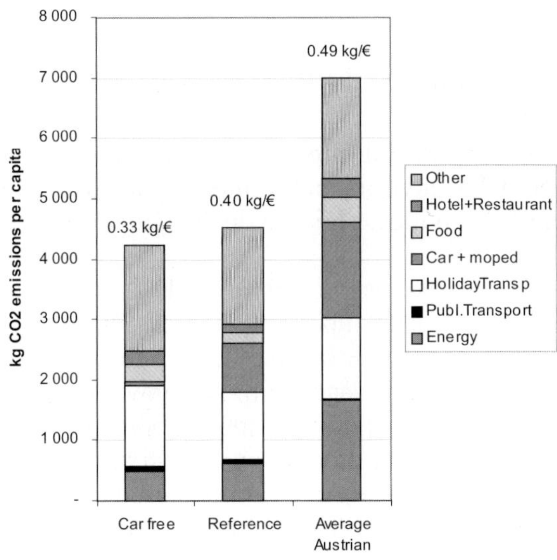


**Fig. 3.** Comparison of the per-capita CO2 emissions in the car-free housing project in Vienna with a nearby reference project and the Austrian average

The results indicate that the $CO_2$ emissions from land transport of the car-free settlement are lower by a factor of 4 than those of the reference settlement. The difference to the average in Austria is even higher. Of course, ground transport is not the only source of $CO_2$ emissions, especially for urban households. The emissions due to air transport, which is the most important component of holiday transport in Figure 3, are about the same for of the two settlements and the Austrian average. Given the importance of other consumption items not investigated in the survey, the difference in the total CO2 emissions per household is not that large. Figure 3 indicates that the households in the car-free settlements have 8% lower per capita $CO_2$ emissions from all types of consumption, even though their income, and thus consumption, is 15% higher. There is a significant difference, however, to the average Austrian consumer. The two settlements have more families with children than the Austrian average, and hence a lower per-capita income. In addition, there is less need for car travel in an urban area. The use of district heating leads to lower CO2 emissions for heating and hot water. The overall evaluation indicates that the car-free housing project is very successful in reducing the emissions due to ground transportation, but that the emissions of other consumption areas are very important and similar. The overall effect is hence less than one would expect.

## 5 Critical assessment issues

Household environmental impact assessment and associated techniques require estimates for environmental impacts for many different products, product types and activities. This is a challenging task. There are a number of critical issues:

1. HEI assessments make use of hybrid analysis, which combines process LCA data and economic input-output data. Different options exist for combining these two types of analysis. The number of processes covered with LCA data can vary from only covering what goes on in the household to describing many different processes, such as food production, energy and material production, which are critical for households. The need for detail on processes depends on the purpose of the analysis and on the detail available in the IO tables. European IO tables typically have 50 sectors, while the U.S. uses 500 sectors. The integration of IOA and process LCA is under development and awaits standardization. We have not found any investigations on how much detail is required for different purposes of analysis.
2. Assessments often do not correctly assess imports and are generally biased to reflect conditions in industrialized countries. A better understanding of global product networks and environmental impacts arising in developing countries is needed, especially if the analysis is used to influence consumption patterns.

3. Assessments frequently underestimate the emissions connected to transport, because the fuel consumption of airplanes and ships is recorded as 'bunker fuels' and not assigned to any specific purpose.
4. Studies today are often not comparable. There is a need for standardizing assessment methods on everything from household activity classifications to system boundaries, methodological issues for taking into account buildings and other infrastructure investments, and environmental impact indicators.

## 6 Recommendations

We recommend a further development and standardization of life-cycle methods for sustainable consumption in a transnational project which includes the application of these methods in a number of local case studies. Such a development could include standard reporting formats and benchmarks. More detailed recommendations have been submitted to the European Commission and are available on our web site.

## Acknowledgements

This paper summarizes the FESCOLA project financed by the European Union's sixth framework programme through grant NMP-ct-2003-505281. The study of the car-free housing project was funded by the Austrian National Bank and the Society of Non-Traditional Technology in Japan. Tania Briceno, Willi Haas, Eirik Haukland, Klaus Hubacek, Katarina Korytarova, Michael Ornetzeder and Christian Solli contributed to various parts of the project.

## References

1. United Nations General Assembly, 2002, World Summit on Sustainable Development: Plan of Implementation, UN Division for Sustainable Development: Johannesburg.
2. Hertwich, E.G., 2005, Lifecycle Approaches to Sustainable Consumption: A Critical Review, Environmental Science & Technology 39/13:4673-4684.
3. Bullard III, C.W. and R.A. Herendeen, 1975, Energy Impact of Consumption Decisions, Proceedings of the IEEE 63:484-493.
4. Herendeen, R.A. and J. Tanaka, 1976, Energy Cost of Living, Energy 1:165-178.
5. Rønning, A., K. Magnussen, and I.S. Modahl, 1999, A life over 365 days: CO2 balance and human life-cycle, OR 04.99, Østfold Research: Fredrikstad, Norway.
6. Frischknecht, R., N. Jungbluth, and M. Nauser, 2002, Embodied greenhouse gas emissions in Switzerland: A case study for the trade with products from the energy and food sector, ESU Services: Ulster, Switzerland. p. 21.

7. Pachauri, S., 2004, An analysis of cross-sectional variations in total household energy requirements in India using micro survey data, Energy Policy 32/15:1723-1735.
8. Wier, M., M. Lenzen, J. Munksgaard, and S. Smed, 2001, Effects of household consumption patterns on CO2 requirements, Economic Systems Research 13/3:259-274.
9. Vringer, K. and K. Blok, 1995, The direct and indirect energy requirements of households in the Netherlands, Energy Policy 23/10:893-910.
10. Norland, I.T., E. Holden, and W.M. Lafferty, 2005, Consumption of Energy and Transport in Urban Households: The role of urban planning vs. 'green consumerism' in promoting sustainable consumption, pp-139-160 in Sustainable consumption: The contribution of Research, Oslo.
11. Brower, M. and W. Leon, 1999, The consumer's guide to effective environmental choices. New York: Three Rivers Press.
12. Fukushima, Y. and M. Hirao, 2004. "EcoLife" as a Navigator for Consumers. Pp. 113-120 in Proceedings, The Third International Workshop on Sustainable Consumption. Tokyo, Society for Non-Traditional Technology.
13. Peters, G.P. and E.G. Hertwich, 2006, The importance of import for household environmental impacts, Journal of Industrial Ecology 10/2.
14. Schor, J.B., 2005, Prices and quantities: Unsustainable consumption and the global economy, Ecological Economics 55/3:309-320.
15. Munksgaard, J., K.A. Pedersen, and M. Wier, 2000, Impact of household consumption on CO2 emissions, Energy Economics 22/4:423-440.
16. Schipper, L., F. Unander, S. Murtishaw, and M. Ting, 2001, Indicators of energy use and carbon emissions: Explaining the energy economy link, Annual Review of Energy and the Environment 26:49-81.
17. Unander, F., I. Ettestol, M. Ting, and L. Schipper, 2004, Residential energy use: an international perspective on long-term trends in Denmark, Norway and Sweden, Energy Policy 32/12:1395-1404.
18. Weber, C. and A. Perrels, 2000, Modelling lifestyle effects on energy demand and related emissions, Energy Policy 28/8:549-566.
19. Rood, G.A., J.P.M. Ros, E. Drissen, and K. Vringer, 2003, A structure of models for future projections of environmental pressure due to consumption, Journal of Cleaner Production 11/5:491-498.
20. Duchin, F., 2005, The sustainable consumption of food: A framework for analyzing scenarios about changes in diets, Journal of Industrial Ecology 9/1-2:99-114.
21. Takase, K., Y. Kondo, and A. Washizu, 2005, An analysis of sustainable consumption by the waste input-output model, Journal of Industrial Ecology 9/1-2:201-220.
22. Hertwich, E.G., 2005, Consumption & the Rebound Effect: An Industrial Ecology Perspective, Journal of Industrial Ecology 9/1-2:85-98.
23. Fritsche, U.R., 2002, Nachhaltige Stadtteile - die Rolle des Warenkorbs der Konsumenten, in Nachhaltiger Konsum: Auf dem Weg zur gesellschaftlichen Verankerung, G. Scherhorn and C. Weber, Editors. Ökonom Verlag: Munich. 335-342.
24. Brohmann, R., U. Fritsche, S. Hartard, M. Schmied, C. Schmitt, C. Schoenfelder, N. Schuett, et al., 2002, Sustainable Districts on Urban Areas: Material Flow Analyses as an Instrument for Evaluation - End Report (German), Oeko-Institut.
25. Briceno, T., G. Peters, C. Solli, and E.G. Hertwich, 2005, Using life cycle approaches to evaluate sustainable consumption programs: Car-sharing, IndEcol Working Paper 2005/1, Norwegian University of Science and Technology, Industrial Ecology Program: Trondheim.
26. Hertwich, E.G. and M. Ornetzeder, 2005. The Environmental Benefit of Car-free Housing: A Case in Vienna. pp. 161-170 in Proceedings, Sustainable Consumption: The Contribution of Research. Oslo.

# A business-oriented approach to the product life cycle

G. Molcho and M. Shpitalni
*Laboratory for CAD and LCE, Dept. of Mechanical Engineering, Technion, Haifa, Israel, shefi@technion.ac.il*

Abstract: Cultural and public attitudes as well as new business approaches and life cycle concepts have impacted and will continue to impact not only matters of consumption and use but indeed almost every aspect of the product life cycle. Ecological awareness, globalization and excess manufacturing capacity along with intensive competition are the main factors that have led to what is no less than a revolution in design and manufacturing. This revolution is enabled by improved information technology. This paper explores and briefly analyzes the forces driving the significant changes in the market that have affected not only design and manufacturing but rather the entire product life cycle. Moreover, it identifies the expected changes and business opportunities in manufacturing due to new legislation requiring manufacturer accountability during the entire life cycle.

Keywords: Life Cycle Engineering, Services and Products, Business

## 1 Introduction and motivation

Population growth, industrialization, resource depletion and globalization [1] have generated a need to create a sustainable society in order to protect the environment and natural resources for future generations. From the global or macro-economic point of view, this is only logical. Yet from an operational or micro-economic point of view, it is proving difficult to initiate such strategies. Indeed, to do so fundamental structural changes are required in products as well as in organization and production technologies, and the economic benefits involved in such changes are either uncertain or associated with risks [2]. Life cycle engineering must therefore aim at economically and responsibly dealing with our limited resources [3], ensuring ecological excellence while at the same time securing the economic interests of industry and creating employment.



*D. Brissaud et al. (eds.), Innovation in Life Cycle Engineering and Sustainable Development,* 17–32.

In order to achieve the required balance between cultural and public attitudes and requirements on the one hand, and the industry demand for profit maximization on the other, every aspect of the product life cycle must be analyzed and optimized. The supporting processes, tools and technology must be researched and developed, and creative solutions should be sought. For example, the solution of selling services rather than products may be one type of bridge that allows migration to a more environmentally conscious industry. However, even this type of solution requires reeducation of both users and manufacturers due to the psychological and cultural consequences.

## 2 Problem statement

Three contradictory elements form the basis of today's revolution in design and manufacturing:

*The drive for more new products and larger varieties, motivated by competition and customer demand*

Human need motivates the demand for products. If this need were reduced, fewer products would be produced. But human demand is not the target of environmental engineering as we know it. In fact, environmental engineering targets the supply side only, i.e. the solutions engineered to satisfy demand, rather than the demand itself or the growing population 4. Reducing demand by changing consumption patterns and life style to reflect a less materialistic consumer culture may well be one of the routes to lower environmental impact in the future. Yet from the average global perspective, the increase in consumption we see today and can expect in the coming decades is generally materialistic, thus leaving the task of reducing environmental impact to manufacturers and government [4].

*The drive by society members for a clean and safe environment*

The environment with its limited natural resources is becoming endangered due to emissions [4]. In face of potential dangers such as climate change, toxic waste, multiple international organizations have formed, and many political efforts have focused on finding solutions and encouraging change. The demand is therefore to evolve to sustainable development, that is, "development that meets the needs of the present without compromising the ability of future generations to meet their own needs" [2]. Environmental constraints enforced by society through legislation need to be incorporated in the modern manufacturing paradigm with the objective of providing more use with fewer resources [3], and having less impact on the environment.

*The drive by industry for maximum profit*

The goal of industrialists is profit maximization. As a result of excess global manufacturing capacity and increased competition, however, a shift in the manufacturing paradigm is required [5]. Hence, Kimura [6] suggests that instead of focusing on "how efficiently we can produce products," we must now consider "how we can avoid producing products while maintaining customers' satisfaction and corporate profits.".

Hence, the main problem posed by the modern manufacturing revolution is as follows. How can we offer more new products, satisfy consumers' changing requirements and respond to frequent, sharp and abrupt changes in the market such that the 'total impact on the environment' is minimized, the usage of each product is enhanced, its life span is extended and the processes and profits involved are beneficial?

The field of product life cycle engineering has recently emerged to cope with this complex problem. In this paper, we describe the product life cycle, discuss the changes expected in modern manufacturing and explore the resulting academic and business opportunities, risks and challenges.

## 3 The product life cycle

The preliminary stage in understanding the product life cycle from the three points of view of customers, the environment and industrialists is to understand the life cycle of a product. This will allow us to pinpoint the places where change can be expected, business opportunities may be found and research and development of processes and tools are required. We present in this paper one common type of product life cycle representative of typical off-the-shelf products with multiple competitive manufacturers, such as refrigerators, televisions, and washing machines. One such example is sufficient to discuss the different aspects of the life cycle; clearly, however, the processes and the order in which they are carried out may vary for each specific product or industry. For example, in a project-based industry, sales precede design, manufacturing and assembly; and manufacturers who sell kits rather than products have no assembly stage. Figure 1 presents an overview of the product life cycle. Here we see that the product life cycle may be divided into three major sections – pre-active life phase, active life phase and End Of Life (EOL) phase. Figure 2 details the pre-active and active life phases, and Figure 3 details the EOL. The stages of each phase are presented, as well as their relevant input and output of materials. Note that the flow of knowledge and information can also be presented on these diagrams by modifying the flow arrows.

It is important to stress that according to new legislation, modern manufacturing will require manufacturer accountability for a product during its entire life cycle, as opposed to the present situation where manufacturers are accountable for their products from design only until the warranty period expires, generally a year from sale. Modern manufacturers will have to develop and implement new models and capabilities for managing and performing all these additional processes for which they will be held accountable, particularly for the EOL processes.

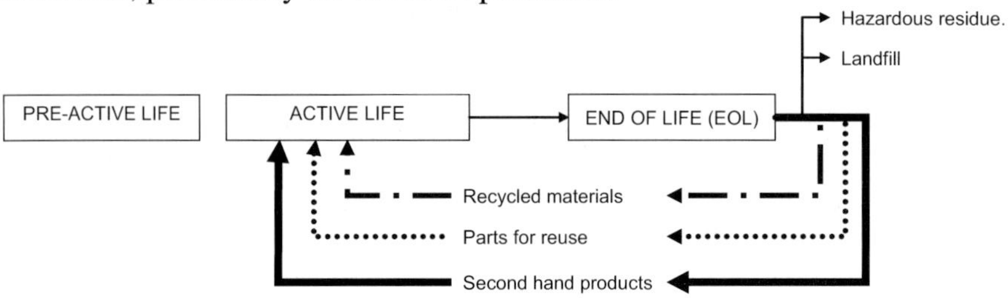


**Fig. 1.** Upper level product life cycle diagram[1]

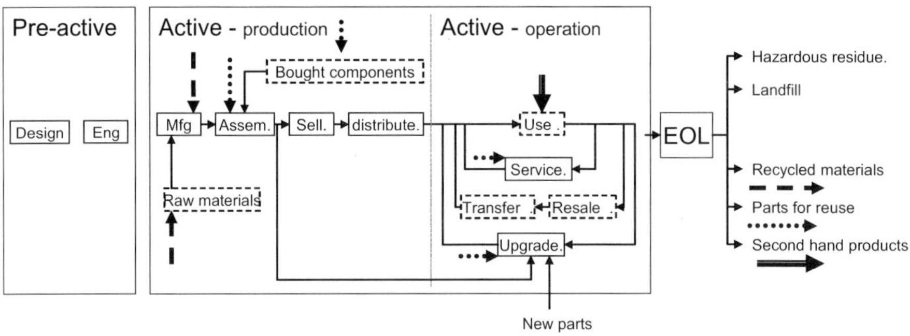


**Fig. 2.** Primary life cycle details

Although manufacturers will ultimately be responsible for all stages of the product life cycle, each stage may be performed in-house or outsourced to sub-contractors. Manufacturers may outsource knowledge or capacity, according to the organization's core competencies and long-term business strategies. Since it seems unlikely that EOL processes will become part of

[1] The flow diagram indicates only the main flow of the product and does not consider all the secondary flows (e.g. products that are rejected after assembly or manufacturing and need to be dealt with).

the core business of most manufacturers, most of these processes will probably be outsourced.

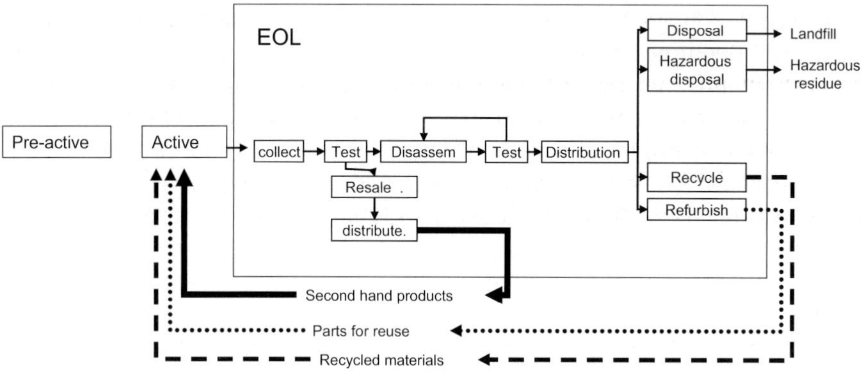


**Fig. 3.** Detailing the EOL phase of the product life cycle

## 3.1 The pre-active life cycle

The pre-active life cycle phase includes two stages, design and engineering. This phase is different from the other two life cycle phases in that (a) it occurs once for a specific product type and is not repeated for each single product produced, and (b) its direct impact on the environment is minimal. It is, nonetheless, considered an integral part of the product life cycle, and is central to achieving the sustainable society target.

### Stage 1: Detailed design

While the design phase (or more specifically the conceptual design phase) has no direct impact on the environment, 75% of the product life cycle cost is determined at this early stage, as is the product's environmental impact. In accordance with the industrial and environmental points of view, the time invested in product design should be expected to increase to allow for innovative designer solutions and thorough verification and validation of the product to ensure a sustainable society on the one hand and the desired profits on the other. In practice, however, the design phase has shortened considerably, due to increased global competition and product demand. Advances in information technologies, such as computing speed, communication and Internet, along with innovations in production systems have dramatically changed processes and engineering practices related to product development [7-9]. Moreover, these advances have also enhanced global manufacturing capabilities to a point far exceeding the transition from excess demand to excess production. As a consequence, competition has dramatically increased. Development cycles for products have been reduced from several years to several months. Consequently, marketing deadlines have become crucial, as most profits are earned within a short period after product release. This demand for improved time-to-market (shortened development cycle) has two main consequences. First, the expected life span of a new product is shorter; and second, design engineers are forced to rely primarily on existing and proven technologies in order to meet these demands, leaving little possibility for research and development. To shorten the design process, designers have begun to reuse good and already proven design modules, to rely on pre-competitive collaboration between organizations or to buy components to be incorporated in their products. The problem of how to introduce new products in order to compete on the market can be approached by reducing the number of manufacturers through mutual acquisitions and by offering more models which use the same modules and components. To demonstrate this point, Table 1 shows how the Volkswagen group offers different models that share the same platform and how market segmentation and brand names enable the group to introduce a variety of new products. We ignore special cars such as the Lamborghini and the Bugati, whose marketing approach is of course different, and focus on the group's mass-produced cars: Audi, Volkswagen, Seat and Skoda [10].

**Table 1.** Partial classification of cars based on labels and market segments

| | SKODA | SEAT | VW | AUDI |
|---|---|---|---|---|
| Label | Value | Sport/young | Main-stream | Premium |
| Supermini | Fabia | Ibiza | Polo | A2 |
| Compact | Octavia | Toledo | Golf/Bora | A3 |

A different yet valid example of introducing new products for competition is the customization of products for a specific individual. The concepts of customization and personalization are continuously growing, thus defining a new competitive field. Customization involves giving the customer set options, while personalization refers to special customer orders (e.g. tailoring of a suit) [10]. Currently the WEEE Directive [11] has no firm rules on product design, other than "Member States shall 'encourage' better design for recycling". Nevertheless, the RoHS (Restriction of Hazardous

Substances) Directive should be referred to, as it bans the use of certain materials after 2007. The EuP (Eco-design requirements for Energy-Using Products), a further drafted directive, will make consideration of product life cycle mandatory at the design stage. Therefore, it would be good design practice to begin incorporating this consideration. Special attention should be given to design of products whose impact on the environment is minimal and whose recovery value is high. The trend toward miniaturization and replacement of mechanical devices (e.g. mechanical timers) by computer-controlled devices aimed at conserving energy and emissions and protecting the environment will, therefore, continue. Development of methodologies and tools dedicated to product design is crucial in the migration process to a sustainable society. Environment parameters such as EOL requirements/possibilities (e.g. modularization), longer life span, less material consumption (in particular hazardous materials) and less energy consumption will have to be studied, improved and incorporated in these improved design tools. Thus a prominent task in life cycle engineering is to improve an organization's knowledge management capabilities to incorporate previous experience in the design process, enhancing cost and functionality, as well as taking into account environmental considerations.

### Stage 2: Engineering

We have separated the Engineering phase from the Design phase to emphasize the contribution of detailed documentation in modern manufacturing. All the Knowledge, Information and Data (KID) created in the design stage needs to be conveyed not only to the manufacturing and assembly stages, as is present common practice (i.e, product file, including technical and manufacturing data), but to all other stages of the life cycle as well. A much broader file is required that incorporates service data (best practices, maintenance data) and EOL KID (disassembly, recovery and disposal KID). Since KID management will be a critical component of success, we emphasize this stage as a cornerstone of product life cycle management.

## 3.2 The active life cycle

The active life cycle may be further divided into two major sections, *production* and *operation*. *Production* includes four stages governed by the manufacturer, manufacturing, assembly, sales and distribution, and *operation* includes two such stages, service (maintenance and repair) and upgrading. Figure 2 presents not only these six stages, but also four additional input processes central to the active life cycle phase (marked by dotted lines): raw material provision, bought product part provision, use and a secondary secondhand market. These additional input processes are part of the central core of the product life cycle but are not governed by the manufacturer. By incorporating these input processes into the model, we allow a comprehensive understanding of the product life cycle while outlining business requirements and opportunities.

### Stage 1: Manufacturing

In the manufacturing stage, parts are produced or purchased to assemble the product. Energy and materials are used and emissions are generated to make the required parts. The environmental impact per produced part is determined according to the materials (type and quantities), the manufacturing processes and the manufacturing facilities. The first two factors are usually determined in the design phase and to some minor extent in the process planning. Today more than in the past, much attention is focused on manufacturing facilities targeted at supplying global market demands and variations. This adaptation is achieved by enhancing the flexibility of manufacturing systems through the introduction of reconfigurable machines and manufacturing systems designed to be efficient, flexible and responsive, in particular when dealing with large part families. It should be noted that integration between design and manufacturing facilities makes a major contribution to the efficiency, flexibility and responsiveness of a system (e.g. modular design). Companies that cannot meet this requirement for flexibility tend to outsource their manufacturing processes. Outsourcing manufacturing processes has multiple advantages and disadvantages. The major advantages include converting manufacturing fixed costs to manufacturing varying costs and reducing the need of major investments in flexible machinery. The downside is loss of flexibility and very often loss of core knowledge.

We expect to see new manufacturing considerations, new processes (with minimal impact on the environment) and new factory and plant configurations with full control and tracing of products by Product State Models [12]. RMS – reconfigurable manufacturing systems – is only one example. These new developments will be based on open control systems, extended flexibility and the need to manufacture new products on demand. At the same time, large companies are expected to act as manufacturing services providers, mainly producing large quantities of basic components (families of parts) at the sub-assembly level for goods manufacturers. The raw material input into the manufacturing stage may be of two types, primary or secondary (recycled). In order to fulfill the vision of a sustainable society, governmental intervention may be required to ensure that manufacturers consume secondary materials, lest they build up to become redundant and therefore end up as landfill. One direct consequence, however, may be an increase in costs of raw material due to EOL costs, and as a result an increase in product price. Contrary to the design phase, which incorporates many knowledge processes, the manufacturing, assembly, sales, distribution and service phases are supported by multiple business process models that detail the desired activities to be performed in a process, their sequence within the process, and the triggering events [17]. Business process modeling may be viewed as an IT-based management technique that allows for detailed examination and specification of the business processes being carried out in the organization [18]. Leading model developers, such as ARIS [16], and leading vendors, such as SAP [13] and Oracle [14], have adopted a taxonomic or ontological view of enterprise business processes, expressing enterprise activity as a structured hierarchy of major functions, main functions, processes and activities. The high level representation of life cycle processes (Figure 2) may be adopted as an alternative ontology or upper level hierarchy in detailed business process modeling and best practice identification. Many of the existing processes will be incorporated in the extended model, but additional processes will be required for EOL processes as well as EOL influences (e.g. forecasting EOL quantities for logistic planning).

### Stage 2: Assembly

Production includes manufacturing and assembly. In the assembly phase, product parts – whether manufactured or purchased – are assembled to create the final product. In modern manufacturing, three prime sources exist for product parts: 1) new parts fabricated in the manufacturing process (in-house or outsourced); 2) new parts bought from an outside source; 3) refurbished parts cleaned and tested from a prior product (in-house or outsourced). Government incentives may be required to ensure that companies incorporate refurbished parts in their products. Moreover, the public will have to be reeducated regarding the reliability of products built of refurbished parts. Including purchased parts in products complicates manufacturer accountability in the life cycle. Hence, clear processes and EOL procedures will need to be developed and implemented to explicate legal aspects of life cycle management. Fundamental changes in the classical assembly process are not expected. Nevertheless, increased built-in flexibility will be required for automated assembly processes to deal with the rapid changes in product design, and assembly processes will be required to consider disassembly requirements, in particular bonding methods.

### Stage 3: Sales

As mentioned above, one creative idea to benefit the industrialists, the environment and the general population simultaneously is to sell services rather than products. The idea is that manufacturers of goods sell their services at the place and time desired by the customer. The operator is the customer, but the owner is generally the manufacturer. Several cases may be distinguished [10]:

1. Expensive products, such as photocopiers or very high quality printers, where customers are offered photocopying and printing services at a certain location or in their facilities: Due to competition with other manufacturers, the manufacturer – who is also the owner of the product – always replaces old products with newer and better ones (Figure 4). Thus, the customers always enjoy the benefits of modern and well maintained equipment. The discarded equipment is then used to upgrade older equipment at other sites, thereby increasing customer satisfaction by continually upgrading equipment while significantly increasing product life span and reducing waste accumulation.
2. Single-use products: In this category, we differentiate between products to be used, such as single-use cameras that are returned with the film and then disassembled, cleaned and reassembled, and products to be consumed, where the package is returned against a minimal charge, such as beverage cans and bottles.
3. Products that are provided "free of charge" or almost free, in order to offer customers services or other products: Examples are cellular phones, razors, cable television converters, razors (specific replacement cartridges are sold), coffee machines, etc. We need to distinguish between home machines that have to provide some fixed functionality, such as a coffee machine for preparing coffee, and products that are in vogue and/or whose functionality is enhanced, as is the case with mobile phones. Products in the second category have a much shorter life span.
4. Home appliances: These home appliances are also owned by the manufacturer, who, for example, must take away a washing machine upon customer request. Since the manufacturers are the owners, they may then disassemble the washing machine and reuse some of its parts in 'new products', therefore gaining added benefit from them. Thus, it can be expected that the average household will use more modern machines (less energy consumption and more reliable), benefiting both the environment and the customers. One question to consider here is whether such appliances will be designed in the same way as traditional appliances are designed today.

This trend of selling services rather than products will go beyond private customers. Consider, for example, the automotive industry, where machine tool builders sell production lines to car manufacturers for manufacturing powertrain components. In the new era, this could be considered a form of selling machining services. Once car manufacturers order a production line, they are vulnerable to changes in market demands, model acceptance, etc. To protect themselves, they must acquire some degree of flexibility by means of reconfigurable machines, some spare production capabilities, etc. But we can take the selling of services approach one step further by letting the car manufacturers buy the powertrain components from the machine tools builders. Namely, the machine tools builders will have to change their business from building machines to providing machining services. The question then becomes: will they build the same machines and the same manufacturing lines for themselves?

### Stage 4: Distribution

No fundamental changes in distribution requirements and processes are expected. Nevertheless, distributors will be at a distinct advantage for capitalizing on the requirement to collect products from the consumer for EOL processing (Figure 3). With minimal alterations their existing framework, capacity and capabilities can be adopted to perform the collection, thus increasing profits considerably with minimal investments. However, additional business processes will be required.

### Stage 5: Service

Whether time-based or condition-based, product servicing (i.e. maintenance, product repair or part replacement) plays an important role in protecting the environment. It considerably elongates the life span of the product and has a much smaller impact upon the environment than producing a new product. From a product life cycle management perspective, the purpose of service is to control the conditions of products so as to provide the functionality required by customers or by society, while keeping the environmental load at a minimum and maintaining appropriate corporate profits [5]. Increasing business opportunities will arise to enhance the product life cycle, such as e-servicing companies aimed at ensuring optimal machine performance, lengthening the life span of the machines and decreasing energy consumption and waste accumulation. Much research is still required to enable such services.

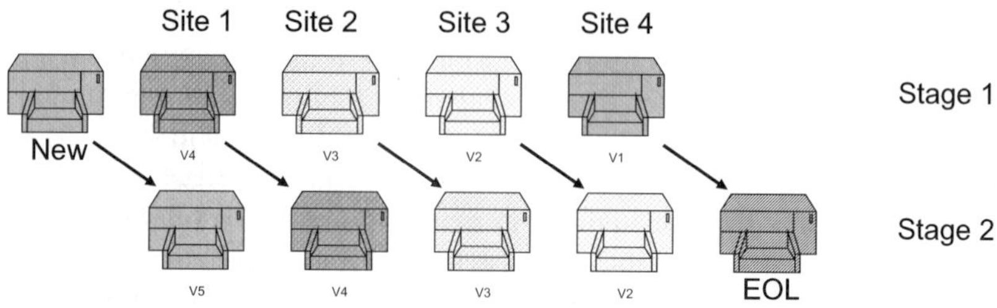


**Fig. 4.** The product life cycle for service-selling

### Stage 6: Upgrading

Often when the performance of a product has deteriorated and new products are available on the market, customers have two available options. The first is to upgrade their product, and the second is to discard the product and buy a new one. Modern manufacturing encourages upgrading since it has less impact on the environment than producing a new product, reduces the need for entailing EOL costs and is usually a lot less expensive for the customer than buying a new product. The functionality of a product may be upgraded by improving its appearance, incorporating new software, or incorporating new components (e.g. replacing the interior of an airplane and providing better seats with new features, or upgrading PC performance to provide increased computing speed or memory size). Increased demand for upgrading has encouraged manufacturers to design module component products that can be incrementally updated and products that can be updated by changes to software alone.

## 3.3 End Of Life phase (EOL)

When the performance of a product has deteriorated and at the same time new products on the market offer much better performance and appeal than the original product, a big gap is created. When this gap exceeds a certain threshold, the customer will generally purchase a new product. At this point the old product, if not sold privately to another customer, enters the EOL phase. Figure 3 presents a more detailed life cycle diagram that highlights new business opportunities and new topics for research and development, both academic and practical. EOL processes have received attention due to new legislative initiatives demanding manufacturer accountability for a product over the entire product life cycle. For example, the European Union's directive on WEEE (Waste Electric and Electronic

Equipment [11]) has been introduced because of concern about the growing volume of electronics products that waste valuable resources and contain toxic substances. The WEEE thus requires a 75% recovery rate for cellular telephones, and batteries, printed circuit boards and liquid crystal displays will be required to be disassembled. Recently, Seliger et al. [3, 15] showed that disassembly and remanufacturing of cellular phones can be a competitive business segment in the European market. Hence, the WEEE directive presents new economic opportunities, in this case by creating a large market for reuse of cellular phones. Each stage in the EOL phase presents a business opportunity resulting from the expected rapid increase in demand due to legislation. Moreover, additional business opportunities will arise to manage and orchestrate the EOL phase, ensuring ongoing cooperation between the relevant parties for KID sharing to increase expected profit gain. It would appear that manufacturers will prefer to outsource the EOL activities. Hence, multiple companies will be served by single suppliers to perform product disassembly, testing, recycling, refurbishment and disposal.

### Stage 1: Collecting, testing and disassembly

The first step of the EOL phase is collecting the products and sorting them. The first set of products to be separated from the collected goods are those that can be sold 'as is', that is, secondhand products that can reenter the active life cycle, such as used cars. Hitachi Construction Machinery Co. is adopting this approach and is expanding its business based on the proper after-sales maintenance of equipment, and also the collection, refurbishment and sales of used equipment. Furthermore, to expand the market for used construction equipment, Hitachi has created a database of used equipment, and a system for selling it overseas via the Internet. Such a concept may be applied to other types of products (home appliances, telecommunication products) when legislation and enforcement thereof come into play. A whole new market for secondhand products is likely to evolve. As a result, a secondary warranty market is also likely to develop in order to encourage requisition of these products (similarly to used car lots, which often offer a six month warranty on the car purchased). The reasoning behind this secondary market evolution is that manufacturers will still be held accountable for their products. Hence, they can benefit economically from reselling a product several times before encountering the EOL costs. The next stages of the EOL phase are disassembly, testing and identification. During these processes, the product is first disassembled, and its hazardous parts that cannot be reused are separated and later disposed of in a controlled manner. Other parts are inspected and categorized for refurbishment or recycling. The idea is of course to recover the maximal value of the product and minimize landfill materials. Disassembly will be one of the major obstacles in achieving the equilibrium between environment, customer and manufacturer. The level of disassembly, the amount of testing and the quantity and quality of products channeled to refurbishing, recycling or disposal are all major economic decisions that will require extensive cooperation and KID sharing, as well as new models and tools to optimize these processes to create win-win situations. Moreover, economic calculations and environmental requirements are likely to clash on many occasions where the expected economic benefits will not cover expected testing and disassembly costs. In such cases, the economic considerations are likely to outweigh the environmental requirements unless governmental intervention targets this conflict.

#### Stage 2: Recycling, refurbishment and disposal

Companies such as Safety Kleen Corp in San Diego or the Escol company in Israel have developed hazardous material disposal facilities. Due to the general consensus of controlled hazardous material disposal and the danger of shipping these materials, numerous local companies provide these services, and the business opportunity they offer has been clear for some time. An added business opportunity that should be encouraged by government funding is hazardous waste *recycling*, which helps reduce or eliminate hazardous waste disposal. Once treated and tested, these materials can then remain in place, no longer hazardous, toxic, or leaching. One such example is Matrix Environmental Products in Arizona, which has successfully treated hydrocarbons, halocarbons, chemicals and a number of heavy metals. Multiple companies for recycling and refurbishment will emerge when legislation comes into play and demand for these services rapidly increases. The need for these services in the near future is beginning to be recognized, and companies such as Nxtcycle (personal computers and monitors), Amber Diagnostics (for medical equipment) and Bruin Electronics (integrated circuits) have seen the potential of reverse manufacturing, refurbishing and recycling, and are already providing such services to organizations.

## 4 Concluding remarks

This paper has explored and briefly analyzed the forces driving the significant changes in the market that affect the entire product life cycle. Business requirements and opportunities have been highlighted, as have the constraints of growing demand, new legislation and increasing global competition. The requirement for extensive research and development was noted as a fundamental component for success in fulfilling the vision of a sustainable society, satisfying the requirements of the three dominant forces in the product life cycle: customers, industrialists and the environment. Maximizing the benefits of the EOL phase is compulsory for economic feasibility and vision fulfillment. Further factors contributing to the target of building a sustainable society are (a) research and development, (b) government intervention, (c) entrepreneurs and economic joint ventures and (d) education. Life cycle engineering focuses on researching and developing technologies, tools and support methodologies over the entire product life cycle that will benefit customers, industrialists and the environment. This includes, for example, technology that enables traceability of products and their specific history for quick identification and classification or technology that reduces recycling and refurbishment costs. In particular, extensive research and development for IT-based tools is required, such as developing e-servicing capabilities. In addition, life cycle engineering involves developing metric systems and measurement capabilities that will allow objective assessment and benchmarking, both for legal and business purposes. Implementation of the new laws in the industry is not trivial due to economic and logistic difficulties. Hence, nonpartial law enforcement, tax regulations and public pressure will probably be required to generate this industrial revolution. Many environmental impact reduction measures have been driven by environmental regulation, for the money required would not have been invested by common economic optimization 4. Ecological consciousness and its accompanying legislation threaten some existing businesses but provide new opportunities for others. Existing businesses may need to redefine, change or extend their main business core, and multiple new ventures will emerge as the "life cycle" concept becomes the backbone of a new industrial culture.

## Acknowledgements

This research was supported in part by the VRL-KCiP Network of Excellence, the Bernard M. Gordon Center for Systems Engineering at the Technion, Grant 2005847, and the Minerva's Schlesinger Laboratory for Life Cycle Engineering.

## References


1. Westkamper, E., Alting, L., Arndt, G., 2000, Life cycle management: Approaches and visions towards sustainable manufacturing, Proceedings from the 7th CIRP international seminar on life cycle engineering, 1-16.
2. Niemann, J., Westkämper, E., 2004, Product life cycle management in the digital age, Intelligent knowledge-based systems: Business and technology in the new millennium, Kluwer Academic Publisher
3. Seliger, G., Muller, K., Perlewitz, H., 1997, More use with fewer resources, Proceedings of the 4th CIRP international Seminar on life cycle engineering.
4. Alting, L., 2004, Architecture of environmental engineering, Proceedings from the 11th PTK.
5. Takata, S., Kimura, F., van Houten, F.J.A.M., Westkämper, E., Shpitalni, M., Ceglarek, D., Lee, J., 2004, Maintenance: Changing Role in Life Cycle Management, Annals of the CIRP, 53/2.
6. Kimura, F., Suzuki, H., 1995, Product Life Cycle Modelling for Inverse Manufacturing, Life Cycle Modelling for Innovative Products and Processes, Chapman & Hall, 80-89.
7. Alting, L., Legarth, J.B., 1995, Life Cycle Engineering and Design, Annals of the CIRP, 42/2:569-580.
8. Bar Cohen, A., 1995, Mechanical Engineering in the Information Age, Mechanical Engineering, 117/12:66-70.
9. Krause, F.-L., Kimura, F., Kjellberg, T., Lu, S.C.-Y., 1993, Product Modelling, Annals of the CIRP, 42/2: 695-706.
10. Shpitalni, M., 2004, Impact of Life Cycle Approach and Selling of Services on Product Design, Proceedings Manufacturing '04 Conference, Poznan, 4-5.
11. P&CEC, 2002, Directive 2002/96 on Waste Electrical Electronic Equipment (WEEE), The Parliament and the Council of the EC, Brussels.
12. Shpitalni, M., Alting, L., Bilberg, A., 1996, Practical Aspects of CALS in Design and Manufacturing of Sheet Metal Products, Annals of the CIRP, 47/1:393-396.
13. SAP, 2004, URL = www.sap.com/business solutions/industries.
14. Oracle, 2002, Business Flow Accelerator standards - demand for discrete manufacturing, PowerPoint presentation.
15. Seliger, G., Franke, C., Ciupek, M., Basdere, B., 2004, Process and Facility Planning for Mobile Phone Remanufacturing, Annals of the CIRP, 53/1:9-16.
16. Scheer, A.W., 2000, ARIS: Business process modeling, Springer.
17. Scheer, A.W., 1994, Business Engineering: Reference Models for Industrial Enterprises, Springer.
18. Vernadat, F.B., 1996, Enterprise modeling and integration: principles and applications, Chapman and Hall.

# Meeting the Climate Change Challenge

## Towards social and technical innovations for a functional society

C. Rynikiewicz

*Lepii-EPE CNRS/Univ. Grenoble - Département Energie et Politiques de l'Environnement (ex IEPE), Grenoble, France,christophe.rynikiewicz@upmf-grenoble.fr*

Abstract: This paper aims at giving quantitative figures on the climate and energy challenge that may change the conditions in which business industry is operating. The material efficiency approach along the whole life cycle of products and services enables to show how to shift away from the unsustainable "business as usual scenario". Indeed, climate policy often focuses on energy systems and price of carbon and technological options and may fail to foster sufficiency solutions compatible with business objectives and a more functional society. A focus on R&D needs and appropriate technological, economical and social framework to enhance the shift to sustainable "product service systems" is needed.

Keywords: Climate Change, Eco-Efficiency, Sufficiency Innovations

## 1 Introduction

Momentum is growing globally such that action to reduce greenhouse gases (GhG) emissions may be required. The UN Convention on Climate Change states that the policy goal should be to limit average global temperature increases to no more than 2°C of pre-industrial levels, which would already have serious impact. Therefore, meeting this climate objective will require a peak in world emissions within a few decades and a strong decrease to stabilise the concentration in atmosphere. A variety of factors related to human activity interact to produce global environmental change : the configuration and operation of population change, technological change, and socio-economic/cultural organization structure the patterns of production, consumption, and impacts throughout the world. Using



*D. Brissaud et al. (eds.), Innovation in Life Cycle Engineering and Sustainable Development,* 33–48.

results of the POLES partial energy equilibrium model, this paper aims at giving quantitative figures on the climate and energy challenge that may call for a material efficiency approach and create the economic conditions for a shift from products to sustainable products-service systems.

Firstly, we provide the basics to understand the emissions profile and on the trend of rising energy prices. Secondly, we propose an adapted framework to analyse the potentials of decoupling growth and $CO_2$ emissions Such new constraints will induce challenge for the co-evolution to a "factor4 emissions reduction world" for the industry in 2050. We explain why a "material efficiency approach" consisting in the optimal choice of materials and eco-design of energy intensive products is of great promise. Climate policy often focuses on energy systems and fails to address systematically the emissions related with material intensive consumption patterns. Finally, we build on the current emerging debate about the contribution of product–service system (PSS) to more sustainable consumption patterns. More research is needed to assess if PSS may enable to link technical and social innovations that are needed for process and system evolution towards emission reductions of a factor 4.

## 2 New imperatives imposed by the climate and energy challenges

### 2.1 Emissions profile and GhG reductions

Climate change is related to a variety of factors associated with human activity and the biosphere engine and Greenhouse Gases (GhG) Emissions are related to a set of assumptions and scenarios on population growth, economic growth, etc.. When tackling these questions, climate models and economics models are generally coupled for long-term strategies. Most scientific work available has thus far assumed that reaching the 2°C target would translate into a long-term greenhouse gas concentration level of 550 ppm $CO_2$ equivalent [1] (which means 450 ppmv $CO_2$ only) in the atmosphere.

Such level of concentration is still subject to uncertainty and new scientific knowledge may become available in the future, specially on the climate sensitivity factor [2]. The German Advisory Council (WBGU) has recommended the stabilisation of $CO_2$ concentration in the atmosphere below 450 ppmv [3].Considering the hypothesis in the Common POLES-IMAGE (CPI) work [4], the resulting global reduction challenge is shown in the figure 1. The baseline describes the development in the main driving forces (population and economic growth) and environmental pressures (energy, industrial and land-use emissions) for the 1995-2100 period with « Technical Change and Policy as usual ». It serves in particular as a benchmark for the assessment of alternative policy schemes.

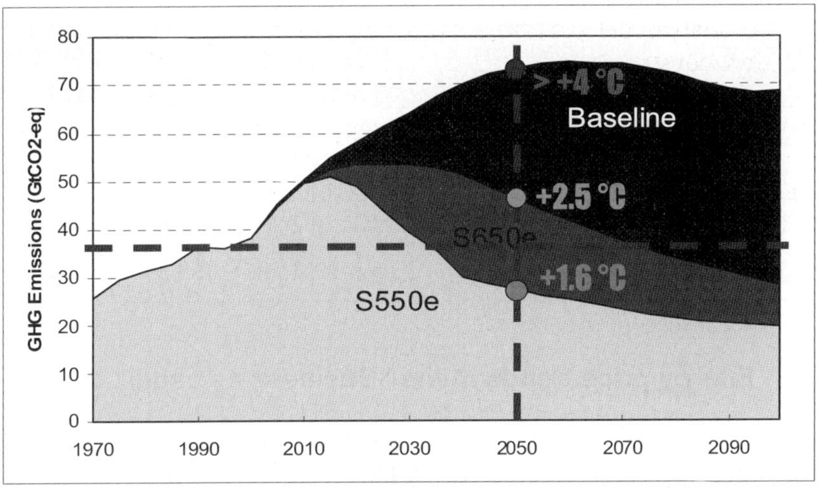


**Fig. 1.** World emissions profile for different $CO_2e$ concentrations in atmosphere
Source: IMAGE2.2 [4]

A recent study "GhG Reductions Pathways" [5] looked at options for a future climate change regime. A concentration level of 550ppm $CO_2e$ would translate into a global reduction of GHG emissions of 15–20 % by the year 2050 compared to 1990 emission levels or by 50–60 % compared to a "business as usual" scenario. The challenge would be particularly important for industrialised countries, as reduction would be important to enable emissions of developing countries to increase. The different GhG emitting sectors have to face this "carbon constraint" and limit their emissions. Environmental economics theory recommends to internalise the GhG externalities. The change in the relative prices of energy and production costs would imply "induced technical change". Even with little constraints resulting from the first application of the EU Trading Scheme and the Kyoto Protocol (-8% for EU in 2012 compared to emissions of 1990), the climate constraint is said to imply an increase in the operating costs.

Preliminary research indicate that 50–75% emissions reduction would lead to much higher prices than those in table 1. Limits of the socio-technical system and the climate change challenge would induce changes in the production, distribution and consumption patterns of steel and other materials. We will describe transitions envisaged in the steel industry from cleaner production to systems innovation Moreover, materials are a central factor in the life-cycle cost of industrial products. Therefore, cutting these costs increases competitiveness and growth.

**Table 1.** Increase in total costs per tonne of finished products assuming full opportunity cost, IEA, 2004 [4]

| Carbon price E/ t $CO_2$ | Electricity | Steel (BOF) | Steel (EAF) | Cement | News-print | Alumin-ium |
|---|---|---|---|---|---|---|
| 5 | 5% | 0.2% | 0.4% | 0.7% | 0.5% | 1.2% |
| 10 | 11% | 0.5% | .08% | 1.5% | 1.1% | 2.4% |
| 15 | 16% | 0.7% | 1.3% | 2.2% | 1.6% | 3.6% |
| 30 | 32% | 1.5% | 2.5% | 4.5% | 3.3% | 7.2% |

## 2.2 Energy price trends in the "Business as usual" case

In the "business as usual" case, energy prices would rise acoording to most foresight exercises of IEA, US DOE . In order to provide some general insights on the trend of energy prices, we use the recursive energy equilibrium model POLES. In the WETO exercise (World Energy Technology and climate policy Outlook) for the EU Commission [5], world final energy consumption is projected to grow by 1,8%/year on average during the 2000–2030 period of the simulation. With the current economic conditions, this reference scenario gives an important share to fossil fuels. The simulation in POLES reflects basically the dynamics and adjustments of world demand and supply.

**POLES Model.** The POLES (Prospective Outlook for Long-term Energy Systems) model is a partial equilibrium model of the world energy system, developed since 1997-1999 under the co-ordination of IEPE and in particular with IPTS [7]. This model is currently used by the European Commission for energy and climate policy simulation. POLES simulates in a recursive year-by-year process the supply and demand of the major energy carriers within the period 2000–2030, and endogenously computes their prices. The main advantages of the model are its comprehensive and detailed structure (regional and sectoral disaggregation), the fact that it provides elements for the endogenisation of technological change and its capabilities for the analysis of technological progress in the context of $CO_2$ abatement policies.

Energy prices are expected to increase as shown in figure 2. For physical reasons related to the acceptable concentration of $CO_2$ in the atmosphere as well energy prices and oil depletion concerns, there is an imperative to shift away from the current technico-economic paradigm [8]. Global abatement policies may even be beneficial to the low-income regions of the world, in particular when environmental cobenefits are taken into account [9]. Two constraints will impact the socio-technical systems: upstream via the fossil fuels resource constraint and downstream via the climate challenge. This reality stresses the need to decouple economic growth and environmental damages.

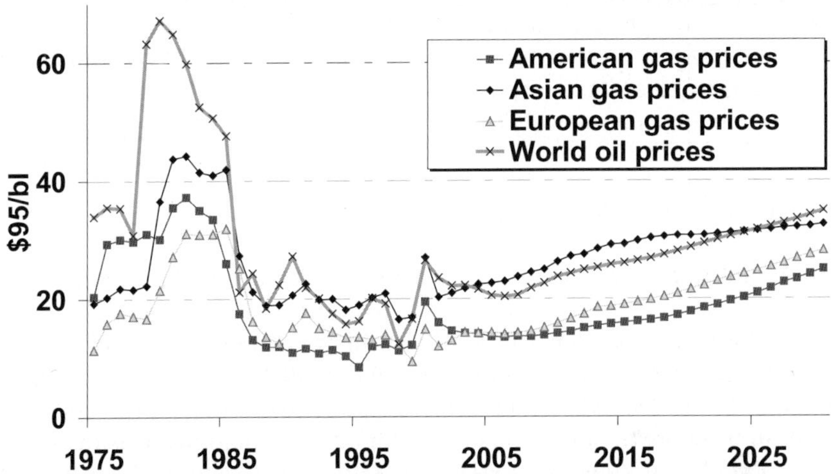


**Fig. 2.** Rising energy prices and past evolution, source POLES, 2003, WETO project [6] (note that foresight does not aim at "predicting" short term events such as oil prices rise)

# 3 Decoupling strategies to shift away from the "business as usual" case

## 3.1 Co-benefits of shifting away from the current technico-economic paradigm

Literature on environment has identified for long the other environmental impacts than GhG emissions. Historically, impacts related to chemicals and toxics, land degradation induced the rising concern about environment and the need for decoupling strategies. Lovins' earlier book on "factor4" argued that numerous technical and organisational opportunities exist for quadrupling resource productivity, enabling a doubling of wealth whilst halving resource use [10]. Economic assessments of climate policies use the Cost-Effectiveness approach, comparing the costs of different strategies for a given target. Finally, even if the issue of comparing the costs and the benefits remains controversial, because of the uncertainties surrounding damage cost evaluation, it is still cheaper to cut emissions than to pay for climate damages. Famous Cline's assessment of climate damages and policy options lead to high carbon prices but the cost benefit ratio ranges between two and four [11–12]. Whatever the future climate regime is and which value human community attributes to climate change risks and damages, there will be a mix of policies and a price attributed to $CO_2$ emissions in order to try to internalise the external costs Then, the question remains how to evaluate the potentials for a systems change.

## 3.2 Assessing a future decoupling potential $CO_2$ and growth

There are widespread demands in society for a decoupling of environmental impact from economic growth. The different possibilities of decoupling $CO_2$ and GDP growth are listed in the equation of figure 3. The ability to realise such decoupling is crucial considering the likelihood of continued economic growth in developed countries and rapid economic growth in many developing countries with high populations. It is therefore of great interest to know how well decoupling have succeeded so far and what potential there is for future decoupling.

Four strategies are listed:
A – "Process innovation" deals with the radical or incremental innovations that would decrease the $CO_2$ emissions per ton of material
B – "Transmaterialisation" implies a recurring industrial transformation in the way that economic societies use materials [13–14]
C – "Dematerialisation"
D – And structural changes

The concept of TMR (Total Material requirement] as defined by WRI or Wuppertal Institute [15]. Total Material Requirement is a compound indicator reflecting all of the physical materials that are mobilized each year to support an economy, including "hidden", non-economic materials such as mineral overburden, processing waste and soil erosion. The TMR includes an aggregate indicator and disaggregated sub-indicators by resource sector. Combined with economic measures (e.g. TMR per unit GDP), the indicator enables to track changes in material intensity and thus tendencies toward dematerialization of an economy (or the reverse see [16]). TMR per unit GDP also provide a tool to compare a country's material economy with those of other countries for a greater understanding of common as well as unique patterns.

$$CO_2 = \frac{CO_2}{T\,steel} * \frac{T\,steel}{T\,materials\,(TMR)} * \frac{T\,materials}{GDP} * GDP$$

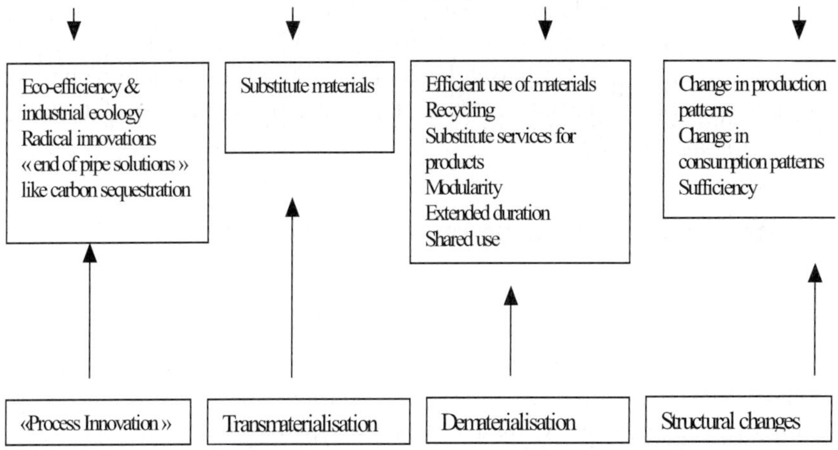


**Fig. 3.** Various strategies to decouple environmental and resource impact from economic growth (adapted from [18–19])

The existing challenge is to evaluate what could be the potential of each strategies. Moreover, the modification of energy prices and the introduction of a carbon value will change the competitive advantages of each material. A fast growing literature focuses on technological change induced by a carbon constraint. However, much of the existing literature deals with energy related emissions and possible projections about future energy systems, $CO_2$ reduction potentials transaction costs, barriers to investment and technological diffusion. Another complementary approach deals with material and energy efficiency. Tackling the problem represents usually a two-thirds reduction of present per capita primary energy use of industrialised countries, implying at least factor of five of energy efficiency improvement and a reflection on material efficiency. The Swiss Board of the Federal Institutes of Technology promoted the vision of a '2000 Watt per capita society' by the middle of the 21st century [17] as early as 1998. Detailed evaluation suggests that this vision is technically feasible within half a century. This approach is of particular interest to elaborate real practical solutions linking the production and consumption patterns.

# 4 Emissions reductions potential revealed by the material efficiency approach

Indeed, considering the important challenge of "factor4", it is still hard to comprehend how economies could evolve towards a much less carbon-intensive path. Any reduction goal compatible with climate stabilisation will have considerable effects on economic activities, markets and behaviours. The demand side (in particular buildings and transportation) will be impacted, via their materials content. Therefore the approach of material efficiency is of great promise.

## 4.1 Relevance of focusing on the material industry

In OECD countries 36% of the primary energy demand is used by industry to manufacture products that are consumed in society. A large part of the energy is dedicated to the production of basic materials used in the products. Moreover, material industry is of particular importance for developing countries in their industrialisation "phase" and it accounts for a major part of the GhG and local pollutants emissions. Progress in cleaner production or "systems innovation" could be adapted and transferred in developing countries enabling "leapfrogging".

## 4.2 Emissions potentials with a life cycle approach

Historical data provide examples how energy efficiency has improved in the industry, thanks to process innovation as described in figure 3. However, industry experts from many sectors say that after 2010 the necessary emission reductions require major technological changes, as the improvement of existing processes will not be sufficient. Several EU research projects focus on cleaner technologies or radical innovations [20]. In the following part of the paper, attention is paid to other solutions. The traditional engineer's approach is highly disputed by some economists : with existing technologies there are potentials of emissions reductions, which would have a positive effect on the profitability of the firm [21–22].

Numbers in table 2 and 3 shows that energy can be saved by more efficient use of materials.

**Table 2.** Final Energy use in the EU-15 Industry (Eurostat)

| | Final Energy use 1990 (Mtoe) | Final Energy use 2000 (Mtoe) |
|---|---|---|
| TOTAL INDUSTRY | 266.01 | 271.68 |
| Iron and steel | 56.04 | 51.49 |
| Glass, pottery and building materials (cement included) | 36.65 | 34.88 |
| Chemicals | 50.29 | 45.33 |
| Food, drink and tobacco | 22.02 | 25.02 |
| Engineering and other metals | 27.35 | 24.51 |
| Paper and printing | 17.99 | 31.12 |
| Textile, leather and clothing | 8.75 | 8.45 |
| Non-ferrous metals | 10.83 | 10.44 |
| Ore extraction | 2.99 | 2.68 |
| Other industries | 20.64 | 34.96 |

**Table 3.** $CO_2$ emissions reduction strategies for Western Europe, compared to autonomous development

| | Cost Euros/t $CO_2$ | Strategy | Potential (% of $CO_2$ emission) |
|---|---|---|---|
| Renewables | 0 - 1 000 | Substitution / prevention | 10 - 25 |
| Nuclear energy | 0 - 100 | Substitution | 10 - 25 |
| $CO_2$ capture and disposal | 0 - 100 | Substitution | 10 - 25 |
| Fossil fuel switch | 100 - 500 | Substitution | 10 - 20 |
| Energy conservation / increased efficiency | 100 - 1000 | Prevention | 10 - 25 |
| Waste heat utilisation | 100 - 1000 | Process integration | 5 - 10 |
| Materials recycling/reuse | | Process integration | ? |
| Dematerialisation | | Prevention / Process integration | ? |

Emissions associated with the demand for energy-intensive materials could be reduced :

- A - by more efficient use of these materials (by improving their design or material properties). Ecodesign has been a promising and growing field of research, which delivers part of the solution, and is particularly interesting to facilitate the collect and sorting of materials for recycling. However, ecodesign objectives may fail to account for absolute limits of the global ecosystem.

- B - by increased recycling or substitution of those materials by less energy-intensive or biomass based materials. The "industrial ecology" central idea is to optimise the flow of materials and energy between different industries and in that way to propose new "industrial metabolisms". It derives partly from a desire to see societies endogenise these impacts through new models of economic development and conceptualisations of societal 'progress'. Zero waste and 3R (reduction, recycling and recovery) approaches have become common concepts and they are often included in the strategic policy of several companies, who view the environmental issue as a priority as much as more traditional aspects concerning productivity, production cost cuts, etc. Like the Kalundborg case, these loop–closing activities slowly developed over time as firms identified and characterised waste sources and sinks.
- C - by a shift from products to services. A change in the products and services sold implies however different institutional frameworks regarding property, liability and fiscal system. We will discuss this option in part 5.

These three stages from end of pipe to product redesign and system changes illustrate the industrial transformation process which is arbitrarily estimated to take place along time scales on the order of 10 to 25 years. Figure 4 from the Industrial Transformation Science Plan [25] illustrates the relation between various response modes, the time scale, and the geographic scale involved.

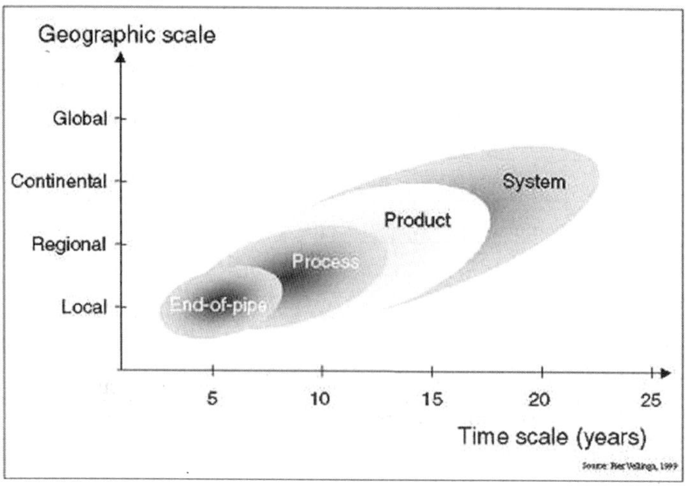


**Fig. 4.** Societal responses to the issue of environment (source [25])

Systems change is a combination of technical change and societal change. We will introduce the product service system (PSS) approach which is believed to be a promising step for transition to more sustainable consumptions. Our question is also whether this approach would contribute to a change of scale in the climate policy arena.

## 5 A logical enhancement of ecodesign : product service system

The precedent parts of this paper have dealt with innovation and technology as tools for achieving higher resource efficiency in society. Here, we would like to focus on solutions coherent and compatible with business strategies that go beyond eco-efficiency and process optimisation.

A critical attribute of ecodesign is that the product is still at the forefront. Value creation and material use are then still intimately coupled. The following table shows the type of business strategies that would help increased resource efficiency with different thinking in business practices. However, increasing material efficiency does not lead necessarily to an absolute decrease in material consumption.

**Table 4.** Resource efficiency and business strategies in the Service Economy

| | **Types of business Strategies** | |
|---|---|---|
| **Increased resource efficiency:** | **closing material loops:** technical strategies | **closing liability loops**: commercial / marketing strategies |
| reducing the VOLUME of the resource flow: | **Eco-products**<br>Dematerialized goods<br>Multifunctional goods | **Eco-marketing**<br>More intensive utilization of goods<br>Shared utilization of goods<br>Selling utilization instead of goods |
| reducing the SPEED of the resource flow: | **Re-manufacturing**<br>Longer utilization of goods<br>Long-life goods<br>Service-life extension of goods | **Re-marketing**<br>Away-grading of goods and of components<br>Re-marketing services<br>New products from waste |
| reducing VOLUME *AND* SPEED of the resource flow | **Technical system solutions**<br>Krauss-Maffei PTS plane<br>Transport system<br>"skin" strategies | **Systemic solutions**<br>Lighthouses<br>Selling results instead of goods<br>Selling services instead of goods |

## 5.1 Introducing the sufficiency paradigm

Necessary emission reductions of a factor4 or 10 require major technological changes and also that innovation arises out of a more integrated arena. An expert group, commissioned by the EU DG research Commission in 2001 addressed the issue of what type of Research, Technology Development and Innovation policies and actions would support the move to a competitive and sustainable European production system in the period to 2020. The group developed an integrated view of competitive and sustainable production. This view links production technology, technologies in products 'in use' whether as artefacts and materials, to the socio-technical systems in which they are embedded. It also argues that purposeful change and innovation in socio-technical systems involves the participation and collaboration of many actors in the networks that surround these systems. The report argues that two types of (complementary) strategies are to be followed: efficiency and sufficiency [27]. Sufficiency is based on the notion of moving from selling product (with its material throughput philosophy) to providing performance, managing the material content of products together with their asset value. The challenges are to help in constructing new ways to meet social needs [28].

**Table 5.** Definitions of PSS in the literature [32–33]

| | |
|---|---|
| The configuration (quantity and quality) of products and services supplied to meet the demand for well being may be described as a product-service mix or product service combination | Manzini (1996), Goedkoop (1999) |
| A Product-Service System is defined as "a marketable set of products and services capable of jointly fulfilling a user's need" | Goedkoop/ van Halen/ Riele/ Rommens (1999) |
| Through servicizing for the consumer, consumption shifts from purchase and use of a product to purchase of a service | White et al (1999) |
| Sustainable services and systems ideally fulfil customer/ consumer demands with the lowest negative impact on natural and social environment are profitable strategies for companies and can be continued over a long period of time | Tischner (2000) |
| PSS is a system of products, services, supporting infrastructure that is designed to be competitive, satisfy customers need's and have a lower environmental impact than traditional business models | Mont (2001) |

## 5.2 Product Service System as an enhancement of ecodesign

The idea of shifting from products to services is now more than 40 or even more years old [25]. In the last decade, it has resurfaced and a growing literature deals with theoretical concepts and practical examples. A more systematic perspective on the combination of products and services is needed. The provision of use is at the forefront and its aim is to satisfy consumer's needs to increase. It is also consistent with current emerging notions of functional society [30–31]. However the definition is still in construction as summarised in table 5. Innovative products or services can clearly increase resource efficiency without adverse effects on functionality or usefulness. More research is still needed to strengthen the market for PSS and see if it is a way to enable more "aggressive" climate protection strategies.

## 5.3 The contribution of PSS to climate protection objective and new areas of research

For some time now, on the business side, a (limited) number of companies position themselves as service providers but quite independently from their environmental aspirations. In the first part of the paper, we indicated that climate protection imperative and energy price trends may change the operating conditions of companies. The question is then to which extent a carbon price and the institutional arrangements may facilitate the creation of PSS activities. Moreover, it is important that social and technical solutions emerge to enable the implementing of the climate policies. Therefore, it is important to systematically address the contribution of PSS to $CO_2$ reduction. Moreover, an appropriate framework has to be designed so that economic instruments may deliver benefits.

**Table 6.** Six main transitions and challenges

| Focus | Service transformation | |
|---|---|---|
| Design | from planned obsolescence | to sustainable product design |
| After sales support | from short term guarantee | to comprehensive after-sales support |
| Form of contract | from ownership | to eco-leasing |
| Mode of consumption | from individual consumption | to collective consumption |
| Need | from dependence | to reduced need |
| Sales revenue | from output maximisation | to least cost supply |

Finally, it has to be investigated whether the climate objectives may influence values in the European Community for example that may facilitate the transitions that are needed at the consumer's level to opt for PSS [34]. Table 6 synthesise six main transitions which can deliver environmental benefits [35–36]. Indeed, PSS implies a change in thinking about categories of ownership and consumption. Sufficiency strategies seem to be more dependent on the choices of final consumers than eco-efficiency strategies and usually include sharing and pooling of products. Consumers are hesitant towards alternatives of consumption without ownership, such as sharing and renting [37].

## 6 Conclusion

Cleaner Production via incremental or radical solutions, Industry Ecology and Life-Cycle thinking are the basis for circular economy approach. The traditional approach in the climate policy arena has been decoupled for too long with material efficiency approach. As the carbon constraint may induce huge change in the conditions to run business, the challenge is to assess the design and contribution of sufficiency strategies. An appropriate framework is needed to foster the development of these activities and enable new comers to contribute to the solution.

## Acknowledgements

We would like to thank ARCELOR and ADEME for a PhD grant and the POLES team for their work on the model. Discussions with Prof Daniel Brissaud were particularly fruitful. All responsibility for the content rests with the author.

## References

1. Eickhout B., Den Elzen M.G.J., Vuuren D.P. van, (2003), "Multi-gas emission profiles for stabilising greenhouse gas concentrations, Report 728001026, RIVM, Netherlands.
2. Hare B., M. Meinshausen, (2004), "How much warming are we committed to and how much can be avoided? ", PIK report n°93.
3. Grass I., Schellnhuber H.J., et al., (2003), "Climate Protection Strategies for the 21st Century : Kyoto and Beyond", WBGU, Berlin.

4. CNRS/LEPII-EPE, RIVM/MNP, ICCS-NTUA, CES-KUL, (2003), "Greenhouse Gas Reduction Pathways in the UNFCCC process up to 2025". http://europa.eu.int/comm/environment/climat/studies.htm
5. Reinaud J., (2004), "Industrial competitiveness under the European Union Emissions trading Scheme", information paper, IEA, Paris.
6. European Commission, (2003),."World Energy, Technology and climate policy Outlook 2030", EUR , Brussels.
7. Criqui P., Kouvaritakis N., Soria A., Isoard F., (1999), "Technical change and $CO_2$ emission reduction strategies: from exogenous to endogenous technology in the POLES model", in Le progrès technique face aux défis énergétiques du futur, colloque européen de l'énergie, AEE, Paris, 473-488.
8. Freeman C., (1992), The Economics of hope : essays on technical change, economic growth and the environment, Pinter, New York, 243 p.
9. Jochem E, Madlenee R., (2003), "The forgotten benefits of climate change mitigation: innovation, technological leapfrogging employment, and sustainable development", OECD, ENV/EPOC/GSP(2003)16/FINAL.
10. Weisacker E., Lovins A.B., Lovins L.H, (1997), Factor four : doubling wealth, halving ressource use, Earthscan, London.
11. Cline W.R, (2004), "Meeting the challenge of global warming", Copenhagen Consensus challenge paper, Environmental Assessment Institute.
12. Newell R., Pizer W., (2001), "Discounting the benefits of climate change mitigation", Pew Center for Global Climate Change, Arlington, USA. http://www.rff.org/Documents/RFF-Resources-146-discount.pdf
13. Hurdelbrink R.J., (1991), "Implications of materials shifts", the New Material Society, Vol. 3, US Governement Printing office, Ch 6.
14. Labys, W., (2001), "Transmaterialisation", Handbook of industrial ecology, Edward Elgar.
15. Adriaanse A., et al., (1997), Resources flows : the material basis of industrial economies, World Ressources Institute, Washington.
16. Bringezu S., (2003), "Industrial ecology and material flow analysis: basic concepts, policy relevance and some case studies", in Bourg, D. and Erkman, S. (eds.), Perspectives on Industrial Ecology, Greenleaf, Sheffield; UK.
17. Jochem E., Spreng D., Favrat D., Von Rohr R.P., Hungerbühler K., Wokaun A., Zimmermann M., (2002), "Steps towards a 2000 Watt Society. a White Paper", Input paper to an international workshop on September 2002, Zurich.
18. Ehrlich P.R., Holdren J.P., (1971), "Impact of Population Growth." Science 171, pp. 1212-17
19. Holmberg J., Azar C., Sten K. , (2002), "Decoupling. Past trends and prospects for the future", Environmental Advisory Council Ministry of the Environment of Sweden, Chalmers and Göteborg University.
20. programme ETAP http://europa.eu.int/comm/environment/etap/ or EU Steel Technological Platforms and ULCOS (Ultra Low $CO_2$ Steelmaking) program.
21. Worrell E., Levine M.D., Price L.K., Martin N.C., Van den Broek R., Blok K., (1996), "Efficient Use of Energy and Materials: Progress and Policies", a report by the Secretary-General of the United Nations, vol. 3, n° 20.
22. Pacala S. and Socolow R., (2004), "Stabilization wedges: Solving the carbon problem with technologies available today." Science, n° 305, pp. 968-972.
23. Okken P.A., and Ybema J.R., (1995), "Three divergent scenarios with 80% $CO_2$ reduction", in Energy technologies to reduce $CO_2$ emissions in Europe : prospects, competition, synergy, IEA/OECD, Paris.

24. Ayres R.U., Ayres L.W., (1996), Industrial Ecology : Towards closing the materials cycle, Edward Elgar, Cheltenham, UK / Brookfield, USA.
25. Vellinga P. and Herb N. (eds.), (1999), "Industrial Transformation Science Plan", report, International Human Dimensions Programme (IHDP), Bonn.
26. Gianrini, O., and Stahel, W.R., (1989/1993) The Limits to Certainty, facing risks in the new Service Economy, Kluwer Academic Publishers.
27. European Commission, (2001), "Sustainable production : challenges and objectives for EU research policy", EUR 19880, Brussels.
28. Ehrenfeld J., (2001), "Designing 'Sustainable' Product/Service Systems", Presented at Ecodesign 2001, Tokyo, December 15.
29. Becker G.S., (1962), "Irrational behaviour and economic theory", Journal of Political Economy, vol. 70,pp. 1-3.
30. Stahel W.R., (1995), "Product stewardship : increased competitiveness due to a higher resource productivity and a system design", ISWA Conf., Vienna.
31. Stahel W.R., (2003), "The Functional society: the service economy", in Bourg D. and Erkman S. (eds.), Perspectives on industrial ecology, with a foreword by Jacques Chirac, President of France, March, 384 p.
32. Manzini E., (1996), "Sustainable product services development, Pioneer industries on sustainable service", in Workshop organised by UNEP-WG-SPD in the INES Conference, 22-25 August, Amsterdam, Netherlands.
33. Montt O., (1999), "PSS : shifting corporate focus from selling products to selling product service system, a new approach to sustainable development", AFR report 285, Lund University.
34. Tukker A., Tischner U., (2003), "First report of PSS review", Suspronet Report.
35. Oosterhuis F., Rubik F., Scholl G., (1996), Product policy in Europe : new environmental perspectives. Kluwer, London.
36. White A.L., Stoughton M., and Felg L., (1999), "Servicizing : the quiet transformation to extended producer responsibility", Tellus Institute, Boston.
37. Behrendt S., Jasch C., Kortman J., Hrauda G., Pfitzner R. and Velte D., (2003), Eco-service development: reinventing supply and demand in the European Union, Greenleaf Sheffield.

# Assessing product life cycle strategies in the Japanese market

Y. Ogushi, M. Kandlikar and H. Dowlatabadi
*Institute for Resources, Environment and Sustainability, Liu Institute for Global Issues, University of British Columbia, 229-1855 West Mall, Vancouver, BC, Canada V6T 1Z2, mkandlikar@sdri.ubc.ca*

Abstract: It is becoming increasingly clear that in order to reduce lifecycle impacts, manufacturers need to go beyond product design considerations alone. Important factors to be considered also include nontechnical factors such as business models, market attributes and consumer behaviour. This paper provides a comparative case study of five categories of products in the Japanese market—photocopiers, household appliances, disposable cameras, personal computers and automobiles—and shows how these factors affect lifecycle outcomes. Based on the case studies, we draw general conclusions about the design of appropriate lifecycle strategies.

Keywords: Lifecycle management and design, Reuse/Remanufacturing, Consumer products

## 1 Introduction

In recent years there has been an emerging focus on post use recovery options for consumer goods. The many reasons for the rising interest in recovery include economic benefits for manufacturing firms and consumers; an increased demand for greener products; and legislation on Extended Producer Responsibility (EPR) in Europe, East Asia, and the United States that requires manufacturers to be responsible for appropriate recovery or disposal of manufactured goods. With the passage of the recent recycling laws, Japan, a global centre for innovation in manufactured consumer products, is now at the forefront of innovation in recovery options. Technical factors such as product and system characteristics and design are important determinants of lifecycle strategies [1]. Also, product factors such as the dynamics of value of a product are important factors to consider when formulating a lifecycle strategy [2]. Nevertheless, as manufacturers



*D. Brissaud et al. (eds.), Innovation in Life Cycle Engineering and Sustainable Development,* 49–64.

attempt to reduce lifecycle impacts of products, it is becoming clear that they need to go beyond product and system design considerations alone. Strategies that minimise lifecycle impacts require careful consideration of nontechnical factors such as business models, consumer attitudes, and market attributes. It is the complex interaction of these factors that leads to different post use outcomes, which in turn help in the formulation of strategies that maximise the effectiveness of post use recovery. For example, some products are more amenable to a strategy of remanufacturing and reuse, while other products may be more easily recycled. This paper examines the 'lifecycle strategies' for consumer products in the Japanese market, and provides a framework for understanding why post use recovery of consumer products differs from product to product. More specifically the paper will:

- Develop case studies of five consumer products in the Japanese market.
- Use the case studies to explore how technical and nontechnical factors conspire to produce different post use outcomes.
- Develop an understanding of the general conditions under which different and appropriate lifecycle strategies emerge for different products.

## 2 Lifecycle options: technical vs. non-technical factors

Figure 1 provides a representation of the hierarchy of post use recovery options [3]. The higher up the process in the hierarchy the more environmentally friendly it is. For example, the performance, reliability and appearance of a remanufactured item are as good as 'new', but reprocessing it may take far less energy and inputs than a newly manufactured good. Recycling involves recovering the material value of the product but destroys its function. Thus, remanufacturing generally has a lower environmental impact than recycling, although a case-by-case analysis may be required to evaluate the precise amount of this difference.

**(Most environmentally preferable)**
Reuse components / refurbish assemblies
Remanufacture
Recycle materials
Incinerate for energy recovery
Disposal as waste
**(Least environmentally preferable)**

**Fig. 1.** Hierarchy of post consumption lifecycle options

The hierarchy appears to be robust for many applications [4], and is borne out by lifecycle assessment studies which demonstrate that reuse tends to have a lower environmental impact than disposal [5], although there is a need for carefully evaluating increased demand for new goods by growth of the secondhand market [5] and optimum number of times of re-use [6]. However, as we demonstrate in this work, recovery options for different products depend on variety of factors specific to the technical design, manufacturing and consumer use of a product. Consequently, appropriate post use recovery outcomes for different products may not follow the simple hierarchy. Rather the appropriate outcome might emerge from a lifecycle strategy that combines technical and nontechnical aspects. A lifecycle strategy would integrate understanding of technical aspects such as product design with broader aspects including nontechnical ones such as business models and systems for post use collection and recovery [7]. Formulating business strategies is upstream of the process of formulating a lifecycle strategy. This includes determining the basic concepts of the product, target users, model of service delivery (e.g. sale or lease), the volume of production and so on. Lifecycle options need to be evaluated to meet the defined objectives by business strategy. Lifecycle options include not only choices made with regards to the post consumption stage such as 'reuse' and 'recycle' but also those that influence other life stages such as design change to prolong life and repair of a product. Table 1 shows lifecycle options categorised by product stages [7]. The term cascading implies that recovered parts or material are used for products other than the original. For example, using an engine from a used car for a motorboat is a cascading reuse, while using recycled plastic from plastic containers for plant pots is cascading recycle.

**Table 1.** Lifecycle options of products

<table>
<tr><th>Product stages</th><th colspan="3">Lifecycle options</th></tr>
<tr><td rowspan="6">Pre consumption stage</td><td>Business strategy change</td><td colspan="2">Model of service delivery (e.g. sale to lease)</td></tr>
<tr><td rowspan="5">Design change</td><td>Reuse</td><td>Reusable design (e.g. design for easy disassembly)</td></tr>
<tr><td>Recycle</td><td>Recyclable design (e.g. use of recyclable materials)</td></tr>
<tr><td colspan="2">Longer life design (e.g. selection of materials)</td></tr>
<tr><td colspan="2">Upgradeable design</td></tr>
<tr><td colspan="2">Repairable design</td></tr>
</table>

| | | | | |
|---|---|---|---|---|
| | Manufacturing | Reuse | Use reused parts | |
| | | Recycle | Use recycled material | |
| **Consumption stage** | Continuation of usage | Maintenance | | |
| | | Repair | | |
| | | Upgrading | | |
| | Transfer of ownership | Resale to second hand market | | |
| **Post consumption stage** | Recovery | Reuse | Reusing parts/units | Closed loop (remanufacturing) |
| | | | | Spare parts |
| | | | | Cascading |
| | | Recycle | Material recycle | Closed loop |
| | | | | Cascading |
| | | | Chemical recycle | Closed loop |
| | | | | Cascading |
| | Disposal | Heat recovery | | |
| | | Simple disposal | Incineration | |
| | | | Landfill | |
| | | | Other | |

## 3 Case studies

In order to capture the full spectrum of factors influencing lifecycle outcomes, we examine five different product case studies for copiers, home appliances, computers, disposable cameras and automobiles. Data on case studies was collected through literature review (primarily from companies' websites) and the results of an electronic survey sent to five copier manufacturers and five home appliance manufacturers.

### 3.1 Copiers

Copiers are a good example of a successful closed-loop product lifecycle. The central strategy for recovering copiers is remanufacturing. Remanufacturing is the process of restoring a nonfunctioning, discarded, or traded-in product to nearly new condition, giving the product a second (or third, or fourth) life [8]. Manufacturers now ship copiers that consist of reused parts at a high rate. For example, five types of Ricoh's remanufactured copiers use 87% (weight basis) of reused parts [9]. Other major manufacturers also use reused parts in their new products at more than 70% [10–12]. Although remanufacturing is a primary focus, material recycling is also carried out for the parts that are hard to reuse. The materials for such parts are either closed-loop recycled into the same parts or down cycled into different parts or products [13]. Manufacturers report the recycling rates of copiers—defined as the reused and recycled mass divided by the collected mass—exceeds 90% [14–16]. The factors that lead to remanufacturing are described below.

### Business characteristics

A majority of copiers are leased rather than sold to business users. Although leasing of copiers started as a marketing tool [17], this business model gives manufacturers a substantial advantage to remanufacture their products. First, manufacturers can recover their products through lease returns without much effort. Once collection routes for leased products are established, there is little chance for used products to 'leak out' of the loop. 97% of products are recovered through leasing companies [18]. Lease period is usually three or five years. Second, manufacturers can plan product design ahead because they know the timing and quantity of products coming back to manufacturers. In response to our survey, one manufacturer makes 'a 10-year plan of product design, identifying which units and parts of which product a future product will have.

### Supporting technologies

Technology needed to support remanufacturing includes the following: disassembly of products; diagnosis of parts so used parts are appropriately reused, repaired or recycled; identification of the remaining lifetime of parts; and cleaning used parts. Manufacturers are striving to develop these technologies [10, 19–20]. Cost incentives may be a reason for developing technologies to promote remanufacturing. Copier manufacturers have adopted 'environmental accounting' systems, which calculate costs for improving the environment and the benefits derived from such improvements. Expenses include research and development cost, costs for recycling, energy saving and antipollution measures, and labour cost related to the improvement of the environment. Benefits include saved energy and material of the company and users, and improved profitability. Although they do not have a separate account for remanufacturing, the total benefits exceed total costs [21–22]. There is likely to be some cost incentives for the development of technologies and systems for remanufacturing. Copier manufacturers have come to be regarded as environmentally conscious companies. In an annual survey on environmental performance conducted by Nihon Keizai Shimbun (a Japanese business newspaper) in 2003, Canon was ranked first, Fuji Xerox on the fourth and Ricoh on the sixth among 599 manufacturing companies [23]. These companies are regularly competing on the top list of this survey. Remanufacturing of copiers is one of the ways by which companies can improve their environmental performance, and these companies are striving to improve recycling rate for better performance. Competition over environmental performance to enhance the manufacturers' public image may also be a driver for research and development.

#### Product attributes

A copier is outdated and replaced at the end of the lease period (three to five years) or even before, while parts are still usable at the time of replacement. In other words, the physical lifetime of parts is longer than the value lifetime of a product, and this allows manufacturers to remanufacture products using reused parts. Manufacturers now attempt to make parts last longer to enable remanufacturing [24].

#### User attributes

There are two major conditions among user attributes for the closed-loop lifecycle scenario—user acceptance and predictable use. Users have good acceptance of remanufactured products, as long as the quality of a product is assured. In particular, government offices are willing to accept remanufactured products, according to one manufacturer. Most users are business users, and their usage patterns are predictable. In other words, most products are used in a manner that is within the range of manufacturers' assumptions about usage patterns.

### 3.2 Home appliances (designated four appliances)

Home electric appliances are products whose lifecycles are likely to be greatly influenced by regulation. The Home Appliance Recycling Law, enforced in 2001, set the recycling schemes of four designated appliances including fridges, air conditioners, televisions and washing machine. Under this law, manufacturers are responsible for recovering used products. Although used products are more likely to return to manufacturers after the implementation of the law, the central recovery strategy appears to be aimed at material recycling, not reuse or remanufacturing unlike copiers. For example, used televisions are disassembled at a manufacturers' facility and then recyclable resources are carefully recovered manually or using high-tech machine. Panel glasses and funnel glasses (containing lead) are separated, crushed into cullet and recycled into new glasses for new televisions [25]. Metal parts go to steel manufacturers and refinery to be recycled into new material. Electric circuit boards go to refinery to recover precious metals. The following factors are contributing to the formulation of material recycling lifecycle strategy evidenced in household appliances.

#### Business characteristics

Most home appliances are sold rather than leased. Once the ownership is transferred to users, it is impossible to control the usage; manufacturers have no way of knowing when a user disposes a product, and whether usage patterns of a product are as expected.

#### Supporting technologies

Material recycling requires the technologies of disassembly and separation. Manufacturers focus on research and development of such technologies at their recycling centres. For example, Matsushita Eco Technology Center (METEC), which started its operation at the same time as the enforcement of the Home Appliance Recycling Law, filed 68 patent applications including the high speed separation of CRTs, normal temperature crushing of compressors, high grade separation of copper and aluminium from heat exchangers, and the whirlpool sorting of blended plastics [26]. All these technologies exemplified above are separation and recovery technologies, not those that facilitate reuse and remanufacturing.

#### Product attributes

The average period of usage of regulated appliances is nine to eleven years [27]. Technological changes take place for this period to the extent that parts from used products are no longer reusable for new products, according to one manufacturer. For example, CRT displays recovered from used televisions are no longer useful for rebuilding televisions using recent technologies such as liquid crystal and plasma display panels. In addition, the long lifetime of products and rapid technological changes lead to heterogeneity of collected parts, which makes reusing parts difficult.

#### System attributes

Major appliance manufacturers have constructed facilities dedicated for recycling of products by the implementation of The Home Appliance Recycling Law. The law also defines collection routes; it obligates retailers to collect used products and return them to manufacturers.

### User attributes

Failure is the most common reason for consumers disposing their appliances—the rate of consumers who replace appliances due to breakdown is 60% for fridges, 69% for air conditioners, 72% for televisions and 76% for washing machine [28]. According to one manufacturer, the remaining lifetime of parts in used products is not long enough to be reused when a product is disposed. Users do not necessarily use products in a predictable manner, making it difficult to forecast the remaining lifetime of parts.

### Other lifecycle scenarios

Other emerging lifecycle strategies after the implementation of the law include prolonging lifetime of products by rental business and through second hand market. Some manufacturers and large retail stores started rental business of appliances targeting short term users such as students and workers away from home for a limited period. The Home Appliances Recycling Law gives an incentive to rental businesses because it forces consumers to pay recycling expenses at the time of disposal. (2,520–4,830 yen [29], US$23–44 at the exchange rate of 1US$=110yen) Rental users are exempted from paying such recycling fee in addition to paying less than owning. Rental users also get the benefit of not having to repair rental products; Toshiba Techno Network provides free repair during rental period, using their existing service network. Rented products are expected to be inspected and repaired, and to flow into second hand market after the rental period [30]. The scale of rental market is still insignificant compared with the sales market. The number of rental appliances is estimated in the order of tens of thousands [31], while the domestic shipment of these four appliances is four to eight million units per year [32]. Consumers' obligation to pay the recycling fee also provides an incentive to the second hand market, but the scale of the second hand market is still insignificant. The number of sales of second hand washing machine was approximately 160,000 in 2001, while that of new ones was 4.06 million [32–33].

## 3.3 Computers

The value of computers declines quickly despite its long physical lifetime. This means that consumers may dispose or replace a product far earlier than its physical lifetime is reached. On the other hand, the demands for second hand computers are steadily increasing. In 2001, the market scale of second hand computers was 830 thousand units (about seven percent of new products) and rising [34]. Reusing computers is becoming a popular lifecycle option. Cleaning, deleting data in storage devices and re-installing operating system software are the normal procedures before a used computer is resold. Even if a computer can be reused, it will lose its value after a certain period of time as current software is no longer compatible with its capabilities or it reaches its physical lifetime, and needs to be disposed. The legislation requires that used computers to be collected and recycled by manufacturers. Used computers are first disassembled by hand into their constituent parts [35]. Reusable parts (e.g. HDD, power unit etc.) are recovered at this stage, followed by inspection. Remaining parts are crushed and separated into materials such as glasses, plastics and metals before being handed over to recyclers. Material recycling is the major method used for recycling crushed computers. In sum, the lifecycle strategy of computers can be said to be a 'cascading.' A product is reused while it has use value, but it is materially recycled once it loses use value or reaches physical lifetime. In other words, the ownership of a computer is transferred in such a way that users' demand and the performance of the computer match until the computer loses use value. Once a computer is outdated to the extent that no user wants to use it, it is materially recycled. A case study by IBM Asset Recovery Center shows that product resale value significantly drops in four years and levels off afterwards, while parts value and material value do not follow this trend [36]. The current trend of cascading lifecycle scenario is likely to be consistent with this study. The following factors are relevant to cascading lifecycle strategy, in particular, the transfer of ownership within the value lifetime of a product.

### Product attributes

In general, the design of computers has little flexibility of reusing parts. Reusing computer parts are limited to certain modules and units, while reusing products is more widespread. In the past, computer hardware was rapidly outdated as new software requiring higher hardware specification was developed. More recently hardware has been so advanced that average users have not needed to upgrade their computers in order to take full advantage of newly developed software. A two-year old computer can sufficiently serve such users. The dynamics of hardware and software advancement led to an increase in demand for second hand products [37].

### System attributes

The legislation is the major driver of recycling postuse computers that have no market value. The Law for Promotion of Effective Utilisation of Resources set the scheme of recycling used computers from business users and personal users. The law defines the responsibility as follows:

manufacturers are responsible for collecting and recycling used computers, and users pay the expense for recycling.

### Market attributes

There is a strong second hand market for used computers, as mentioned above. The legislation defines users' responsibility for paying recycling expenses. The typical recycling fee is 3,150yen (US$29) for a desktop and laptop computer and 4,200yen (US$38) for a CRT monitor [35]. These fees are included in the price of products sold after the legislation enforcement in October 2003, but not included in the price of products sold before. (This only applies to personal users.) Users pay the fee at the time of disposal if the fee is not prepaid. Therefore, the obligation to pay the recycling fee at disposal works as an incentive for users to sell a product to the second hand market as long as a product has a market value.

## 3.4 Disposable cameras

Disposable cameras were first introduced in the mid 1980s and have been widely available for over a decade. When they were first released in 1986, there was no recycling system and they tended to be disposed after a single usage. Four years later, owing to consumer pressure, manufacturers started to build recycling systems and renamed their products as 'film with lens', emphasising that the product was no longer disposable [38]. Now, more than 90% of its parts are reusable and the recycling rate (defined in 3.1) exceeds 95%, according to the two major manufacturers [39–40]. The lifecycle strategy of disposable cameras is remanufacturing which is very similar to that of copiers. Products are recovered from customers through camera shops and dealers. After being classified, a disposable camera is disassembled into modules and parts. These modules and parts (e.g. flash module, shutter unit, battery etc.) are inspected and reused in new products. Covers are materially recycled into the same products or downcycled into other products. Manufacturers strive to improve recyclability at design stage; for example, only recyclable materials are chosen and modular design of component parts are common. The major factors that contribute to the remanufacturing of disposable cameras are as follows.

### Supporting technologies

Manufacturers developed their own recycling facilities where used products are classified, disassembled, tested, and repaired if necessary. The technology seems to be so matured that over 90% of parts are reused.

### Product attributes

The time of usage is short. Theoretically, it is not likely that parts reach their physical lifetime before the products are recovered. According to the experiment conducted by Hokkaido University, a battery in a disposable camera had the capacity to power the flash more than 200 times, while the number of exposure is typically less than 40 [38].

### System attributes

A buyer is expected to bring the product to a shop for film developing. The nature of the product makes it easy for manufacturers to collect used products. With regard to the recycling system, when disposable cameras were first released, recycling system was not in place. Around 1990, public and media regarded disposable cameras as a symbol of disposable age, and criticised the wasteful nature of the product. Critical public opinion helped to redirect manufacturers toward recycling and reuse [38]. Three major manufacturers collectively started to build recycling systems in 1991.

### User attributes

Users accept remanufactured products as long as quality is assured.

## 3.5 Automobiles

The lifecycle of automobiles is far more diversified than the products discussed above. The lifecycle scenarios include the reuse of products through the second hand market, reuse of parts, material recycling and prolonging life by maintenance and repair. On the other hand, closed-loop recycling is not a popular lifecycle option for automobiles. The lifecycle scenarios are similar to those of computers in that reuse before reaching lifetime and material recycling at the end of life are the main scenarios, but largely different in product lifetime and reuse of parts. The existing recovery system, market and regulation play a big role in determining the lifecycle scenarios of automobiles. The following is the outline of how the existing system, market and regulation affect the lifecycle.

### Existing system and market

There had been no regulation with regard to recycling end-of-life vehicles (ELVs) until very recently. This means that there had not been unified measures to recover ELVs. Historically, recovery of used parts from and recycling of ELVs were totally an emergent property of the market for such material. ELVs are usually handed to dismantlers through dealers, insurance companies and garages. Although dismantlers play an important role in recycling ELVs, the details of dismantler activities are not well known [41]. It is said that there are around 5,000 dismantlers in Japan handling about four million ELVs every year. Dismantlers vary in origin, size, facility and scope of work, and there is no standard business model of a dismantler. In general, dismantlers have the following main functions: [41]

- Collect residual oil, gas and CFCs from ELVs.
- Remove reusable (marketable) parts from ELVs. Parts that have no market value are left as scrap, to be shredded at a later stage.
- Clean, repair and rebuild (wherever necessary) used parts before sale. ('Production' of used parts.)
- Separate valuable metals and sell them to steel refiners and smelters.
- Hand over scraps to shredding factories. Under the current market condition, dismantlers pay about 7,000yen (US$64) per tonne to shredding factories.

The business of dismantlers is vulnerable to market conditions. The main source of profit used to be scrap metals sold to steel refinery and smelters, but due to the plummeting price of metal scraps in the 1980s, the main source of profit shifted to used car parts sold to users and garages [41]. Parts are demanded from users with a damaged vehicle or one with a failed component. Each dismantler develops a marketing strategy of used parts. (e.g. establishing brand name, networking with other dismantlers to extend marketable parts, specialising in certain types of parts etc.) The money flow from dismantlers to shredding factories also started in 1980s when the expense for final disposal increased due to the scarcity of available land. In this way, the market largely defines the business of dismantlers. This indicates that the lifecycle scenario of vehicles largely depends on market conditions. For example, the demand for used parts determines the level of disassembly for a given make and model of the ELV, regardless of reusability of parts. On the other hand, manufacturers of vehicles have had little participation in the historic recycling process of ELVs. Without the participation of manufacturers, closed-loop systems such as those of copiers are not likely to take place because considerations need to be given at design and production stages.

### Regulation

New regulation is coming into force in recycling of ELVs. The End-of-Life Vehicles Recycling Law was implemented in January 2005. It is aimed at the reduction of automobile shredder residue (ASR) that goes to scarce landfill sites. Under this law, manufacturers and importers are responsible for collecting and recycling ASR and airbags, and collecting and destroying CFCs. Expenses for recycling ASR and airbags are paid for by car owners at the time of purchasing new cars or at the time of inspection of cars purchased before the implementation of the law. The target recycling rate (including thermal recycling) is 95% by 2015, while the recycling rate before the enforcement of the law is around 80% [42]. Although the End-of-Life Vehicles Recycling Law adopts the EPR principle, its design assumes taking advantage of the existing infrastructure. Manufacturers are only responsible for recycling ASR and airbags, and destroying CFCs. The remaining of the recycling system will be left to the existing mechanism and the market. Closed-loop systems are unlikely under such circumstances.

# 4 Summary and conclusion

## 4.1 Interactions of factors on lifecycle outcomes

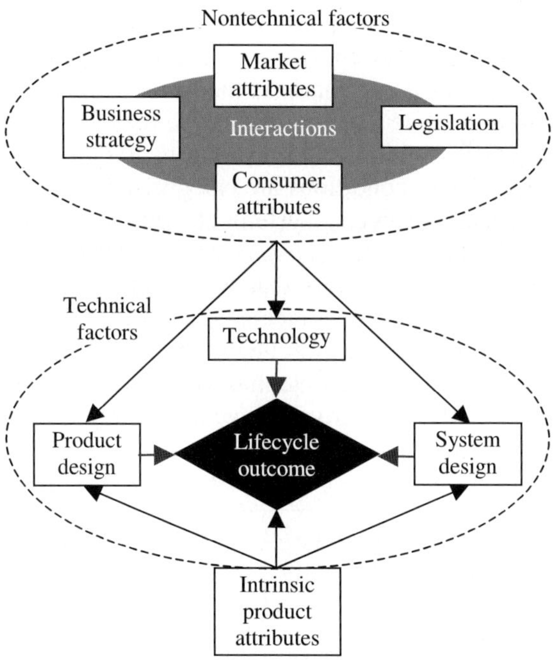


**Fig. 2.** Both technical and nontechnical factors affect lifecycle outcomes

Figure 2 shows how lifecycle outcomes are influenced by technical and nontechnical factors. It is well known that lifecycle outcomes are determined by technical factors such as product design, system design and supporting technologies – usually reflected in design. As demonstrated in the case studies, these technical factors are influenced by exogenous nontechnical factors such as: business strategy, legislation, consumer behaviour and market attributes. The pattern of interactions and the magnitude of the influence of these nontechnical factors vary depending on products. In addition, intrinsic product attributes are also reflected in design, or in some cases, constrain it. For example, a disposable camera is destined to a film developer mainly by an intrinsic attribute of the product rather than design, and this attribute is considered and reflected in product and system design.

## 4.2 Summary of case studies

Table 2 summarises conditions and driving factors of different lifecycle outcomes derived from case studies of recycling practices involving five products. Lifecycle strategies were categorised, and the products investigated in the case studies fit in the following categories: Closed-loop manufacturing: copiers and disposable cameras, Cascade recycling: computers and automobiles, Simple material recycling: appliances.

Note that there is an overlap between cascade recycling and simple material recycling; end-of-life scenario is the same while the transfer of ownership occurs within value lifetime in cascade recycling. The table summarises major lifecycle options and conditions required to achieve each lifecycle strategy. These conditions are categorised into four major areas—technology, product attributes, system attributes and market attributes. Dots indicate the condition is necessary to form such a lifecycle scenario. Table 2 also shows major drivers that were identified in the case studies. Most of these drivers are nontechnical and amenable to direct management through regulation and business strategy. Overall, the table indicates how management practices have the potential to be harnessed in management of product design to achieve superior lifecycle outcomes.

**Table 2.** Summary of case studies of five products

| | **Closed-loop manufacturing** | **Material recycling** | | **Major possible drivers** | |
|---|---|---|---|---|---|
| | | **Cascade** | **Simple** | **Drivers** | **Objective** |
| **Product examples** | Copiers<br>Disposable cameras | Computers<br>Automobiles | Appliances | | |
| **Major lifecycle options** | | | | | |
| Remanufacturing | • | | | | |
| Product reuse | | • | | | |

| | | | | | |
|---|---|---|---|---|---|
| Parts reuse | | • | | | |
| Material recycling | | • | • | | |
| **Technology** | | | | | |
| Disassembly | • | • | • | R&D | Compliance to regulation<br>Cost incentive<br>Competition |
| Separation | | • | • | | |
| Diagnosis | • | • | | | |
| Cleaning | • | • | | | |
| **Product attributes** | | | | | |
| Designed for disassembly, separable products, units and parts | • | • | • | R&D | Compliance to regulation<br>Cost incentive<br>Competition |
| Physical lifetime is longer than value lifetime of a single user | • | • | | R&D | Compliance to regulation<br>Cost incentive<br>Competition |
| **System attributes** | | | | | |
| Collection and recovery system | • | • | • | Regulation<br>Business strategy | Less environmental impacts |
| Controlled ownership | • | | | Business strategy | |
| **Maket attributes** | | | | | |
| Highly managed usage | • | | | Business strategy | |
| User acceptance of recovered products | • | | | | |
| Strong second hand market | | • | | Regulation (user cost) | Less environmental impacts |
| Demand for recycled materials | | • | • | Market conditions | |

## References


1. Rose, C.M., Ishii, K. and Masui, K., 1998, How Product Characteristics Determine End-of-Life Strategies, http://mml.stanford.edu/.
2. Kato, S. and Kimura, F., 2003, Typology of Recycling Business Using Lifecycle Costing (in Japanese), The Japan Society of Precision Engineering 2003 Spring Semestral Conference Proceedings p.202.
3. Graedel, T.E. and Allenby, B.R., 2003, Industial Ecology Second Edition, Education Inc., Pearson, New Jersey.
4. Faruk, A.C., Lamming, R.C., Cousins, P.D., Frances, E.B., 2002, Analysing, Mapping and Managing Env. Impacts along Supply Chains, J. of Ind. Ecology, 5/2:13-36

5. Thomas, V.M., 2003, Demand and Dematerialisation Impacts of Second-Hand Markets, Journal of Industrial Ecology 7/2:65-78.
6. Okamura, S., Morikuni, T. and Okino, N., 2003, Environmental effects of physical life span of a reusable unit following functional and physical failures in a remanufacturing system, Int. J. of Production Research, 41/16:3667-3687.
7. Manufacturing Science and Technology Centre, Inverse Manufacturing and Lifecycle design (in Japanese), www.mstc.or.jp.
8. Hauser, W. and Lund, R.T., 2003, The Remanufacturing Industry: Anatomy of a Giant, Department of Manufacturing Engineering, Boston.
9. Ricoh website (a), www.ricoh.co.jp/ecology/product/09.html.
10. Fuji Xerox (a), www.fujixerox.co.jp/release/2003/1002_dc507md.html.
11. Canon website (a), http://web.canon.jp/ecology/eco2004/p17.html#p17.
12. Sharp website, www.sharp.co.jp/corporate/rd/journal-73/pdf/73-10.pdf.
13. Ricoh website (b), www.ricoh.co.jp/ecology/history/Presen/PRE4.html.
14. Ricoh website (c), www.ricoh.co.jp/ecology/report/pdf2004/43-47.pdf.
15. Canon website (c), http://web.canon.jp/ecology/report/eco2003.pdf.
16. Fuji Xerox website (b), www.fujixerox.co.jp/release/2000/0914_haiki.html.
17. Fishbein, B.K., McGarry, L.S. and Dillon, P.S., 2000, Leasing: A Step Toward Producer Responsibility, INFORM, New York.
18. METI (the Ministry of Economy, Trade and Industry) website, www.meti.go.jp/report/downloadfiles/g11026g24j/pdf.
19. Canon website (c), http://web.canon.jp/ecology/eco2003/p59.html.
20. Ricoh website (d), www.ricoh.co.jp/ecology/greenreport/05/003.html.
21. Ricoh website (e), www.ricoh.co.jp/ecology/account/index2003.html.
22. Canon website (d), http://web.canon.jp/ecology/eco2004/p51.html.
23. Nikkei website, www.nikkei-r.co.jp/kankyo/2003/seizo/s_ranking.htm.
24. Canon website (e), http://web.canon.jp/ecology/eco2004/p31.html.
25. METEC website (a), http://panasonic.co.jp/eco/metec/index.html.
26. METEC website (b), http://panasonic.co.jp/eco/metec/qa.html.
27. Economic and Social Research Institute, Cabinet Office, Government of Japan website (a), www.esri.cao.go.jp/jp/stat/menu.html#shohi.
28. Economic and Social Research Institute, Cabinet Office, Government of Japan website (b), www.esri.cao.go.jp/jp/stat/shouhi/0403shouhi.html.
29. Association for Electric Home Appliances website, www.rkc.aeha.or.jp/.
30. Environmental GOO, http://eco.goo.ne.jp/offer/files/rensai/re2/0203.html.
31. EIC net website, www.eic.or.jp/library/pickup/pu030417.html.
32. www.env.go.jp/doc/toukei/contents/es15/es150335.xls.
33. IBPC (International Business Promotion Center) Osaka Network Center, www.ibpcosaka.or.jp/network/trade_japanesemarket/others/used_goods03.
34. MIC Research Institute Ltd. website, www.mic-r.co.jp.
35. Personal Computer 3R Promotion Centre website, www.pc3r.jp/home.html.
36. Grenchus, E., Keene, R. and Nobs, C., 2000, Composition and Value of Returned Consumer and Industrial Information Technology Equipment, Proceedings of the IEEE International Symposium.
37. PC view website, www.pc-view.net/Outsourcing/031031/.
38. Hokkaido University website, www.hps.hokudai.ac.jp/Education/4-4.html.
39. www.fujifilm.co.jp/corporate/environment/pdf/report/ff_env_2003_012j.pdf.
40. http://konicaminolta.jp/about/environment/activity/report/konica/pdf/02/part12.pdf.
41. Takeuchi, H., 2004, Car Recycling: Now and Future of Vein Industry (in Japanese), Toyo Keizai, Tokyo.
42. Nikkei Ecology, 56, p.41.

# Applications of service engineering methods and tool to industries

T. Sakao, Y. Shimomura, M. Lindahl and E. Sundin
*Mitsubishi Research Institute, Inc./The University of Tokyo, Japan, sakao@ race.u-tokyo.ac.jp, Research into Artifacts, Center for Engineering, The University of Tokyo, Japan, Department of Mechanical Engineering, Linköping University, Sweden*

Abstract: Servicification of economic activities could lead to dematerialization. The authors have been carrying out fundamental researches on a novel engineering paradigm to deal with services, service engineering. This paper demonstrates the effectiveness of service engineering methods and tool to support service design processes through applications to two service examples in industries' real operation. In an application to a hotel service, redesign options were obtained including an introduction of a sun-lighting facility and a cash-back system for not used towels. Thus, the methods and tool were proved to be effective to support designers.

Keywords: Service, Design, CAD, Application

## 1 Introduction

We should reduce the production and consumption volume of artifacts to an adequate, manageable size to solve environmental problems although we do not want to make our quality of life lower than now. Consequently, we must aim at qualitative satisfaction rather than quantitative sufficiency, and thus the decoupling of economic growth from material and energy consumption [1]. This is a new paradigm, which is free from the mass-production paradigm of 20th century. To achieve this paradigm, products should have more values, supplied largely by knowledge and service contents, rather than just materialistic values. This dematerialization of products requires the enrichment of service contents. To this end, we need a novel engineering method to evaluate services and to design the services. This novel engineering is called service engineering, which includes design methodology of both products and services. This novel engineering



*D. Brissaud et al. (eds.), Innovation in Life Cycle Engineering and Sustainable Development,* 65–83.

differs in the definition of value from Value Engineering [2], where value is defined as function over cost. Conventional design methodology [3–4] also deals with only functions of the object rather than satisfaction of the consumers by the functions. This paper aims at demonstrating the applicability and the practicality of the service engineering methods and the tool [5–7] to support service design through two case studies on services in industries' real operation. One is provided by an Italian company in the accommodation industry [8], while the other is by a Swedish manufacturer in the logistics industry. In this paper, Chapter 2 explains problems in current service design and existing researches briefly first. Next, Chapter 3 describes the methods and tool for service engineering. Chapter 4 and 5 explain the results of two applications. Chapter 6 and 7 describe discussions and conclusions.

## 2 Problems in current service design and existing researches

### 2.1 Problems in current service design

Traditional processes of designing services normally consist of analyzing customers, designing adopted products, and market testing [9]. It is engineers that are requested to carry out designing physical products to be utilized in the services after marketers specify the values to be provided by the service. Since a product design process in general consists of conceptual design, embodiment design, and detailed design [4], traditional service design processes as a whole follow the flow depicted Figure 1(a). However the service design methods following this flow have the problems below.

1. Decoupling of working agents. Engineers in charge of designing physical products have difficulties to obtain the enough information on the values to be provided to the final customers, since the results from analyzing customers are not fully transferred from different agents, namely marketers.
2. Decoupling between products and services. Engineers normally end up with the activities to identify the physical structures of the product that are expected to play the predefined roles, therefore they often fail to design what the customers really need. Namely, there exists a problem that effective services cannot be designed, because the process for identifying the provided values and that for generating the product functions stand separately.

In order to solve this problem, the flow depicted in the upper half of Figure 1 (b) is requested for the processes to identify the service values and the product functions:

1. Those two processes are achieved by an identical agent. In addition, engineers are supposed to carry this out in service engineering, since it is relatively easier than marketers.
2. Product functions are identified through recursive processes of analyzing customers and conceptual design, so that they will meet customers' requirements.

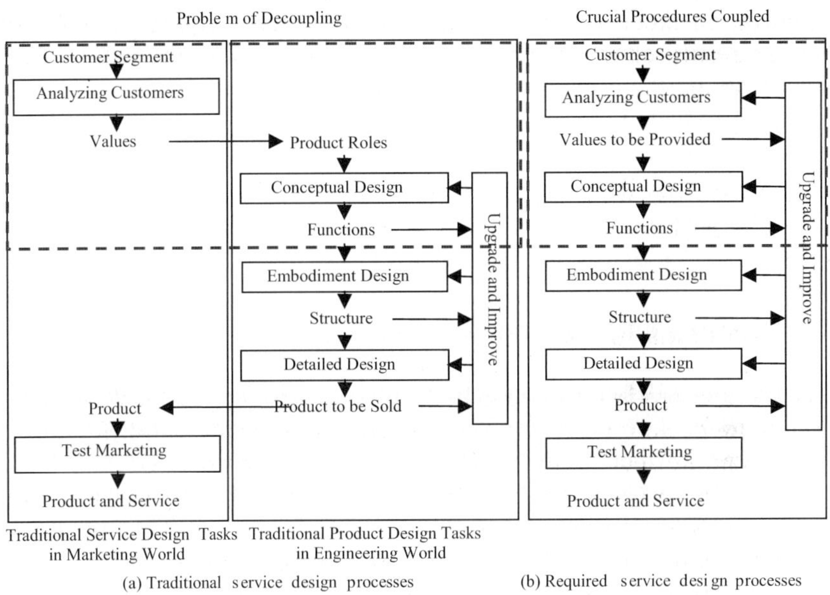


**Fig. 1.** The traditional and required service design process

## 2.2 Existing researches of service design

First, existing researches on service design and development are reviewed. Shostack [10] and Ramaswamy [9] have proposed methods to design and manage services properly mainly for marketers in the service industry. It should be noted that the main designed object is the activities of the working people at the sites. Among them, Ramaswamy discusses the method to design the process after the service constituents are classified into product, facility, and process. Bullinger et al. [11] discuss the ways to manage organizations, human resources, and Information Technology systems after they recognize services as not only designed object but also the objects for Research and Development. Meier [12] proposes a life cycle based service design method in cases that an adopted physical product is given. Next, researches on PSS (product service system), which are relatively new trial activities in the manufacturing sector, are summarized. Most of the existing researches [13-15] have analyzed customer values, economic costs, or environmental impacts through the product life cycles, but have not tackled design issues. Exceptionally, Weber et al. [16] have proposed a scheme to design and develop PSS by adding an element of a customer, who has a functionality to feedback to the PSS requirements from product characteristics in their model of traditional product design. However, differences in details between customers' requirements for PSS and those for product design are not given, nor the design method. As reviewed above there exists no method to design services and associated physical products based on the inputs of customers' requirements for the services in any research in those two different fields above.

# 3 Methods and tool for service engineering

## 3.1 Service engineering

A *service* is defined as an activity that *a provider causes, usually with consideration, a receiver to change from a state to a new state that the receiver desires*, where both contents and a channel are means to realize the service [17]. Hence, a service receiver satisfies with just contents, which are any of material, energy, and information. A service channel is used to transfer, amplify, and control the service contents. In this definition, artifacts can change from contents to channels and vice versa. Artifacts have their own functions, behaviours, and states, and therefore they can be designed with conventional CAD systems. Since both service channels and service contents also have the above-mentioned characteristics, we introduce the similar methodology into the design of services. The term "service" is defined. Then what is *service engineering*? It is a design methodology to increase the value of artifacts and to decrease the load to the environment by reasons of focusing service. Note that service engineering has both analytical and synthetic aspects. In this paper, we discuss both. Service engineering aims at intensifying, improving, and automating this whole framework of service creation, service delivery, and service consumption. To increase the total satisfaction of receivers, we can improve functions and/or quality of both channels and contents. Traditionally, engineering design has aimed to improve only function. A better function of a new product, we believed, makes consumers satisfied. In service engineering, however, not only the functions of artifacts but also meaning of contents must be matched to the specifications given by receivers. Then the satisfaction of receivers increases.

## 3.2 Service modeling

### The outline of a service model

Let us define the design procedure of a set of services. In the argument regarding the design process [18–20], it is widely accepted that design might be a search for a physical structure matching the required function. The design of a service is also a search for both physical and non-physical structure, but it differs from the conventional design in terms of evaluation. Conventional design regards mainly the performance of the channels; it does not consider the state change of the receiver except for a happiness that comes with owning the products. The design of a service is based on the degree of satisfaction with the state change of the receiver. Therefore, it is necessary to express state changes by means of the received contents.

### Receiver's state

Change of a receiver is represented by a set of *receiver state parameters* (RSPs). Thus, a service can be represented by a set of RSPs necessarily and sufficiently. Since a RSP consists of quantitative values, including Boolean logic and multi-value logic, we can compute any comparison between two RSPs. In addition, we introduced a new assumption that all RSPs are observable and controllable. This assumption has been unproven with human receivers because we have not had a reliable method to measure the consumer behavior. The RSPs change by received contents. Hence, in this research, we assume that contents consist of various functions, whose name is Function Name (FN), whose operating objects are Function Parameters (FPs) and whose effect is represented by Function Influences (FIs). Since both channels and contents are described by the functions with FPs and FIs, RSPs also belong to functions. As the receiver's states may change with respect to supply of contents, RSPs can be written as functions of contents. Parameters expressing contents are called *content parameters* (CoPs). In the same way, the parameters of channel, which make the flow of CoPs change and thus influence RSPs indirectly, are called *channel parameters* (ChPs). Hence, in this research, we assume that both contents and channels consist of various functions. These parameters create a network with one another. We studied several examples and chose different sets of parameters; some of the examples are from daily life such as restaurant, travel agents, and laundry. Other studies are from manufacturing sectors such as disposable camera, copiers and elevator maintenance service. The details are not discussed in this paper but we need to point out that the selection of contents within various parameters is subjective. It seems the greatest reason that services have not been dealt with in engineering issues.

### Flow model

One of the typical services is a travel agent, which arranges and purchases various tickets on behalf of customers. Contents are different from the tickets, even if the tickets are delivered to the customers. In this way, services can be delivered through complex multiple structures consisting of various go-betweens. The intermediate agents have double characteristics of a receiver and a provider, which is represented by a symbol of smiling circle "intermediate agent" as illustrated in Figure 4. Providers can be defined as an agent whose receiver side (left half) inside of the agent is hidden by the system, and in the same way receivers do as that whose provider side is hidden. When we focus on the relationship between a receiver and a provider, many intermediate agents exist among them. In this research, we call the sequential chain of both contents and channels 'a flow model of a service'.

### View model

Receiver state parameters (RSPs) change according to how the receiver evaluates subjectively the received contents. An RSP is linked to several CoPs because the receiver evaluates the contents as mentioned in Section 2. The CoPs may be supported by several ChPs existing in the chain of several agents in the flow model. A view model is defined as a tree of CoPs and ChPs with a single RSP at its root. The view model is expressed in a directed graph that consists of nodes representing parameters (CoPs and ChPs) and arks representing their relationships. In the current implementation, both positive and negative influence can be added. An example is shown in Figure 5 and 7. A channel and its contents are expressed by FNs as lexical expressions and FPs. FPs, influencing RSPs directly, are recognized as CoPs, and those connecting indirectly to RSPs are ChPs. The body of the function is expressed in the function influence (FI). The view model illustrated visually the relationships among the parameters (RSP, ChP, and CoP) in the graph of connected nodes. Thus, it helps the improvement of RSP by changing FPs.

### Scope model

Practical services have very complicated structures of intermediate agents, connected to one another on an infinite chain. Therefore, it is necessary to specify the effective range of the service from an initial provider to a final receiver as illustrated in Figure 4. In comparison to the view model in which a single RSP is expressed, the scope model can deal with all RSPs within the providers and the receivers. In other words, a scope model handles multiple view models, namely multiple RSPs. Thus, it helps designers to understand the real activities between the two.

## 3.3 Service explorer

We have developed a prototype system of a computer-aided modelling tool for service design, which is called ***Service Explorer.*** Designers can describe services, and register them in a database. They can operate the service in the following ways:

- Express a service model following the definition mentioned-above.
- Edit the models: reroute arcs among function units, change attributes of function units, and so forth.
- Evaluate the total service by means of assigning each value of the components.
- Search suitable service components in the database such as analogous services and partly-related services.

The current version of Service Explorer was developed using Java (Java2 SDK, Standard Edition 1.4.1) and XML version1.0 in the Microsoft Windows XP Home Edition environment.

### Model building

To model services on the Service Explorer, we register services in the database as scope models, and edit, evaluate, search services. A typical procedure is as follows:

- Define various RSPs and then express each view model of RSP.
- Generate a scope model by means of binding multiple view models of a receiver and of selecting their initial providers.
- Edit function parameters in each view model, and change the arcs between function units.

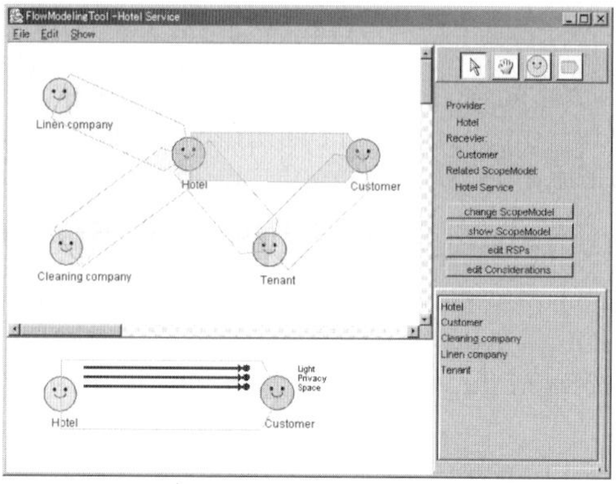


**Fig. 2.** A Screenshot of a scope model panel

Note that a service is a scope model and that it consists of the following elements: (a) provider and receiver of the service, (b) a set of the receiver's RSPs, (c) the structures of the functions. On editing a scope model, we can use an integrated interface of Service Explorer, where (a) and (b) can be edited on the scope model panel (Figure 2) and (c) on the view model panel (Figure 3). We can simply switch these panels on the same window.

### Model evaluation

It is one of the greatest problems in Service Explorer to determine the influence weight of each parameter to RSPs. Thus, we also implemented 'service evaluation module' based on AHP (Analytic Hierarchy Process) [21] and QFD (Quality Function Deployment) [22], and performed on a scope model [23].

- The influence weights of RSPs are computed numerically by AHP method according to bilateral comparisons between parameters.
- The importance weight of function parameter, considering the weights of RSPs previously obtained. That is, which function parameter, we consider, is more important for satisfying the receiver's RSPs. These weights are given as numerical value by QFD method.

A service receiver judges a service by unifying several criteria of many different providers in the following way: (1) make the structure of a service as a chain of agents that exist in the service; (2) determine the sub-services as scope models, which include each provider; (3) edit and adjust the structure and function parameters so as to improve RSPs; (4) clarify the whole structure of a service by editing the view models and reunifying the scope models. A unified evaluation of the multiple RSPs has not been discussed yet. It should be noted that we discuss the *value* over the *cost* on a service from a viewpoint of a final receiver at present. Modeling and evaluating that from a viewpoint of a provider that sends contents and channels to various different receivers via various agents is one of the future works. This may be also useful for designers who work for manufacturing industries.

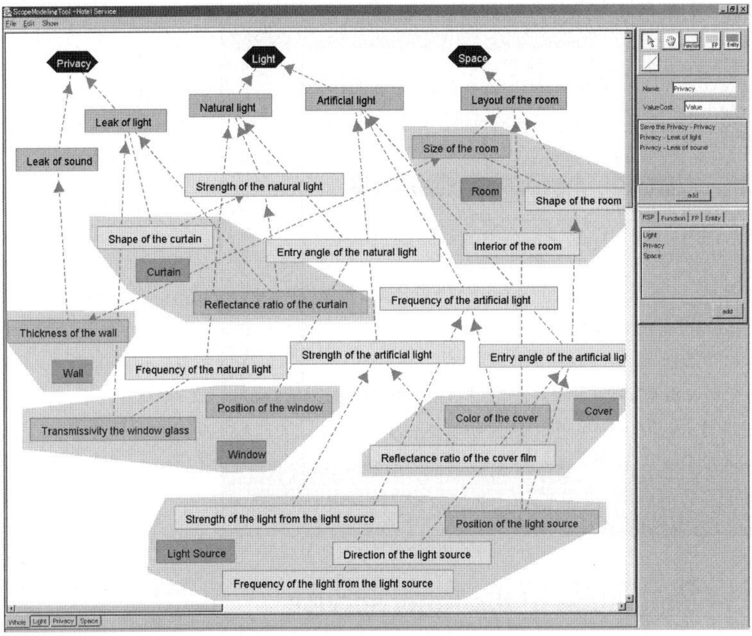


**Fig. 3.** A Screenshot of a view model panel

## 4 Application to a hotel service

This chapter reviews an application to an Italian hotel service based on a literature [8].

### 4.1 The target hotel service

As a target of the service modelling, a hotel service in Abruzzo region, Italy was selected. This hotel has three-star certification, and is characterized by the various efforts to reduce the environmental impact. The hotel management regards the environmental consciousness as attractiveness to the guests, wants to reduce the environmental impact and develop a new attractive service for the guests.

**Table 1.** Results of the on-site survey (partial)

| RSP | Weight |
|---|---|
| Freshness of towels and bed linen | 9.8 |
| Clean rooms | 9.7 |
| Correct billing | 9.5 |
| Rooms free of unpleasant odours | 9.5 |
| Indoor air pollution level | 9.4 |
| Everything in working order in room | 9.4 |
| Comfortable bathrooms provided with various amenities | 9.4 |
| Room ready at arrival | 9.4 |
| Comfortable beds | 9.4 |
| Prompt, competent and attentive staff | 9.3 |
| Quiet rooms | 9.3 |
| Friendly and polite staff | 9.3 |
| Fast and efficient check-out | 9.2 |
| Adequate heating and/or air-conditioning | 9.2 |
| Comfortable and spacious rooms | 9.1 |
| Efficient booking | 9.0 |
| Well-furnished and attractive rooms | 9.0 |
| Quality of food and beverages at breakfast | 9.0 |
| Friendly welcome at arrival | 9.0 |
| Reduced release of pollutants into the environment | 8.9 |
| Good lighting (both natural and artificial) | 8.8 |
| Fast and efficient check-in | 8.8 |
| Reduced waste generation | 8.7 |
| Use of materials from renewable resources | 8.6 |
| Energy saving | 8.6 |

## 4.2 Investigation of the hotel service

For the sake of the modelling and evaluation of the hotel service, an on-site survey about the service quality to the guests in the hotel from 15th, April to 19th, May was carried out. For each RSP of the service (such as "freshness of towels and bed linen" and "clean rooms"), three criteria were given. The first criterion is the importance of the factor, which gives the importance of each RSP. The second criterion is the evaluation of the service in the target hotel. This indicates the present service level of the target hotel. The third one is the evaluation of a competitor hotel, which asks the best service quality of the similar hotel which the customers have ever used. This may give us the gap this hotel should fill in. During the above-mentioned time span, 1,233 questionnaires were delivered to guests, and 116 filled-in questionnaires were collected (corresponding to a 9.4% coverage ratio). Table 1 shows the average weights obtained for the various service quality requirements (because of space limitation, only those with the highest weights are listed). It should be noted that the weight was supposed to be given from 1 to 10. A line with white characters refers to environmentally related issues, though this information was not given to the guests. In addition, the actual hotel service was investigated by interviewing with the workers and the owner of the hotel. Based on the result of the investigation, the model of the hotel service was described.

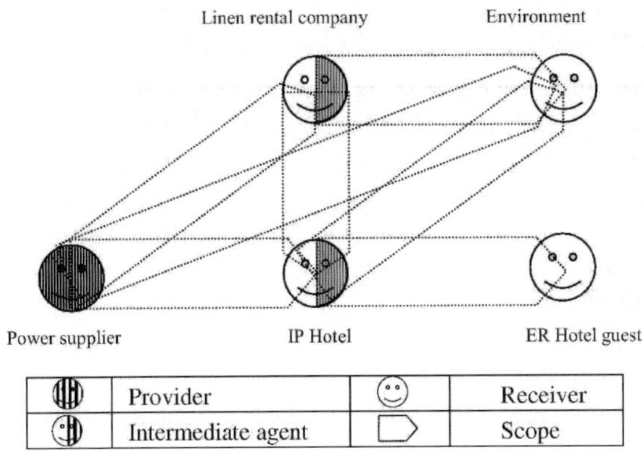


First, the service agents which participate in the hotel service were modeled: the hotel, the hotel guests, a bed linen rental company, a power supplier, and the environment (Figure 4). The agent 'environment' is a virtual agent to evaluate the environmental impact caused by each agent's activity. Second, the RSPs of the hotel guests were deployed (for example, an RSP "cleanliness of the room". Third, based on the investigation of the hotel service, the function structures for each RSP were developed. They describe the details of the hotel service from the point of each service receiver. Figure 5 shows just a small part of the developed model of the service.

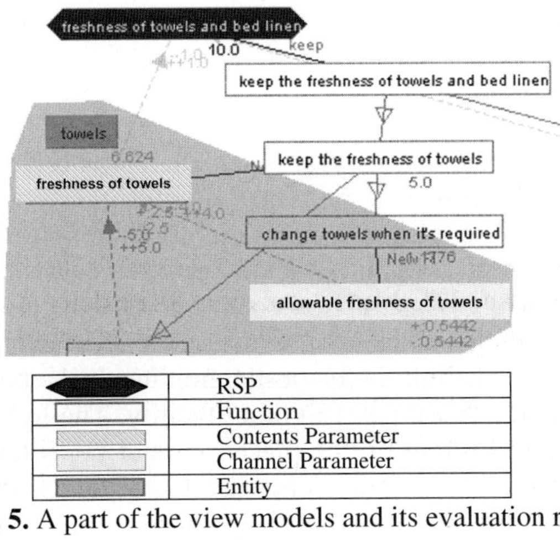


**Fig. 5.** A part of the view models and its evaluation result

## 4.3 Evaluation of the hotel service

By using the importance of each RSP obtained from the on-site survey and a relative weighting between the receivers, the hotel guests and the environment (3:1), given by the service designer, the importance of each RSP, the importance of each CoP and ChP were calculated. In Figure 5, the number under each node represents the importance of those parameters. In addition, the authors made some evaluations of plural service receivers by putting the relative importance on each receiver. For example, the evaluation of the service of the hotel to the hotel guests and the environment was made to compare the importance of the environmental effect and the satisfaction of the hotel guests.

## 4.4 Improvement proposals for the hotel service

Designers are able to investigate redesign options to obtain the higher satisfaction of the hotel guest with the less impact on the environment according to the described realization structure and the relative importance of the elements of this service. This section explains the examples for the obtained concrete redesign options on this service.

### Proposal 1: Window shielding films

Special films put on the window to shield light, an improvement proposal, would allow the guests view the outside. It probably results in more heat's entering the room than in the case of a shutter, the total customer satisfaction may be increased depending on the weights on the view of the outside and the energy consumption for the temperature control.

### Proposal 2: A sun lighting system

According to the RSP weightings obtained, it was revealed that the importance of RSP "adequate light" for the hotel guests was rather small (3.0), while the RSP "energy consumption" for the environment was quite large (10.7). Therefore, a lighting facility with the equivalent or even less quality and less energy consumption could improve the total quality of the service. A sun lighting system [24], which gathers the sunlight outside the shelter and canalize the light to the destination inside the building, can be a concrete measure for that type of lighting facility. The light volume of the sun lighting system to light up the place may not be as large as a normal light source. In addition, it works better in the corridors and the bed rooms, which have small importance of their brightness.

### Proposal 3: "Cash-back per unwash" system

Currently, the hotel give choices for the guests to order to wash the towels used in the bath rooms or not. This system can give more satisfaction level to several RSP. They include "reduced release of pollutants into the environment" in the scope models between the hotel and the guests, between the hotel and the environment. "Cash-back per unwash" system for the towels would also give a chance for the guests to decrease the economic cost. Thus, this type of charging system would bring benefits to more types of RSP.

### Proposal 4: Various goods rental

The view model of RSP "comfort of the bath room" describes various amenities in the bathroom influence the RSP. Other goods provided by the hotel could improve the quality of the hotel service. What hotel guests use during their business trip stay includes shirts, jackets, neckties, bags, and personal computers. Women may need cosmetics, too. The hotel can rent a variety of those useful things to their guests. This improvement option would delete or reduce the degree of the negative RSP "cost of transporting things" of the guests, since they do not have to bring some of their own belongings with them and just rent them on-site. In addition, this option can bring a totally new RSP "joy of using something new" even though it gives negative influence on the existing RSP "sense of property". "Joy of using something new" in this case is meant as joy to try something they never or seldom own.

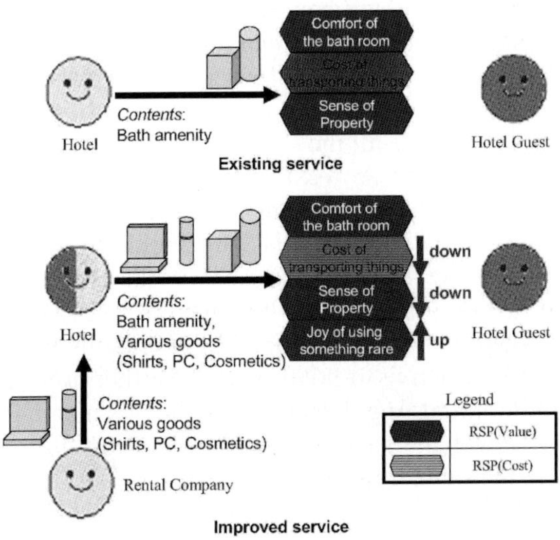


**Fig. 6.** Description of the proposal 4

The examples of that RSP include women's enthusiasm to exchange one type of cosmetics to another, and excitement to wear a jacket with a remarkable brand name. It should be emphasized that this type of RSP used to be out of the focus of the hotel service. If the hotel runs the rental service by themselves, it would cost too much probably. Thus, to avoid the too much cost, it is reasonable that the hotel receives the rental service by another agent, a rental company. It should be emphasized that in this option, several new RSPs for the hotel guests were added by setting a new provider so that the total satisfaction of the service might be increased (Figure 6).

#### Proposal 5: Room customizing system

The hotel provides any type of guest with a standardized set of equipments such as a desk and a sofa in a room. This is a good solution from the viewpoint of mass production with a low cost. However, service engineering pays more attention to a variety of customers' ways of satisfaction. An option to fill in the gap is a system to customize the equipments depending of a guest's needs. For instance, a business trip guest would be able to use a larger desk instead of a comfortable sofa.

### 4.5 Feedbacks from the hotel company

The feedbacks on the outputs from the case study were obtained from the hotel company as follows. According to the hotel company, all the proposed improvement options are new to them. Actually, they started to investigate the feasibility of the light shielding window films (proposal 1) and "Cash-back per Unwash" system (proposal 3). They also started to collect the detailed product information of the sun lighting system (proposal 2). However, they recognize all the various goods renting service (proposal 4) would not necessarily be so feasible. One reason is a renting is not so familiar in the Italian culture at present. Another reason is that Italian people do not like to wear a "standardized" suits or shirts. In addition, the target hotel is located in Abruzzo, where most of guests travel for a short period and a short distance. This means that the cost of transporting goods is not so large. On the other hand, in a hotel in international cities like Rome, this service can be attractive to hotel guests. Furthermore, cosmetics renting service is quite interesting, because a cosmetics manufacturer can utilize a hotel as a channel and a hotel can reduce the cost of preparing various types of goods at the same time.

# 5 Application to a logistics service

## 5.1 The target logistics service

The logistics service investigated is provided by one of the major actors in the global market in the field, and has already taken actions towards increased service content and remanufacturing of used forklift trucks.

## 5.2 Investigation of the customer requirements

In order to investigate the customer requirements, three different data collection methods were used; questionnaires, interviews and literature review. The logistics company's pamphlets and reports were examined, and external company related reports were also reviewed. In order to gather the rest of the required information, semi structured interviews were conducted. Finally, a questionnaire was used in order to gather the customer needs and requirements on the logistics service. Table 2 lists the requirements included in the questionnaire. They are categorized into the five groups. A "fleet" means a set of equipments such as forklift trucks prepared to meet the user needs.

**Table 2.** Customer requirements for the service

| **0. Basic Requirements** |
|---|
| Making the logistics operation more efficient. |
| Confidence for your service provider regarding quality of the provided logistics system. |
| Importance of the logistics system for your ability to satisfy your customers. |
| Importance of the logistics system for your companies other operations, e.g. production. |
| Improving your company's environmental image by using this service dematerialization. |
| Low price for the logistics service |
| Predictable price for the logistics service |
| Possibility to get rid of the responsibility of the devices for the logistics operation. |
| **1. Contract Preparation** |
| Fast validation of logistics needs |
| Optimized logistics needs |
| Short time for identifying your logistics needs |
| Short time for making contract proposal |
| **2. Fleet Delivery and Installation** |
| Short time between signing the contract and the fleet delivery |

| |
|---|
| Reliability in fleet delivery |
| Fast fleet system installation |
| **3. Fleet system operation, maintenance and reinstallation** |
| Logistics reliability |
| Fast time for upgrading configuration |
| Flexible fleet system configuration |
| Frequent validation of your logistics needs |
| Low energy usage during logistics operation |
| Fast support |
| Accurate support |
| Environmental benefits of the logistics service operation |
| Use of materials from renewable resources |
| Easy error reporting |
| Fast error reporting |
| Fast corrections of malfunction |
| Easy detection of malfunction |
| Fast detection of malfunction |
| Fast reaction on malfunction |
| Short time for getting spare parts |
| Support of validation of future logistics needs |
| **4. Fleet system take-back** |
| Fast take-back of previously used devices, such as old trucks and batteries. |
| Possibility to get rid of /sell earlier used logistics equipment |

## 5.3 Modelling the logistics service

The authors are building view models for the service. Figure 7 describes partially a view model for an RSP for this service, which corresponds to "Fast fleet system installation" listed in Table 2. The RSP in this case, the time for delivering the ordered fleet, is linked to the functions such as "decrease time for fleet preparation".

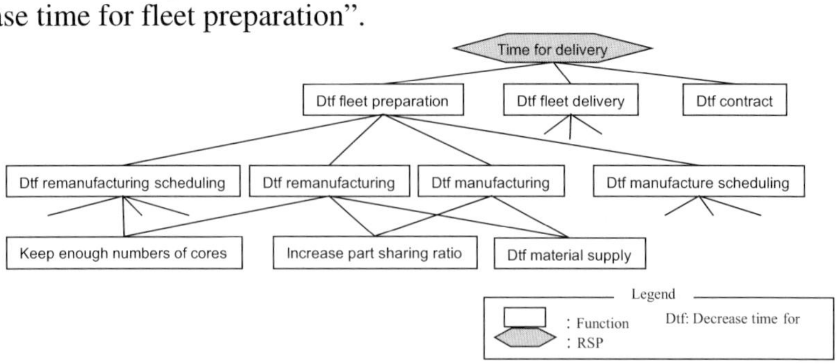


**Fig. 7.** The view model in part for an RSP "Time for delivery"

The investigation described in 5.3 nor the modelling of the service partially shown in 5.4 is not yet finished. After completing these tasks, the authors will generate redesign options for the service with the help of Service Explorer.

# 6 Discussion

## 6.1 Service modelling and evaluation methods

Firstly, the actual service including the activities of the related service agents was found to be able to be described with the service modelling method in the case of the hotel service. This model indeed provided service designers the structured information of the service and opportunities to explore improvement ideas. For example, the relationship between the RSP "comfort of the bathroom" and the amenity resulted in the proposal 4 as described in 4.5. Secondly, the service evaluation method was proved to offer the improvement focus with numerical importance values. The improvement focus in the proposal 2 is one of the examples. In addition, the service can be evaluated not only from one receiver's perspective but also the perspective of the multi agents. However, it isn't possible to evaluate the degree of improvement from the viewpoints of service quality or environment. This needs another evaluation method.

## 6.2 Comparison with other methods

The proposed service modelling and evaluation method is compared below to an existing related work, Quality Function Deployment for Environment (QFDE) [25]. QFDE is a methodology to support ecodesign developed by incorporating environmental aspects into QFD [22]. QFD is a methodology used in a planning stage to analyze required functions and the structure to realize a designed object. QFDE is in principle applicable to services, since the applicability of QFD to service has been already verified. The proposed method is superior to QFDE as followings.

1. It deals with a structure of all the agents participating in the service providing which is a source of requirements on a target service, while QFDE can only treat those requirements in the form of a list.
2. The proposed method presents much more design parameters than QFDE. They include the structure of the agents and RSP. On the other

hand, RSPs are given and the agents are not described in QFDE. The proposal 4 is an outcome taking advantage of this feature.

## 6.3 Service modelling and evaluation tool

Service Explorer, which was used for the service modelling and evaluation in this research, was able to describe the structures of the service and the importance of the service elements. This visualized information could help service designers consider the improvement of the service. However, the current version of Service Explorer doesn't have a functionality to propose improvement ideas. As a support system to derive improvement options with reasoning logic, Universal Abduction Studio (UAS) is being studied [26].

## 6.4 Customer category

Although only an average value was used in these case studies, redesign options specific to a customer category could be generated. Actually, some typical types of the hotel guests were observed in the results of the on-site survey, and some of their requirements were different from one another.

# Acknowledgements

This research was partially supported by VINNOVA, Swedish Agency for Innovation Systems. We give special thanks to Mr. Watanabe, and Mr. Hara for their effort on the implementation of Service Explorer.

# References

1. Tomiyama, T., 1997, A Manufacturing Paradigm Toward the 21st Century, Integrated Computer Aided Engineering, 4:159-178.
2. Miles, L.D., 1971, Techniques of Value Analysis and Engineering, McGraw-Hill, New York.
3. Rodenacker, W., 1971, Methodisches Konstruieren, Springer-Verlag, Berlin.
4. Pahl G., Beitz, W., 1988, Engineering Design: A Systematic Approach, Springer-Verlag, Berlin.
5. Arai, T. and Shimomura, Y., 2004, Proposal of Service CAD System -A Tool for Service Engineering-, Annals of the CIRP, 53/1:397-400.
6. Sakao, T., Shimomura, Y, 2004, A Method and a Computerized Tool for Service Design, International Design Conference, Dubrovnik, 497-502.

7. Sakao, T., Watanabe, K., Shimomura, Y., 2003, A Method to Support Environmentally Conscious Service Design Using Quality Function Deployment . In Proceeding of Eco Design, Tokyo, IEEE, CD-ROM, 567-574.
8. Watanabe, K., Shimomura, Y., Sakao, T., Raggi, A., Petti, L., 2004, Application of a Service Modeling Tool to Hotel Industry, The Sixth International Conference on Eco-Balance, Japan, 495-498.
9. Ramaswamy, R., 1996, Design and Management of Service Processes, Addison-Wesley, Reading, MA.
10. Shostack, G.L., 1981, How to Design a Service, in Donnelly, J.H. and George, W.R. (eds.), Marketing of Services, American Marketing Association.
11. Bullinger, H., Fahnrich K., Meiren, T., 2003, Service engineering – methodical development of new service products, International Journal of Production Economics, 275-287.
12. Meier, H., 2004, Lifecycle-based Service Design for Innovative Business Models, Annals of the CIRP, 53/1:393-396.
13. Goedkoop, M.J., van Halen, C.J., te Riele, H. R., Rommens, P. J., 1999, Product Service Systems, Ecological and Economic Basics, VROM.
14. McAloone T.C., Andreason, M.M., 2004, Design for Utility, Sustainability and Social Virtues, Developing Product Service Systems. In Proc. Design 2004, 1545-1552.
15. Berg, C., Tukker, A. (eds.), 2004, Proceedings of the Second International SusProNet Conference "Product Service Systems: Practical Value".
16. Weber, C., Steinbach, M., Botta C., Deubel T., 2004, Modeling of product-service systems based on the PDD approach, In Proc. Design 2004, 547-554.
17. Tomiyama, T., 2001, Service Engineering to Intensify Service Contents in Product Life Cycles, Proc. of EcoDesign 2001, IEEE, 613-618.
18. Suh, N.P., 1997, Design of Systems, Annals of the CIRP, 46/1:75-80.
19. Umeda, Y., Ishii, M., Yoshioka, M., Shimomura, Y., Tomiyama, T., 1996, Supporting Conceptual Design based on the Function-Behavior-State Modeler, AIEDAM, 10/4:275-288.
20. Shimomura, Y., Yoshioka, M., Takeda, H., Umeda, Y., Tomiyama, T., 1998, Representation of Design Object Based on the Functional Evolution Process Model, J. of Mechanical Design, ASME, 120:221-229.
21. Saaty, T.L., 1980, The Analytic Hierarchy Process, McGraw-Hill.
22. Akao, Y., 1990, Quality Function Deployment, Productivity Press.
23. Sakao, T., Watanabe K., Shimomura, Y, 2003, A Method to Support Environmentally Conscious Service Design Using Quality Function Deployment, Proc. of Eco Design 2003, Tokyo, IEEE, 567-574.
24. http://www.gel.civil.nagasaki-u.ac.jp/text/method/met50/met50.html
25. Masui, K., Sakao, T., Kobayashi, M., Inaba, A., 2003, Applying Quality Function Deployment to Environmentally Conscious Design, International Journal of Quality and Reliability Management, 20/1, 90-106.
26. Takeda, H., Sakai, H., Nomaguchi, Y., Yoshioka, M., Shimomura Y., Tomiyama T., 2003, Universal Abduction Studio -Proposal of A Design Support Environment For Creative Thinking In Design-. In Proceedings of ICED03.

# End-of-Life Strategies

# Towards self-disassembling products

## Design solutions for economically feasible large-scale disassembly

J.R. Duflou, B. Willems and W. Dewulf
*Department of Mechanical Engineering, Katholieke Universiteit Leuven, Belgium*

Abstract: From an economic point of view the productivity associated with commonly available disassembly methods today seldom makes disassembly the preferred end-of-life solution for massive take back product streams. In consequence systematic reuse of parts or components, or recycling of pure material fractions are normally not achievable. Economic models demonstrate that the efficiency of disassembly operations should be increased an order of magnitude to assure the competitiveness of ecologically preferred, disassembly oriented end-of-life scenarios. Using fully automated disassembly techniques does not allow to overcome this efficiency gap if not combined with innovative reversible joints. Enhanced disassembly principles, in which external trigger signals allow to simultaneously reverse the action of multiple fasteners, forms a promising approach to low cost mass disassembly. In this paper a state-of-the-art of these emerging techniques is sketched, categorising the fasteners according to their generic applicability and the degree of imbedded automation of the triggerable disassembly activity. A number of perspectives for innovative reversible fasteners are sketched as a contribution to this promising paradigm of self-disassembling products.

Keywords: Active Disassembly, Disassembly Embedded Design, Self-Disassembling Products, Design for Disassembly

## 1 Introduction

The ladder of Lansink principle [1] has long been recognised as the ecologically inspired hierarchy of preference for the end-of-life (EoL) treatment of products. Aiming for the highest levels in this hierarchy as systematic objectives in EoL treatment scenarios, namely product reuse and disassembly with optimal reuse of parts and components, would bring the ideal of a closed loop economy closer to reality. Although this



*D. Brissaud et al. (eds.), Innovation in Life Cycle Engineering and Sustainable Development,* 87–110.

principle is well-known, and in recent year the societal goodwill to adopt such scenarios was noticeable in a number of advanced economies, the present day practice in countries with a systematic take back obligation demonstrates that from an economic point of view these scenarios seldom prove to be the preferred solution. Where the responsibility for systematic EoL treatment of discarded products is assigned to organisations operating according to market principles, minimising the cost/benefit ratio is typically the preferred operational practice. De facto this leads to systematic shredding of discarded products, after minimal de-toxification according to government regulations. Take back organisations, often affiliated with the manufacturing and/or distribution sectors, are typically also not expected to obstruct the sales of new products by supporting alternative, second hand spare part supply circuits or remanufacturing activities. Only when such activities could be made highly profitable, could systematic disassembly be considered a valid option. Modelling of existing discarded product collection and EoL treatment systems allows to analyse the sensitivity of the different product, component and material flows for changes in the involved cost structure. Such a study was conducted with as specific aim to determine the required reduction of the disassembly costs in order to systematically influence the preferred EoL scenario [2]. Assuming that purely economic considerations are determining the choices made by every actor in the logistical chain, and using real cost data available for the Belgian situation anno 2004 as input for the model, the results summarised in figure 1 were obtained. The graph indicates the required disassembly time reduction levels to change the preferred EoL scenario from shredding to partial disassembly and to full disassembly respectively.

In the logic of the applied model, disassembly time reductions are linearly related to personnel costs and form a measure for the efficiency of the EoL treatment method. Simulations were performed for a range of different product categories of which 10 representative product types are shown in figure 1. From these results it can be concluded that influencing EoL treatment scenarios for low value products with a limited functional lifetime is hard to achieve. Extreme efficiency increases would be required to make disassembly a feasible alternative for shredding. However, as the residual value of products and the functional life time increases, the required disassembly time reductions becomes more realistic, with 65 to 87% reductions leading to a preference for systematic disassembly. Realising disassembly methods that could achieve such an efficiency leap is however not evident.

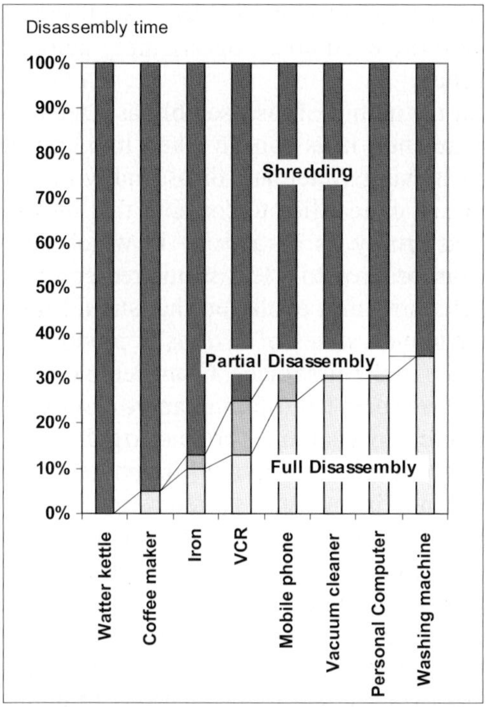


**Fig. 1.** End-of-life treatment scenarios as a function of required disassembly time (100% corresponds to manual disassembly by an experienced operator)

In a pro-active approach the efficiency of disassembly operations may be significantly influenced by an optimisation of the product structure [3–5]. More pragmatic contributions start from the assumption that products in a worldwide economy cannot be influenced, but should be treated as received in well-automated facilities [6–7]. Despite the efforts, it is doubtful whether these individual approaches, can reach sufficiently large efficiency improvements to affect the economically optimal EoL treatment methods. Although observing the limited impact of these efforts could lead to the conclusion that systematic disassembly of large streams of discarded products is fictitious, there are reasons to believe that this should not necessarily be the case. Efforts to automate disassembly activities have typically been oriented towards some kind of reversed assembly. When dealing with conventional fasteners, such as e.g. bolds, screws, rivets or spot welds, the technical complexity of reversed assembly can be huge. Factors such as non-uniformity of returned product models, customer specific maintenance history and deformation or corrosion of the products can make non-destructive disassembly automation an unrealistic task. Since the economic benefits to be expected from such operations are typically outbalanced by the involved costs, disassembly automation often remains an academic exercise.

When however the nature of disassembly as a technical task is reflected upon, some fundamental laws can be taken into consideration to support the believe that achieving systematic disassembly in a highly efficient way should be possible. It can suffice to consider that in principle disassembly, in contrast with assembly, is an activity in which parts and components evolve from a more ordered to a less structured configuration. Such an increase in entropy is a natural evolution that should basically not require a high effort or a complex series of actions if products are not designed to prevent disassembly in the first place. Considerable efforts are spent by designers to prevent products from falling apart in an uncontrolled way during their service life. To guarantee product rigidity, boundary conditions, as can be expected to occur in different life phases, are taken into account during product engineering activities. Identifying environmental conditions that are not expected to occur as functional boundary conditions is however not a common task for designers. Nevertheless such an exercise could help to identify conditions under which a product could be allowed to lose its structural coherence. Products that are typically not subjected to dynamic forces during their functional life could, for example, very well be designed to fail quickly due to fatigue when submitted to an intensive load cycle in an EoL treatment plant. During such a treatment different joints could be disconnected in parallel rather than as consecutive tasks. Furthermore multiple products could be processed simultaneously and in a highly automated way, effectively helping to achieve the required cost reduction for economic disassembly activities.

In recent years a number of fastener systems have been proposed that fit into this framework of externally triggerable reversible joints. They are characterised by the 1-to-n principle: a single external trigger allows reversal of multiple fasteners simultaneously. Depending on the fastener design this connection reversal can be accompanied by an active disassembly action that allows to effectively remove parts or components from the product structure. Due to the reversal of multiple connections in a single action, substantial efficiency increases can be expected, effectively making disassembly an economically viable alternative for shredding.

In the remainder of this paper methods that can contribute to such new generations of fasteners are reviewed. They form a tool box that can be called upon for designers working in a pro-active DfE context.

## 2 State-of-the-art of one-to-many disassembly

Within the group of fasteners suitable for enhanced disassembly leading towards self-disassembling products, two categories can be distinguished according to their generic applicability. They are labelled *Disassembly Embedded Design* and *Active Disassembly* and are described in more detail in the following sections.

For the *Disassembly Embedded Design* category, product-specific separation features are embedded in the product structure during manufacturing. These features are unique for the product, and typically cannot be used in other applications without modification. They are characterised by the fact that they require physical contact of the disassembly tool with the fastener to initiate the disassembly process. In order to increase the process efficiency (Section 1) and to unfasten different connections with a single disassembly action, fasteners should be linked to each other. Despite the opportunities they generate, these techniques have several drawbacks. Since the solutions are design specific and the fasteners have to be integrated into a product wide solution, considerable amounts of time and expertise are required from the designer. Moreover, due to physical restrictions, the number of connection that can be coupled and simultaneously unlocked is limited. Finally, as long the disassembly process involves physical contact, no substantial efficiency increase can be guaranteed. Renz [8] pointed out that the disassembly time can be divided into handling time (tool usage, product movement, segment disposal) and search time (localisation and identification without simultaneous handling). Field trials pointed out that the searching time covers approximately 30% of the total disassembly time. Therefore, eliminating the searching time and the time needed for reaching out and positioning of tools, is required to guarantee a substantial disassembly time reduction.

For the *Active Disassembly* category, innovative fasteners, generically applicable in different situations, are developed. By means of innovative materials or structures a set of connections is created for which an external trigger or a combination of triggers initiates the unfastening process. The concepts are developed such that they are applicable as modular connections in a wide range of other products. Therefore they are considered as generic for certain functional requirements. Since they are activated by external triggers, no physical contact is required. This implies that an unlimited number of connections can be unfastened with the same disassembly action, without requiring internal links, as long as the joints are positioned in the influence field of the trigger signal. An overview of techniques, announced in recent years, that enable the shift towards self-disassembling products, is offered in table 1.

**Table 1.** Classification of innovative fasteners for self-disassembling products

| Trigger principle | Disassembly embedded design | Active disassembly |
|---|---|---|
| **Mechanical** | Mechanical force (Braunschweig 2003) Pneumatic force (Braunschweig 2003) | Pneumo-elements (Neubert 2000) |
| **Thermal** | Heat-activated joining structures (Nishiwaki 2000) (Li 2003) | SMA (Chiodo 1997,1999, 2002) (Sakai 2003) Freezing elements (Neubert 2000) |
| **Chemical** | | Hydrogen storage alloy (Suga 2000) Soluble nuts (Neubert 2000) |
| **Electro-magnetic** | Electro-magnetic field (Braunschweig 2003) (Klett 2003) | |
| **Electrical** | Heating wire (Masui 1999) (Klett 2003) Piezoelectricity effect (Braunschweig 2003) Self-activating SMA (Jones 2004) | |

## 2.1 Disassembly embedded design

An early attempt towards Active Disassembly can be found in Masui et al. [9]. Masui introduced the idea of embedding a separation feature inside a product during manufacturing and activating it at disassembly. The concept was applied on a CRT monitor: a nickel-chrome wire is embedded in the joint between funnel and screen, which are made out of recycling-incompatible glass types (figure 2).

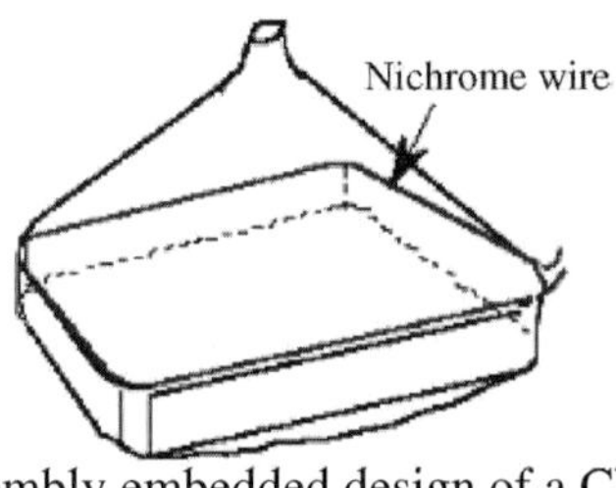


**Fig. 2.** Disassembly embedded design of a CRT monitor

An electric current through the wire, causing a distributed Joule effect, induces thermal stresses in the CRT tube. This results in separation by fracture along the joint between the different glass types when the allowable stress level is exceeded. Although this concept was able to solve a complex disassembly problem, questions about the implementation cost and the efficiency of the concept arose, due to its product specific nature. No similar examples of this technique were found in literature.

Klett et al. [10-12] presented a similar concept of using a heating wire in order to activate disassembly. They developed a bolt with a locking feature realised through a plastic band. A heating wire is positioned between the plastic band and the bolt (figure 3). When an electric current flows through the wire, the heat cuts the plastic band, causing the bolt segments to leave their position and unlocking the connection. Klett et al. [12] developed a similar nut and bolt with a locking feature realized through a set of balls blocking the relative movement between parts (figure 4). The system can be unlocked by pulling the cap, covering the set of balls, in the opposite direction. This movement can be initiated by an electromagnet, which pulls the cap in the required direction.

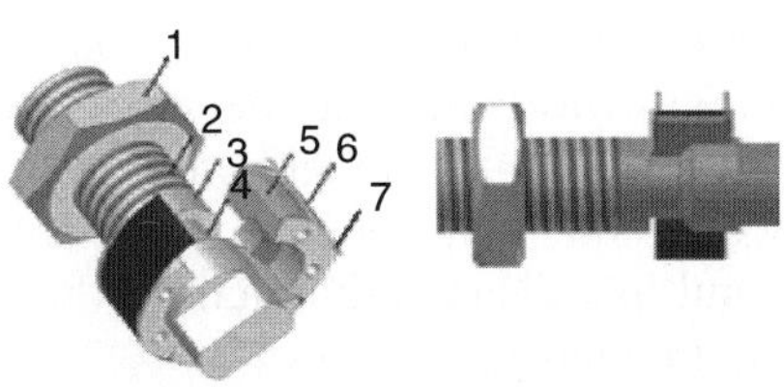


1) nut 2) thread 3) joining feature 4) groove 5) segment 6) plastic band 7) heating wire

**Fig. 3.** Disassembly embedded nut and bolt, triggered by an electric current

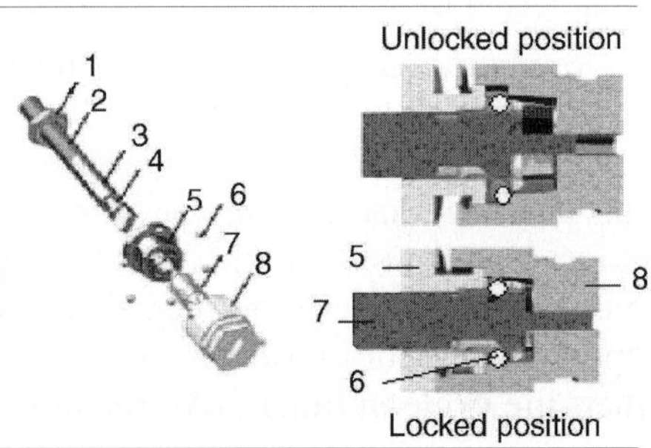


1) nut 2) thread 3) rod 4) groove 5) ball holder 6) balls 7) cage 8) cap

**Fig. 4.** Disassembly embedded nut and bold triggered with an electromagnet

Braunschweig [13–14] explored a similar concept of using electromagnets as trigger for disassembly. By attaching a magnet anchor on the flexible part of a snap fit, the near presence of an electromagnet initiates the disassembly action (figure 5). This concept offers a solution for the unsnapping problem of unreachable snap fits. However, other problems such as the identification or localisation of fasteners in the product remain. Therefore this technique is categorised under Disassembly Embedded Design. A second concept of disassembly embedded product design described by Braunschweig [14] makes use of mechanical force. By means of an internal, mechanical coupling of different joining elements, like snap fits, a central unlocking mechanism is constructed: exerting a mechanical force at one location causes the unfastening of all individual snap fits (figure 6). This concept is of particular interest for large industrial systems.

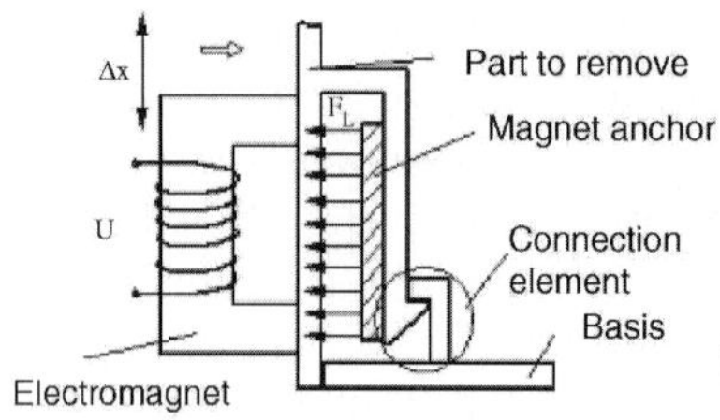


Fig. 5. Disassembly embedded snap fit triggered by an electromagnetic field

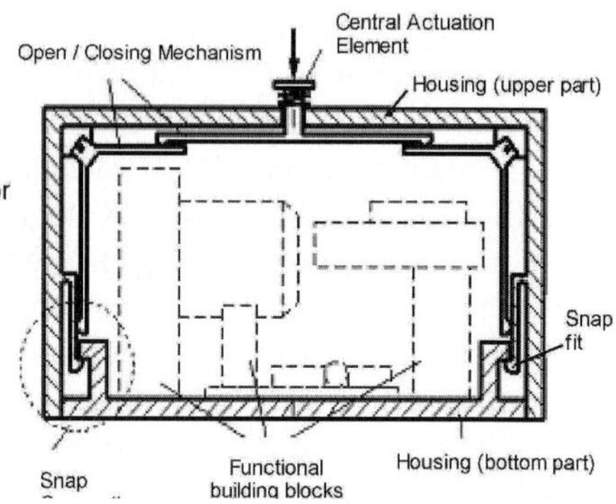


**Fig. 6.** Disassembly embedded product design with mechanical coupling

Another concept is the use of piezo elements that force a shape change of adjacent snap fits when an electrical tension is applied over the piezo crystals (figure 7). Since one piezo element is needed for every snap fit, this concept has severe cost implications. Moreover, at the end-of-life stage of the product, the question arises how to efficiently remove the piezo elements from the snap fits. However, this concept could offer some opportunities for products that require multiple disassembly actions during their life cycle in function of repair or maintenance.

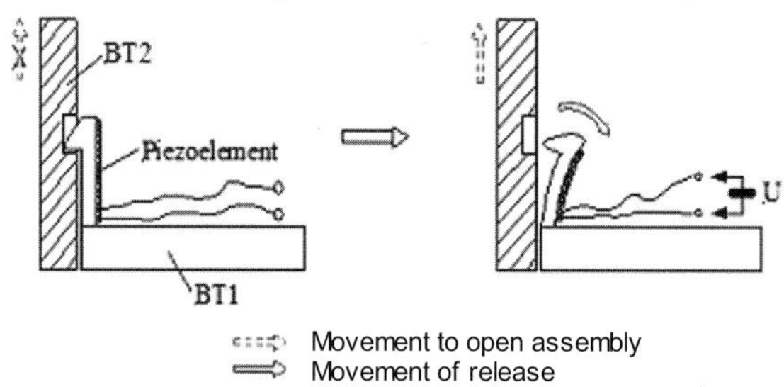


**Fig. 7.** Disassembly embedded snap fits using Piezo element based actuation

Finally, Braunschweig [14] described a pneumatic system to disassemble snapped joints. The product is designed such that the flexible part of the snap fit is connected to a flexible area in the housing of the product. A special disassembly tool is used to create a cavity next to the flexible area. When compressed air is driven into the cavity, this area will move into the product, while unsnapping the snap fit (figure 8). In this configuration, problems of visibility, safety and producability arise.

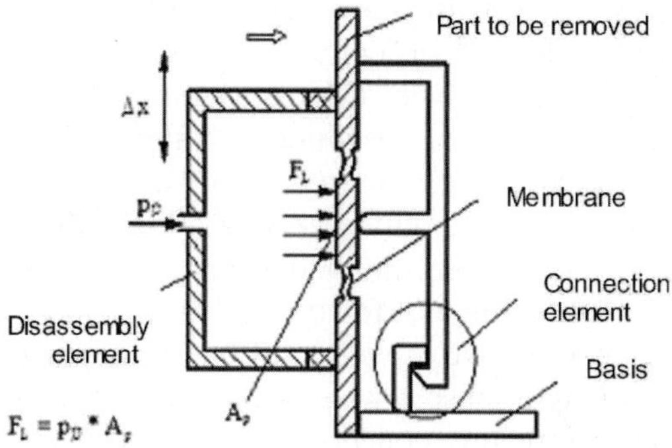


**Fig. 8.** Disassembly embedded snap fit, pneumatically activated

In Nishiwaki et al. [15] and Li et al. [16] the use of product specific integral joining structures was proposed, where the snapped joints are unfastened by the application of localized heat sources, generated through non-uniform electric current. These heat sources will cause localised thermal expansions of materials, which in turn results in a deformation of the whole structure and unsnapping of the joining elements (figure 9). Topology optimisation techniques were used to generate the optimal design structure to obtain the required disassembly effect. This concept has several advantages over the existing thermal actuator types. Firstly, any material can be used to make this kind of fasteners, as long as a sufficiently high thermal expansion coefficient is guaranteed. Secondly, a design structure based on the temperature distribution pattern is less sensitive to design modifications than the other embedded actuation concepts. And finally, due to the specified actuation time and the required accuracy of the heating location, undesired actuation is avoided, which reduces safety problems during use.

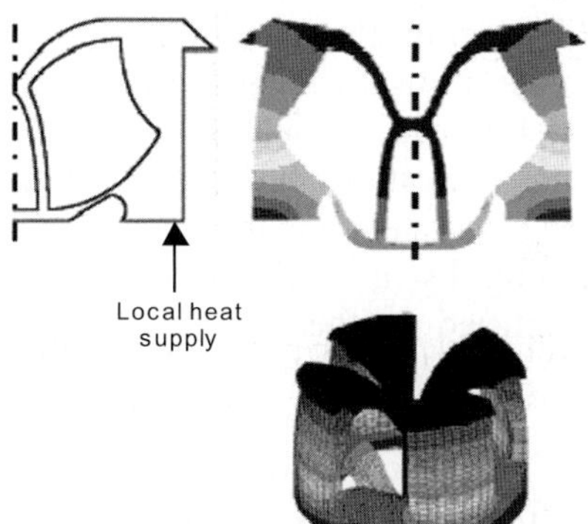


**Fig. 9.** Heat-activated joining structures for disassembly embedded design

Jones et al. [17] describe the next step towards autonomously disassembling products by using, electrically self-powered systems. The authors propose the use of a muscle wire (Shape Memory Alloy (SMA) wire) that is triggered by the residual energy found in batteries of discarded products. After triggering the wire contracts in such a way that joining elements are unfastened (figure 10). Since only a very small initiation signal is required and the product itself delivers the actual disassembling energy, this concept is labelled as autonomous. However, if a substantial disassembly time reduction has to be reached, several fasteners in the product should be connected with a single muscle wire. This requires a well-adjusted, dedicated design. Therefore this concept is categorised as Disassembly Embedded Design.

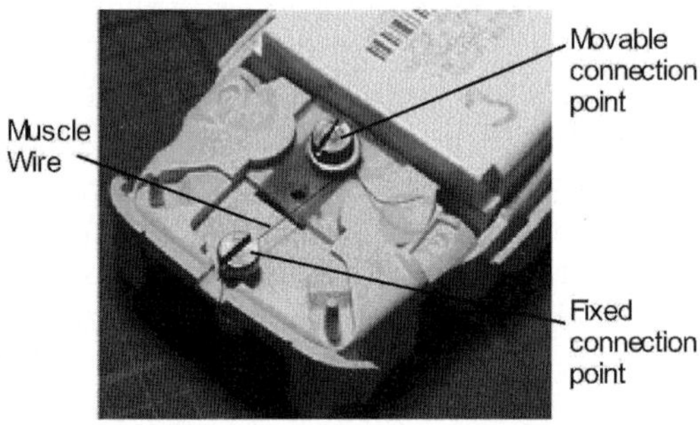


**Fig. 10.** Autonomously disassembling mobile phone: the battery is released after the residual energy in the battery is used to activate the muscle wire

The Disassembly Embedded Designs techniques described above can be regarded as a fist step in the direction of more efficient disassembly processes. Moreover, in some cases innovative solutions resorting under this category can solve very complex disassembly problems, as shown in the separation of the CRT monitor glass fractions. Nevertheless, this category of solutions has a number of disadvantages. First, the features embedded in the product are unique for each application, and consequently they cannot be transposed to other applications without considerable redesign effort. Moreover, the number of connections that can be linked in order to allow simultaneous unfastening by a single trigger is restricted. Since this characteristic was identified as the key driver in reducing disassembly effort (Section 1), disassembly embedded design solutions fail in this respect. Furthermore, the extra costs during manufacturing are not negligible, which puts a question mark above the economic viability. For all these reasons, a shift towards more generic fastener types is essential.

## 2.2 Active disassembly

For Active Disassembly, innovative though generic fasteners are used: well-standardised fasteners can be used in different situations or products. Moreover, the number of connections that can be unfastened simultaneously by application of a single trigger is in principle unlimited. When the energy to initiate the disassembly mechanisms is negligible, these actively disassembling products can be regarded as autonomous. Little research has been performed in the field of Active Disassembly techniques yet. The most advanced research is based on Shape Memory Materials (SMM), which change shape when a trigger temperature is exceeded. During the production, a fastener made from SMM, is trained to remember an initial state, which differs from its shape in use. When the product is heated or cooled down during the EoL-phase, the fastener is able to return to its initial shape. For example, screws can turn into simple pins when triggered by an external heat source, thus losing their clamping behaviour. Chiodo et al. [18-23] applied this technology for the disassembly of a range of products among which different types of telephones. These products were designed with shape memory snap fits. When the product is heated, all snap fits are deformed thus releasing the individual components (figure 11). Subsequently, the product is dismantled. Thus far, for mobile telephones, total unfastening times of 14 seconds have been reached, which is 5 to 10 times faster compared to the average time for manual disassembly [21]. Since many products can be dismantled simultaneously during Active Disassembly, the equivalent individual disassembly time reduction is even higher. Different heat sources were investigated, such as activation by hot air, hot liquids, microwaves and induction.

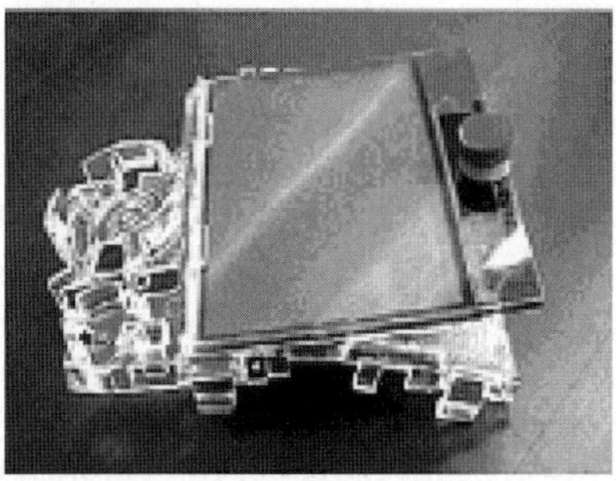

**Fig. 11.** Shape memory snap fit applied in the cover of a mobile display

Sakai et al. [24] investigated similar concepts making use of the shape memory effect and tested them on the joints of a flat screen. Apart from the shape memory parts, Sakai described the use of shape memory screws as well. The screws have a separate head that is fixed on the threaded cylinder by means of a tight fit (figure 12a). By adding heat, the SMA head changes its shape while it releases the threaded part (figure 12b). When using this type of screws, the treaded part remains in one part of the product. Chiodo developed a similar concept with shape memory polymers [23].

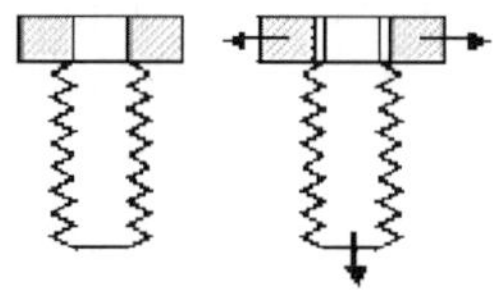

**Fig. 12.** Shape memory alloy screw (Based on [24])

Suga and Hosoda [25] use a chemical reaction as trigger for disassembly actions. A hydrogen storage alloy was presented as a bonding intermediate, which is pulverized when it absorbs hydrogen. When an Al-LaNiAl alloy joint, which was bonded by the surface activated bonding method, is put in an hydrogen atmosphere of 3Mpa at room temperature, the joint fails due to the pulverization of the LaNiAl-layer without applying any external force (figure 13).

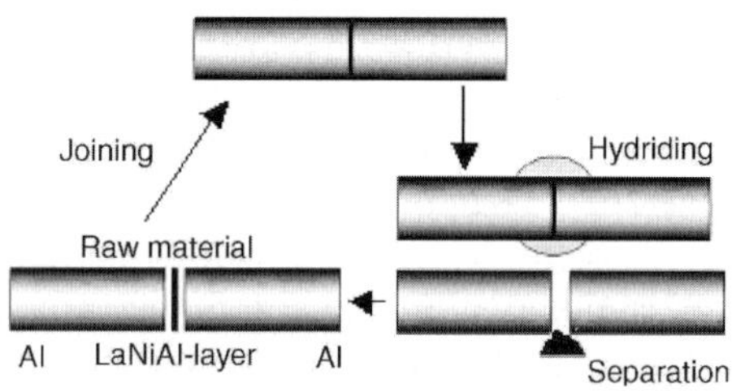


**Fig. 13.** Active Disassembly using hydrogen storage alloys

Finally, Neubert [26] proposed three other working principles for fasteners fitting the Active Disassembly category, making use of pneumatic expansion, the soluble characteristic of materials and the expansion of water when freezing. He investigated the methods theoretically and implemented each technique in a real-life case. The first solution Neubert offers for Active Disassembly is based on pneumatic expansion. He applied this physical phenomenon in so-called "pneumo-elements", which are air-filled, closed, hollow bodies with partial flexible or mobile walls. By increasing the ambient pressure, the air included in the element is compressed. Differences of pressure between the enclosed gas volume and the environment are used to generate a displacement. Air pressure variations, generated in pressure vessels, offer the compressing function. Neubert (2000) examined piston and membrane elements to validate this concept. An example of this concept is given in figure 14. The advantage of pressure based triggers, is that they are more controllable than temperature based triggers. The experienced pressure variations during the lifetime of a product are more gradual and are limited compared to the temperature variations. However, the disadvantages of this technique cannot be neglected. Due to the fact that the release forces of old connections can hardly be predicted, a severe safety factor must be used to compensate for this uncertainty. Also the accuracy of the pneumo-elements is rather low. Secondly, ambient pressure can vary strongly once the limits of the regular use environment are crossed, e.g. in non-pressurized cargo compartments of airplanes. Finally, the pneumo-elements require closed chambers, containing a sufficiently large volume of air. Especially in small products, this takes a considerable amount of space.

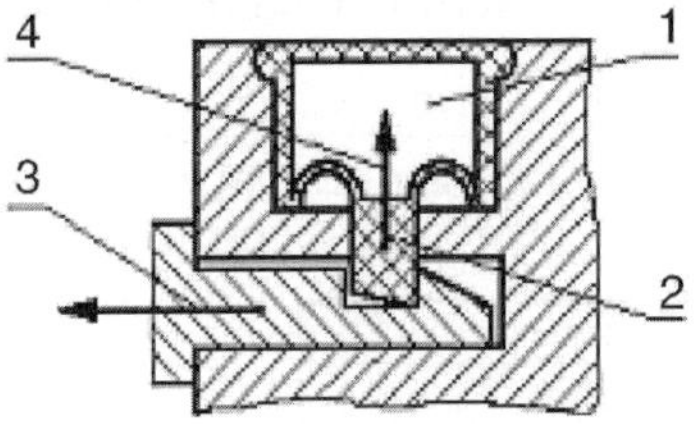


1- Closed air chamber
2- Moving shape element
3- Degree of freedom of the snap fit
4- Degree of freedom of the shape

**Fig. 14.** Example of a pneumo-element

The second concept described by Neubert [26] is the use of water-soluble connections. These connections lose their firmness when exposed to water to the extent that they can only withstand severely reduced forces. It is not necessary that parts are fully dissolved, a reduction of the volume to some extent is sufficient. Neubert describes two alternative base materials: Methyl Cellulose (MC) and Carboxy Methyl Starch (CMS). Methyl Cellulose (MC) is an ether, made of cellulose and methyl chloride or dimethyl sulfate. The molecules composed of glucose units are etherified in an alkaline medium whereby $CH^3$ groups substitute the OH groups. MC is water-soluble with a substitution degree from 1.4 to 2. The soluble characteristic of MC is only valid in cold water (<40°C) and with a pH-level between 2 and 10. The water hardness and the water supply direction have an influence on the solubility speed as well. The lower the Calcium content The softer the water and the more perpendicular the water supply, the better soluble. Carboxy Methyl Starch (CMS) is a paste from potato starch. In the presence of water the viscosity of the CMS material is maximal. The solvable characteristic increases with a higher temperature (>60°C) and is influenced by the pH level: favourable pH-level for fast dissolving is 10. Both MC and CMS are available in powder and granulates that can be used for production input. The characteristics of these materials are to some extent different from the regular polymers. The bending strength of both materials is comparable with those of traditional plastics. However, the stress resistance becomes significantly lower under long time exposure to humid environments. On the other hand, the stress relaxation parameters for these materials have values higher than the regular plastics. Therefore, they are both suitable for flexible elements. Apart from the technical characteristics, also the safety issue is important. Since humid environments can cause degradation of functional elements in a product, this causes obvious safety problems. Therefore this concept is only valid when strict boundary conditions can be guaranteed. The corresponding loosening times, using this technique, are summarised in. These times are not only influenced by water temperature, pH-level and water hardness, but also by the geometry of fasteners. Depending on the screwing length and size, the loosening times will vary as well. Therefore, thin nuts and bolts are to be preferred.

**Table 2.** Loosing times of Methyl Cellulose (MC) and Carboxy Methyl Starch (CMS) immersed in water

| | Worst scenario | Best scenario |
|---|---|---|
| MC | 2 days [1)] | 5.5 hours [3)] |
| CMS | 3.5 hours [2)] | 3.5 minutes [4)] |

[1), 2)] 20°C - contact length - cl 3mm [3)] 20°C - axial water supply-slit - cl 3mm [4)] 60°C - cl 2mm

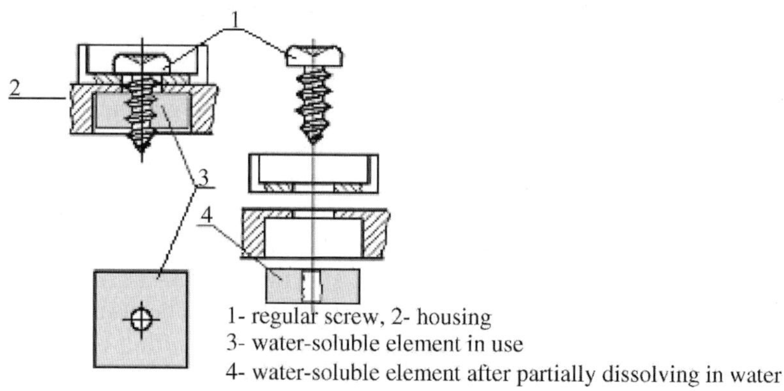


**Fig. 15.** Example of a water-soluble fastener

Neubert [26] describes freezing elements as a third alternative system for simultaneously disconnecting fasteners. Freezing elements make use of the substantial volume expansion of water when freezing, for generating displacement. Nevertheless, the energy efficiency of these systems is rather low (5.2%). The disadvantage of freezing elements is that they require a closed system, which takes a lot of space, especially in small products. Secondly, ice melts again under pressure. For this reason a safety

partially dissolving in water

factor needs to be included when determining the dimensions and actuation temperature of the freezing element. Figure 16 illustrates an example of this concept.

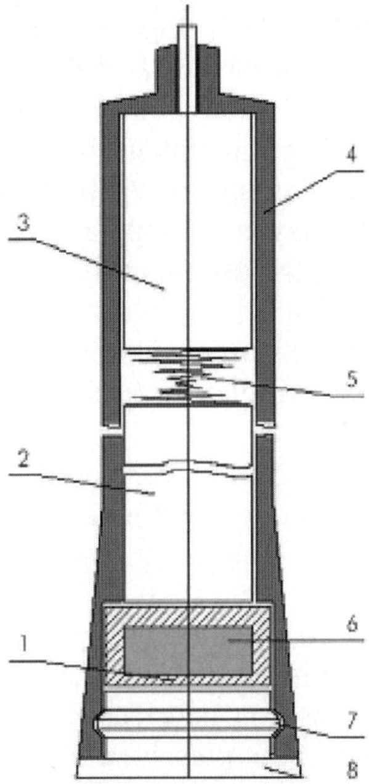


**Fig. 16.** Example of a freezing element as fastener in an electric toothbrush

## 2.3 Evaluation of the current techniques

All proposed techniques described above, are aiming at a reduction of the disassembly time and cost. However, the efficiency increase to be expected from them varies greatly between the different concepts. Table 3 presents an evaluation matrix where the alternative concepts are linked to criteria determining the disassembly effort. These criteria are based on those of the Disassembly Effort Index developed by Das et al. [27] and the Design for Disassembly guidelines listed by Jovane et al. [28]. Using the same procedure as presented in [27], an evaluation score was calculated. The following criteria are taken into account to obtain a single score per fastener type:

**Time.** Since most disassembly activities are manual, the disassembly time is a direct measure of the associated labour cost and hence the profitability of the activity [27]. During Active Disassembly the operator is only performing manual disassembly tasks for a fraction of the total disassembly time, t. Therefore, t is multiplied with a proportional factor, p, reflecting the manual time spend by the operator, to obtain the actual time spend and the corresponding effort score.

**Fixture.** Fixturing requirements have been one the primary impediments to efficient disassembly of large products [27]. Less contact is required, the more degrees of freedom the operator has for disassembling the product.

**Localise-Visible-Access.** Apart from time and fixture, localising and reaching fasteners prior to unfastening remains an important obstacle for efficient disassembly processes. Eliminating the visibility, reachability and direct contact requirements during disassembly operations, and thus reducing the disassembly time, diminishes the disassembly effort to pure unfastening.

**Assembly effort.** Disassembly solutions may not affect the assembly process. Therefore, additional parts or complex assembly actions are to be avoided.

**Level of integration.** Finally, the ability of integrating connections into one system, such that they can all be unfastened with a single disassembly action, is favoured because this causes significant disassembly time reductions.

Based on data found in the related literature an evaluation matrix was completed (see table 3). The relative importance of the five evaluation criteria, as well as the value-score transition matrices for each criterion are listed in table 4. Evaluating the concepts presented above shows that the Disassembly Embedded Design techniques demand a higher disassembly effort than the Active Disassembly concepts, which was expected. But additionally, some of the former techniques are even scoring worse than the commonly used mechanisms. When snap fits are replaced by disassembly embedded fastener types on a one to one basis, no significant efficiency gains can be expected. Only where complex disassembly operations can be substituted by a single disassembly embedded module, as is the case for the CRT with integrated NiCr heating wire, these disassembly mechanisms can offer economically viable alternatives.

**Table 3.** Evaluation matrix of Disassembly Embedded Design and Active Disassembly techniques

| Group | Technique | Time (sec), t | Proportial handling time (%), p | Actial time spend by operator, T= t*p | score, $s_1$ | Fixture | score, $s_2$ | Localise - visible - access | score, $s_3$ | Assembly effort | score, $s_4$ | Level of integration | score, $s_5$ | Total score, S= |
|---|---|---|---|---|---|---|---|---|---|---|---|---|---|---|
| | Regular snap fit | 15 | 100 | 15 | 3 | clamps | 9 | yes | 20 | med | 7 | low | 25 | 64 |
| Disassembly Embedded Design | Heating wire in CRT monitor | 20 | 100 | 20 | 4 | clamps | 9 | yes | 20 | high | 15 | low | 25 | 73 |
| | Pneumatic element in snap fit | 20 | 100 | 20 | 4 | 2 hands | 6 | yes | 20 | high | 15 | low | 25 | 70 |
| | Electro-magnetic snap fit | <5 | 100 | <5 | 0 | 1 hand | 3 | yes | 20 | high | 15 | low | 25 | 63 |
| | Heating wire in nut/bolt | 10 | 100 | 10 | 2 | 2 hands | 6 | yes | 20 | med | 7 | low | 25 | 60 |
| | Heat-activated joining structures | 20 | 100 | 20 | 4 | clamps | 9 | yes | 20 | high | 15 | med | 12 | 60 |
| | Self-activating SMA | 20 | 100 | 20 | 4 | 2 hands | 6 | yes | 20 | high | 15 | med | 12 | 57 |
| | Electro-magnetic nut/bolt | <5 | 100 | <5 | 0 | 1 hand | 3 | yes | 20 | med | 7 | low | 25 | 55 |
| | Piezo element in snap fit | <5 | 100 | <5 | 0 | 2 hands | 6 | yes | 20 | high | 15 | med | 12 | 53 |
| | Mechanical force in snap fits | <5 | 100 | <5 | 0 | 1 hand | 3 | yes | 20 | high | 15 | med | 12 | 50 |
| Active disassembly | Freezing elements | 60 | <10 | <6 | 1 | none | 0 | no | 0 | med | 7 | high | 0 | 8 |
| | Pneumo-elements | 20 | <10 | <2 | 0 | none | 0 | no | 0 | med | 7 | high | 0 | 7 |
| | Soluble nuts | >210 | <10 | >21 | 4 | none | 0 | no | 0 | low | 0 | high | 0 | 4 |
| | Hydrogen storage alloy | >210 | <10 | >21 | 4 | none | 0 | no | 0 | low | 0 | high | 0 | 4 |
| | Shape memory polymers | 20 | <10 | <2 | 0 | none | 0 | no | 0 | low | 0 | high | 0 | 0 |

**Table 4.** Values - score relationships as used in Table 3

| Evaluation criteria | weight% | % |
|---|---|---|
| Time (sec) | 25% | 5% |
| Fixture | 15% | 5% |
| Localise-visible-access | 20% | % |
| Assembly effort | 15% | 5% |
| Level of integration | 25% | % |

| Time (sec) [27] | |
|---|---|
| sec | score |
| <5 | 0 |
| 25 | 5 |
| 50 | 10 |
| 90 | 15 |
| 140 | 20 |
| >240 | 25 |

| Fixture [27] | |
|---|---|
| Type | score |
| none | 0 |
| 1 hand | 3 |
| 2 hands | 6 |
| Clamps | 9 |
| Winch | 12 |
| Automation | 15 |

| Localise-visible-access | |
|---|---|
| Yes/No | score |
| no | 0 |
| yes | 20 |

| Assembly Effort | |
|---|---|
| Level | score |
| low | 0 |
| medium | 7 |
| high | 15 |

| Level of integration | |
|---|---|
| Level | score |
| high | 0 |
| medium | 12 |
| low | 25 |

## 3 New opportunities

Next to the concepts encountered in literature, a number of additional operational principles can be proposed. State-of-the-art design techniques, such as a group brainstorm and TRIZ, were used by the authors in order to enhance creativity and to generate a list of concepts for fasteners suitable for Active Disassembly. Distinction was made between the trigger opportunities used to unfasten connections and the material properties needed to enable disassembly. For example, Chiodo [19] proposed the use of shape memory materials to ensure disassembly. Hereby, "heat" is the trigger and "shape memory material" characteristics are needed to provide the required effect. For the triggers, many principles were identified. From a mechanical point of view, disassembly based on centrifugal forces or on excitation of eigenfrequencies can be imagined. More inspiration can be found in changing material properties due to temperature changes or light/laser (e.g influencing brittleness), thermo-couples, lasers, etc. Finally, (bio-) chemical triggers can be imagined based on PH-level, enzymes, etc. The results of the brainstorm are listed in table 5.

The exploration of various material groups creates opportunities in view of exploiting characteristic material properties. The most elaborate group are the materials sensitive to *electrical triggers*. First there are the electroactive polymers. These materials respond to electrical stimulation with a shape or size change [29–30]. The advantage of these materials is that they are easy to mould, flexible and lightweight. Within this group of electroactive polymers (EAP) distinction is made between the ionic EAP's, that need to be used in aquatic environments, and the electronic EAP's, that are used in dry environments. They are both driven by electric fields but the electronic EAP's need rather high voltages. Vogan et al. [31] present a practical example of material made of a layer of dielectric elastomeric film surrounded by layers of compliant electrodes. The actuator has two stages: on and off. By putting high DC voltage difference on the electrodes these electrodes will move towards each other causing a shape change. Eury *et al.* [32] presented a similar example of electrostrictive polymers. The same triggering effect can be found in electro-active polymer gels [33, 34]. Polymer gels facilitate bending motions initiated by electric fields. To obtain deformation the material should be brought into an electric field. Through the design of the shape of gels, the direction of the motion can be manipulated. Current applications of these materials can be found in robots or artificial muscles. A second group of materials triggered by electrical fields are the electrochemical actuators. Onoda et al. [35] describe that polypyrole, as a doped material with electrons, shows an actuation initiated by an electric field. Examples of applications are not available yet.

Another set of materials can be triggered by *magnetic fields*. Firstly, in analogy with the electrically triggered materials, there are the *magneto-active* materials. These materials show a deformation induced by a magnetic field. The price of these materials depends very strongly of the strain that should be achieved. Commercial graded crystals achieve a strain of $1.6x10^{-3}$. For powders a strain of $6.5x10^{-4}$ can be expected, which limits the applicability of these materials [36]. Secondly there are the *magneto-striction induced shape memory alloys*. They can reach a strain up to 6% (Ni2MnGa) and 1.2% (FePd) due to a rearrangement of the martensite structure induced by a magnetic field of 300kA/m. These materials have two stable phases, similar as the shape memory alloys induced by temperature [36]. The maximal strain is higher than for the regular magneto-active materials and the speed is higher than the shape memory alloys induced by temperature. The large strain and high responsiveness are the main advantages of those materials. However, an important drawback of all magnetic induced materials is the occurrence of a strong hysteresis [37]. The last category comprises the *magneto-rheological* materials. In absence of a magnetic field these materials are in fluid state. But under a strong magnetic field, their viscosity can be increased in milliseconds by more than 2 orders of magnitude, exhibiting solid-like behaviour. Drawbacks are low yield strength, sensitivity to contaminations and demanding high voltages [38].

A last group of promising materials are triggered by *irradiation of light*. Ikehara et al. [39] describe photo-induced phase-transition (PIPT) materials that show bi-stability in their solid phase, affecting structure, optical properties, magnetic properties, etc. The properties of the material are changed by irradiation with light of fixed wavelength, as well as by temperature or external fields. The induced strains, due to the phase transition, were measured at max 2% in one direction, which makes them suitable for mechanical actuation. The main drawback common for all these materials is that they are single triggered which might cause safety problems. In every application the chance of reaching the trigger level during operational conditions should be investigated in order to avoid undesirable disassembly. The introduction of multi-triggered systems can solve this problem when the chance on simultaneous occurrence of both trigger levels is negligible. Moreover, for some materials high investments or training is required in order to obtain the triggering effect. This might put question marks next to the economic viability of some methods and needs to be thoroughly evaluated.

**Table 5.** Non exhaustive overview of concepts for Active Disassembly

| *External Trigger Mechanism* | *Possible trigger principle* | *Possible effect* |
|---|---|---|
| *Mechanical Force* | ▪ Centrifugal force<br>▪ Acceleration<br>▪ Water jet | **Deformation**<br>▪ Elastic deformation<br>▪ Plastic deformation<br>**Material failure**<br>▪ Erosion<br>▪ Splintering<br>▪ Breakage<br>**Function failure**<br>▪ Removal of blockage elements |
| *Vibration* | ▪ Mechanical vibration<br>▪ Sound waves<br>▪ Water waves | **Material failure**<br>▪ Destruction after reaching eigen frequencies<br>**Function failure**<br>▪ Loosening of tight fits |
| *Pressure* | ▪ Pressure variation (air / water) | **Deformation**<br>▪ Elastic deformation<br>▪ Plastic deformation<br>**Phase transition**<br>▪ Melting<br>▪ Evaporation<br>▪ Sublimation |
| *Electrical* | ▪ Electric current | **Deformation**<br>▪ Elastic deformation<br>▪ Shrinking<br>▪ Expansion<br>**Phase transition**<br>▪ Melting |
| *Chemical reaction* | ▪ Reagent in surrounding atmosphere (pH-level, hydrogen, $H_2O$, …)<br>▪ Submerging in reagent | **Deformation**<br>▪ Shrinking<br>▪ Expansion<br>**Material failure**<br>Changing material properties<br>▪ Corrosion<br>▪ Dissolving<br>▪ Pyrolisis<br>▪ Pulvarisation |

| | | |
|---|---|---|
| ***Thermal reaction*** | ▪ Joule effect<br>▪ Radiation (Laser, infrared rays, ...)<br>▪ Microwaves<br>▪ Submerging in hot water tubs | **Deformation**<br>▪ Elastic deformation<br>▪ Plastic deformation<br>▪ Shrinking<br>▪ Expansion<br>**Phase transition**<br>▪ Melting<br>▪ Evaporation<br>▪ Sublimation<br>**Material failure**<br><u>Changing material properties</u><br>▪ Creep<br>▪ Brittleness<br>▪ Viscosity change<br><u>Material breakage</u><br>▪ Thermal shock effect<br>▪ Diverse material expansion coefficient<br>▪ Inverse material expansion |
| ***Magnetic field*** | ▪ Presence of electromagnet (Magnetising vs. Demagnetising)<br>▪ Magnetic Ray Interference (MRI) | **Deformation**<br>▪ Elastic deformation<br>▪ Plastic deformation<br>**Phase transition**<br>▪ Solid to liquid<br>**Function failure**<br>▪ Attraction vs. Repulsion |
| ***Light radiation*** | ▪ UV-radiation | **Material failure**<br>▪ Surface corrosion<br>▪ Pulvarisation due to material property changes<br>▪ Brittleness |
| ***External Trigger Mechanism*** | ***Possible trigger principle*** | ***Possible effect*** |
| ***Biological action*** | ▪ Presence of bacteria<br>▪ Enzyms inducing chemical reactions<br>▪ Bionically designed systems | **Material failure**<br><u>Changing material properties</u><br>▪ Corrosion<br>▪ Dissolving<br>▪ Pulvarisation<br><u>Changing material phase due to eating bacteria</u><br>▪ Melting<br>▪ Evaporation<br>▪ Sublimation |

## 4 Conclusions

From an economic point of view the productivity associated with commonly available disassembly methods today seldom makes disassembly the preferred end-of-life solution for massive take back product streams. In consequence systematic reuse of parts or components, or recycling of pure material fractions are normally not achievable. Enhanced disassembly principles, in which external trigger signals allow to simultaneously reverse the action of multiple fasteners, forms a promising approach to low-cost mass disassembly.

In this paper a state-of-the-art of these emerging techniques is sketched, categorising the fasteners into Disassembly Embedded Design and Active Disassembly techniques. It is obvious that the described methods do not form an exhaustive list of techniques that could fulfil the criteria for self-disassembling products. Most presented concepts are still in an early stage of development and need further development and validation to prove their effectiveness. From a technical perspective a number of complementary joining techniques can be expected to emerge that can offer solutions for products operating in different functional environmental conditions. A more fundamental question that requires attention from policy makers is how appropriate ecotax systems can be conceived to link the cost and benefits associated with products designed for disassembly. As long as benefits of more efficient disassembly activities do not directly affect the product designers and the manufacturers, the incentives for implementing a systematic Design for Disassembly strategy will be low. The self-disassembly solutions reviewed in this paper however indicate that the technical obstructions making disassembly an unrealistic undertaking are gradually evaporating. This should pave the way for policy makers to redesign the take back obligation systems and to influence the design optimisation criteria used by product manufacturers, for example by introducing appropriate tax (reduction) systems.

## 5 Acknowledgements

This paper is based on research funded by the Institute for the Promotion of Innovation through Science and Technology in Flanders (IWT-Vlaanderen).

## 6 References

1. Nederlandse Tweede Kamer, Motie Lansink, Vergaderjaar 1979-1980, 15-800, nr.21.
2. Willems B., Dewulf W., Duflou J., 2004, Can Large-Scale Disassembly Be Profitable? A Linear Programming Approach to Quantifying the Need for Disassembly Time Reductions, Submitted in International Journal of Production Research.
3. Meerkamm H., 2003, Integrated product development as a key to sustainable products, Proc.ICED '03, Stockholm, Sweden, 19-21 August, Paper No.1673.
4. Boothroyd G., Alting L., 1992, Keynote Paper: Design for Assembly and Disassembly. Annals of the CIRP, 41, pp. 625-636.

5. Harjula T., Knight W., Boothroyd G., Rapoza B., 1996, Design for Disassembly and the environment, Annals of the CIRP, 45, pp.109.
6. Seliger G., Basdere B., Keil T., Rebafka U., 2002, Innovative Processes and Tools for Disassembly. CIRP Annals Manufacturing Technology, San Sebastian, pp. 37-40.
7. Scholz-Reiter, B., Scharke, H. and Hucht, A., 1999, Flexible robot-based disassembly cell for obsolete TV-sets and monitors. Robotics, Computers and Integrated Manufacturing, 15, 247–255.
8. Niewohner J., Renz R., 2003, Electronic Waste-quantification of the disassembly time, Proceedings of Colloquium e-ecological manufacturing, Berlin, p.111-113.
9. Masui K., Mizuhara K., Ishii K., Rose C., 1999,Development of products embedded disassembly process based on end-of-life strategies, Proceedings of the EcoDesign '99: First International Symposium on Environmentally Consious Design and Inverse Manufacturing, Tokyo, Japan, December 10-11, pp. 570-575.
10. Klett J., Blessing L., 2004, Selection and modification of connections, Proceedings of the 8th International Design Conference, Dubrovnik, Croatia, pp.1283-1288.
11. Klett J., Blessing L., 2003, Methodische Konzeption demontageunterstutzender verbindungen, Proceedings of the Design for X Symposium, Neukirchen, Germany, pp.71-78.
12. Klett J., Cosiglio S., Blessing L., Seliger G., 2004, Systematic development of easy-to-unlock connections and a flexible unlocking tool, Proceedings of the Global conference on sustainable product development and life cycle engineering, pp.965-968.
13. Braunschweig A., 2003, Flexibel Automatisierte Demontage - Flexibel Automatic Disassembly, Konstruktion, September 9, pp.54-58 .
14. Braunschweig A., September 23-25, 2004, Automatic Disassembly of snap-in joints in electromechanical devices, Proceedings of the 4th International Congress Mechanical Engineering Technologies '04.
15. Nishiwaki S., Saitou K., Min S., Kikuchi N., 2000, Topological design considering flexibility under periodic loads, Structural Multidiciplinary optimisation, pp. 4-16.
16. Li Y., Saitou K., Kikuchi N., 2001, Design of heat-activated reversible integral attachments for product-embedded disassembly, Proceedings of the EcoDesign '01: Second International Symposium on Environmentally Conscious Design and Inverse Manufacturing, Tokyo, Japan, December 11-15, pp.360-365.
17. Jones N., Harrison D., Billet E., Chiodo J., 2004, Electrically self-powered active disassembly, Proceedings of the institution of mechanical engineers - Part B: Journal of Engineering Manufacture, vol. 218 (7), pp.689-697.
18. Chiodo J., McLaren J., Billett E., Harrisson D., 2000, Isolating LCD's at end-fo-life using active disassembly technology: a feasibility study, Proceedings IEEE International Symposium on Electronics and the Environment , San Fransisco, California, USA, May, pp. 318-323.
19. Chiodo J., Billett E., Harrison D., 1999, Active Disassembly Using Shape Memory Polymers for the Mobile Industry, IEEE International Symposium on Electronics and the Environment - ISEE, Denvers, Massachusetts, USA, May 11-19, pp. 151 - 156.
20. Chiodo J., Jones N., Billett E., Harrison D., 2002, Shape memory alloy actuator for active disassembly using 'smart' materials of consumers electronic products, Materials and Design, Vol. 23, pp.471-478.
21. Chiodo J., Harrison D., Billet E., 2001, An initial investigation into active disassembly using shape memory polymers, Proceedings of the institution of mechanical engineers - Part B: Journal of Engineering Manufacture, Vol. 215 (5), pp.733-741.
22. Arnaiz S., Bodenhoefer K., Herrmann C., Hussein H., Irasarri L., Schnecke D., Tanskanen P., 2002, Invited session: Active disassembly using smart material (ADSM) A status report of the ongoing EU project, Care Innovation, November 25-28 Vienna.

23. Wilkes S., 2003, Active Disassembly - the end of electronic life as we know it, Materials World, 11 (11), pp.20-22.
24. Sakai K., Okada H., Tanigawa M., Yasuda T., 2003, Study of auto-disassembly system using shape memory materials, Proceedings of the EcoDesign 2003 Conference: 3rd International Symposium on Environmentally Conscious Design and Inverse Manufacturing, Tokyo, Japan, December 8-11, pp. 504-509.
25. Suga T., Hosoda N., 2000, Active disassembly and reversible interconnection, Proceedings IEEE International Symposium on Electronics and the Environment, San Fransisco, California, USA, May, pp.330-334.
26. Neubert H., 2000, Simultan lösbare Verbindungen zur Rationalisierung der Demontage in der Feinwerktechnik. Dissertation, Technical University of Dresden.
27. Das S., Yedlarajiah P., Narendra R., 2000, An approach for estimating the end-of-life product disassembly effort and cost, International Journal of Production Research, vol. 38 (3), pp. 657-673.
28. Jovane F., Alting L., Armillotta A., Eversheim W., Feldmann K., Seliger G. and Roth N., 1993, Keynote Paper: A Key Issue in Product Life Cycle: Disassembly, Annals of the CIRP, 42, pp.651-658.
29. Bar-Cohen Y., 2002, Electro-active polymers: current capabilities and challenges, Proceedings of the SPIE Smart Structures and Materials Symposium, EAPAD Conference, San Diego, California, USA, March 18-21, pp.1-7.
30. Ashley S., 2003, Artificial Muscles, Scientific American, October, pp.35-41.
31. Vogan J., Wingert A., Plante J., Dubowsky S, Hafez M., Kacher D., Jolesz F, 2004, Manipulation in MRI devices using electrostrictive polymer actuators: with an application to reconfigurable imaging coils, Proceedings of the 2004 IEEE International Conference on Robotics and Automation, New Orleans, Los Angeles, USA, pp. 2498-2504.
32. Eury S., Yimnirun R., Sundar V., Moses P., Jang S., Newnham R., 1999, Converse electrostriction in polymers and composites, Materials Chemistry and Physics, Vol. 61, pp. 18-23.
33. Otake M., Kagami Y., Ishikawa K., 2002, Shape design of gel robots made of electroactive polymer gel, Proceedings of the SPIE Smart Structures and Materials Symposium, San Diego, California, USA, March 18-21, pp. 194-202.
34. Otake M., Kagami Y., Inaba M., Inoue H., 2002, Motion design of a starfish-shaped gel robot made of electra-active polymer gel, Robotics and Autonomous Systems, Vol. 40, pp. 185-191.
35. Onoda M., Shonaka H., Tada K., 2004, A self-organized bending-beam electrochemical actuator, Current Applied Physics, In Press.
36. Pasquale M., 2003, Mechanical sensors and actuators, Sensors and Actuators, Vol. 106, pp. 142-148.
37. Nishi Y., Yabe H., Hagiwara N., Oguri K., Asaka T., Uchida H., Matsumura Y., Uchida H., 2001, Three types of magnetic field operated shape memory effects, Proceedings of the SPIE, Vol. 4234, pp.76-81.
38. Wang J., Meng G., 2001, Magnetorheological fluid devices: principles, characteristics and applications in mechanical engineering, Proceedings of the Institution of Mechnical Engineers, Vol. 215, pp. 165-174.
39. Ikehara T., Tanaka M., Shimada S., Matsuda H., 2001, Optically-driven polymer actuator using reversible photo-induced phase-transition phenomena, Proceedings of the SPIE, Vol.4234, pp.215-222.

# Indicators to measure sustainability of an industrial manufacturing

E. Raizer Neto, M.T. Mariotte and R.T.P. Hinz
*Sociedade Educacional de Santa Catarina, raizerneto@ig.com.br*

Abstract: Most of the industries have mechanisms to measure their economical efficiency through economical and financial indexes, consolidated globally. The same does not happen with the measure of their efficiency in relation to sustainable development. The sustainability indicators foreseen in the ISO series, are difficulty to measure, have high cost for determination of some indicators, they turn impracticable to use for a great part of the world companies and without conditions of being applied by the small industries and services companies. In this work it is proposed 25 indexes, easily measurable, as electric power consumption, water, material balance, mass of liquid, solids and gaseous effluents generated.

Keywords: Indicators, Eco-Efficiency, LCA

## 1 Introduction

All of the industrial or services companies, well structured, measure their efficiency from several indicators consecrated by the use. Among the main ones, we mentioned the profitability, the return on the investment, the participation of the company in the market and profitability. The easiness found to establish the financial and economical indexes, is not found when you intend to measure the evolution of the measures of social and environmental behavior. When the company applies any modification in the productive process aiming reach an environmental requirement or a Clean Production program or LCA, there is need to measure the effects of the decisions taken and their evolution. How to know to choose between two processes, which the more environmentally soundly? A lot of proposed indicators appeared in the last ten years and the main is that foreseen in the ISO 14000 series. However, some of the indicators foreseen in this regulation are complex for their determination and many of them extremely



*D. Brissaud et al. (eds.), Innovation in Life Cycle Engineering and Sustainable Development,* 111–122.

expensive as, for instance, to identify the contribution of the company on to deterioration of ozone's layer. This work, besides revising, to comment on the indicators developed in other research centers, it still presents the indicators which we established to accompany the eco-efficiency measures in a small company of metals heat treatment and a great company of compressors, both of the State of Santa Catarina, Brazil.

## 2 Review

The need to monitoring the industrial activities and development has been motivating many researchers in the creation of sustainability indicators. Daly [1] summarizes three indicators, that for us, are confused with the objectives of the sustainable development:

1. the rate of use of natural renewable resources should not be larger than the rate of renewal of the resource.
2. the rate of use of the no renewable resource does not exceed the renewable substitute's rate.
3. the pollution generated should not exceed the assimilation capacity of the environment.

After `92 the United Nations (UN) has been multiplying efforts to measure the sustainability degree. UN created in 1995 the Work Programme on Indicators of Sustainable Development, with the intention of defining the indicators and their application methodology in five years. In a first attempt, United Nations [2] 134 indicators were presented, what was done with that Scientific Committee on Problems of the Environment [3] (SCOPE) organism of International Council of Scientific Unions (ICSU), it presented a work proposing:

1. an agreement on the terminology to be adopted
2. the identification, definition of a group of indicators with political and analytical relevance, contained in three great groups (environmental impact and environmental and social conditions), as well as the measuring methodology.

In the literature there are several meanings for **indicators**, "a parameter, a value, an index, a relative weight of information etc." Olsthoorn [4], Moldan, Billharz and Matravers [5] Adriaanse, [6] they make an extensive revision, including some discrepancies in relation to the conceptual aspect of the indicators. Wackernagel and Onisto [7] affirm: We are far away of reaching the sustainability. But, how much far away? If we are not capable to measure we don't have as taking decisions¨. This study proposes a methodology and makes the analysis of the sustainability degree in 52 countries that together represent 80% of the population and 95% of world GDP. Other important works present the fundamental concepts on the indicators, among them we mention Berkhout and others [8] they present countless considerations for the establishment of indicators. MEPI - Measuring the Environmental Performance of Industry) looking for an answer for some very important subjects, about the comparison of indicators for different companies from identical industrial activities. They also discuss the hypothesis of Porter and Linde , if these is or not some relationship between the economical performance and the environmental sound. Wehrmeyer, Tyteca and Wagner [9], Ditz & Ranganathan, [10]; Azzone & Noci, [11] Wehrmeyer & Tyteca, [12] present a great revision about the necessary amount of indicators. Gee & Moll, [13]; Wright, Allen, Clift, and Healthy, [14]. present application in very specific examples. This article intends to present some indicators which we adopt in a small company of heat treatment of metals as well as in a company of compressors. These indicators were taken in consideration of values physically easy to measure and their relationship to sustainability.

## 3 Indicators structure

The initial concern was to establish indicators for productive systems. In our attempt to understand initially some hypotheses should be made for the validation of these indicators, what presented to proceed. In this first work we present as boundary condition, the space for the producing unit.

The impact on the sustainable development due to the manufactures has Internal influences, referred to the control volume and external influences, related with the external variables. The main internal influences are:

1. the input of raw material in natural state.
2. source of electric power generation.
3. transformation process.
4. generation of residues and toxicity.

External influences are:

a. industrialized raw material inputs.
b. product usage by the consumer
c. final disposition of the product after reaching the useful life.

There is not a priori, a greater or smaller importance on one above the other. There are cases, for instance, which the industry of the bread-making, whose larger impact is concentrated in the planting tasks, crop and storage of the wheat than in the bread´s production and, where the external influences are preponderant. In these cases, the indicators improvement should also be driven where the influence is preponderant. When the dispersion of these external influences is high, the actions depend much more on government plans than on the producers, properly it would be impossible the application of individualized plans of improvement.

## 4 Source of indicators

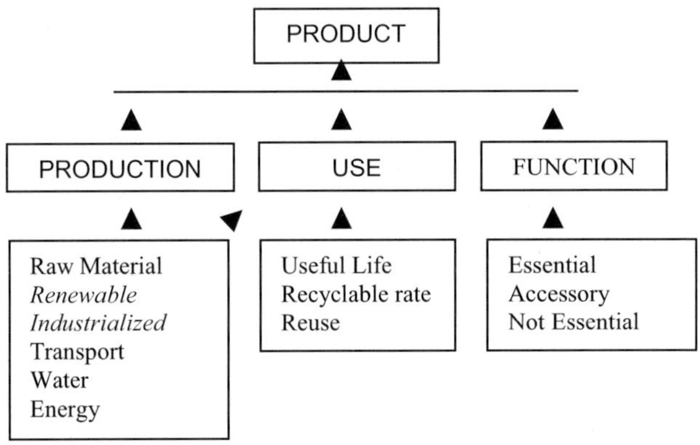


**Fig. 1.** Variables concerned to the product indicator

An indicator represents a tendency and only this. The purpose of an indicator is the establishment of the initial value and their continuous evolution after accomplished modifications. The comparison of a series of the same indicator however, can supply a scene of the factory is a national or world level, as well as to identify how this industry is in comparison with similar one. This will serve, for instance, to indicate which unit operations need to concentrate efforts to reach the wanted degree of sustainability. The indicators established contemplate the following aspects:

PRODUCT
- Production
- Uses
- Function

PROCESS
- Transformation
- Generation of Residues
- Pollution

MANAGEMENT
- Investments in technology
- Investments and operational costs of the waste treatment
- Economical dependence regarding the raw materials
- Dependence regarding the consumer market

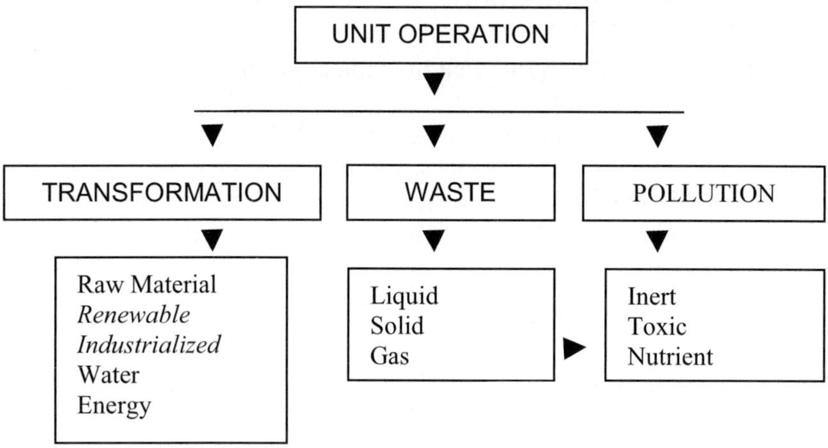


**Fig. 2.** Variables of the process indicators

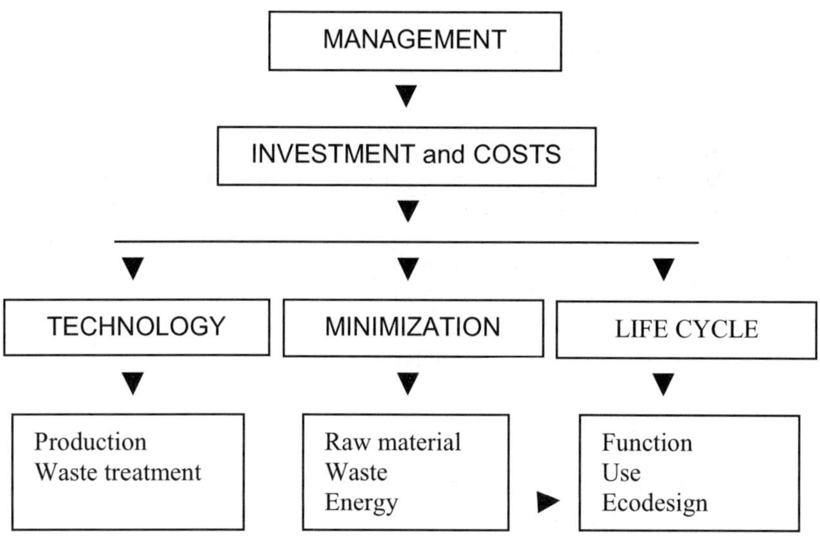


**Fig. 3.** Management influences on indicators

Accomplished the justifications and the conception used, to establish the indicators, will be to proceed their presentation and a brief interpretation of each one and their influence on the sustainability and development of the industry.

# 5 Indicator establisment

We have imposed as a rule that the indicators should have the following properties:

1. dependent on variables easy to measure and on any enterprise scale-up.
2. to use physical greatness as mass, volume, flow, heat etc.
3. when this is not possible coefficients are used where valuation can be objected and improved however of concept universally accepted.

## 5.1 Product Indicators

We have established the following indicators related with the production.
$IMP_D$ = relative Impact to the direct materials used to manufacture the product.
This indicator allows to identify the contribution of each material for the sustainability of the process and globally, allowing to take eco-design actions, minimization and substitution procedures.

$$IMP_{Di} = \frac{m_i}{IMN_i * m_p} \tag{1}$$

Total indicator of raw materials

$$IMP_D = \frac{1}{m_p} \sum_{1}^{n} \frac{m_i}{IMN_i} \tag{2}$$

IMNi -rate of renewal of the material i.

mi - mass of the natural resource i in the mass of the product p

mp - mass of the product p

IMN = rate of renewal of the material.

$$IMN_i = \frac{vu_p}{nr_i} \tag{3}$$

nr - necessary time for the renewal of the material in the nature

vu - useful life of the product.

IMA = relative Impact of the water used in the product

$$IMA = \frac{m_a}{m_p} \tag{4}$$

$m_a$ - mass of water in the product p

$m_p$ - mass of the product p

IME = relative Impact to the consumption of energy.

$$IME_i = \frac{ce_i}{CEN_i * mp} \text{(kwh/kg)} \quad (5)$$

Total impact due to the energy

$$IME = \frac{1}{mp}\sum_{1}^{n}\frac{ce_i}{CEN_i} \quad (6)$$

The electric power, thermal or mechanics have different origins, in some cases with the combustion of fossil fuels, water in hydroelectric, radioactive material, coal, wood etc. The mass equivalent consumption of the fuel is given for:

$$C = \sum \frac{ce_i}{CEN_I * IEN_I} \text{(kg)} \quad (7)$$

This value represents a consumption of combustible, raw material, that it should be added $IMP_D$.

$ce_i$ - expense with energy of the type i in a period t

mp - mass of products manufactured in the period t

$CEN_I$–unitary cost of the energy i

$IEN_I$ energy produced by unit mass of combustible. (wind, water, nuclear, coal, fuel)

IMT - Impact due to the transport of materials and products.

This indicator, like the previous, also transforms the values of financial expenses in fuel mass equivalent consumed for the transport and it also represents consumption of raw materials and, should be added to $IMP_D$.

$$IMT = \frac{(tm + tp) * 0{,}5}{cd * mp} \quad (8)$$

tm - expense with transport of raw materials

tp - expense with transport of products

mp mass of products transported with the expenses (tm+tp)

cd - unitary cost of the fuel mass

The value of the total impact of raw materials used will be, therefore

$$IMP_T = IMP_D + \frac{C}{IMN_c} + \frac{IMT}{IMN_t} \quad (9)$$

## 5.2 Relative to the use

Most of indicators does not contemplate the raw materials consumed by the user. We thought that the development of a product should take into account this factor. As example, we mentioned the case of the electric coffeepot and extraction coffee (Italian) with direct steam. Figures 4 and 5 represent the flow of these two machines. It is easy to be observed, that steam coffee machine produces smaller amount of wastes. It has larger useful life and it should present a larger sustainability level than an electrical machine. We intend to demonstrate that to the same function we can have many products with different amount of raw material consumed by the end user. Therefore not always when production or process indicator are appropriate does not mean that the product really contributes to the sustainable development.

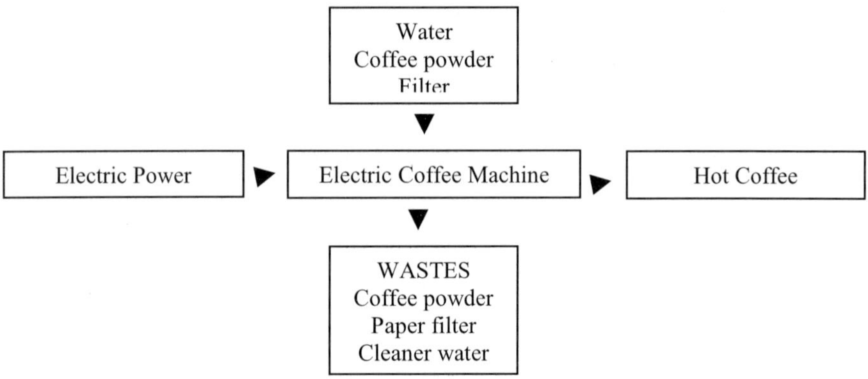


**Fig. 4.** Flowchart of an electric coffee machine

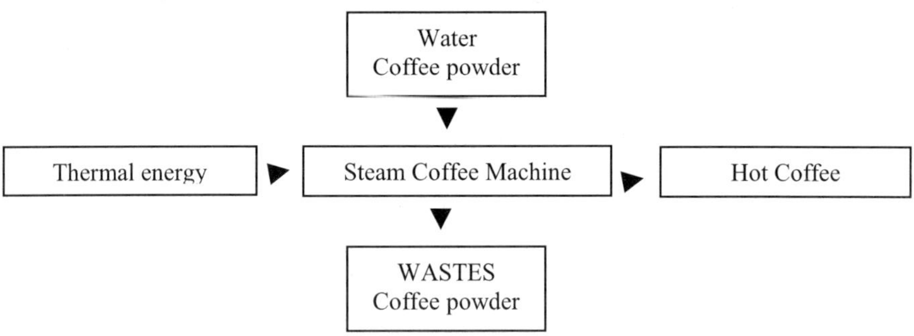


**Fig. 5.** Flowchart of a steam coffee machine

These indicators have the same structure of the previous (series IM_), differentiates only in term of calculation base, raw material and goods produced by the final consumer.

$$IUP_{Di} = \frac{m_i}{IMN_i * m_p} \tag{10}$$

$$IUP_D = \frac{1}{m_p} \sum_{1}^{n} \frac{m_i}{IMN_i} \tag{11}$$

$$IUA = \frac{m_a}{m_p} \tag{12}$$

$$IUE_i = \frac{ce_i}{CEN_i * m_p} \tag{13}$$

$$IUE = \frac{1}{m_p} \sum_{1}^{n} \frac{ce_i}{CEN_i} \tag{14}$$

$$C = \sum \frac{ce_i}{CEN_I * IEN_I} \tag{15}$$

$$IUP_T = IUP_D + \frac{C}{IEN_c} \tag{16}$$

The notation is similar than the production indicator the only difference is: the series IU are related to the consumption and production of the end user.

## 5.3 Indicators of technology

### *Relative to the transformation*

(rate of transformation of the raw material by mass of product)

IPM = rate of raw material (i) transformation. This indicator is indirectly associated to the applied technology.

$$IPM_i = \frac{m_i^T}{m_p * f_i} \tag{17}$$

$m_i^T$ total mass of raw material **i** acquired in a certain time(t) to produce product p

$m_p$ total mass of product p manufactured in certain time(t)

fi fraction of raw material i in the product p

$$IPM = \frac{\sum m_o^T}{m_p} \tag{18}$$

***Relative to the residues and pollution***

It relates the conversion of raw material in liquid, solid or gaseous wastes.

IPL = Indicator of liquid effluent generation

$$IPL = \frac{m_e}{m_a + m_{il}} \tag{19}$$

$m_e$ mass of liquid effluent in a certain time(t)
$m_a$ mass of consumed water in a certain time(t)
$m_{il}$ mass of liquid raw material consumed in a time(t)

IPS = Indicator of the generation of solid residue

$$IPS = \frac{m_s}{m_{is}} \tag{20}$$

$m_s$ mass of solid waste in a time (t)
$m_{is}$ mass of solid raw material consumed in a certain time (t)

IPG = Indicator of the generation of gaseous effluents.

$$IPG = \frac{m_g}{m_{ig}} \tag{21}$$

$m_g$ mass of gaseous effluents in a time(t)
$m_{ig}$ mass of gaseous raw material consumed in a time(t)

## 5.4 Indicators of management

Many times mistakes in administration of an effective technology interfere in an entire process. We put in evidence some indicators which allow to show these cases.

IGP - it indicates the investment evolution in technologies.

$$IGP = \frac{IGN}{IGT} \tag{22}$$

IGN - investments in innovation in a elapsed time(t)
IGT - total investments in technologies in a certain time (t)

IGR = Evolution of Investment in Waste Treatment

$$IGR = \frac{IGW}{IGP} \tag{23}$$

IGW- Investment in Waste Treatment
IGP – Investment in Production

IGM - Evolution of Investment in waste minimization

$$IGM = \frac{IMR}{IGN} \tag{24}$$

IMR – Investment in Waste Minimization.

TGM - Rate of improvements caused by the investment in waste minimization.

$$TGM = \frac{IMR}{VM(m_t - m_{t-1})} \tag{25}$$

VM - value of the material i

$m_t$ - consumed mass of raw material after innovation

$m_{t-1}$ - consumed mass of raw material before innovation

## 6 Conclusion

These indicators are easy to obtain because they are current of the mass and energy balances, already accomplished by most of the companies. We have been applying in two types of industries of Santa Catarina, Brazil, and we did not have difficulties in the attendance of their monthly evolution. Some of them needs to be perfect to best assist the needs of the companies, especially those when operate for production systems for order, without a continuous production line. In the case of the unit of metals thermal treatment there is a need to adapt it to the countless variations of the process in function of the several applied materials. Although they are not still perfect they have shown better than those foreseen in ISO 14040, especially for the difficulty of measurement. Our team continues to research the adaptations which should suffer the indicators previously presented, as well as we are developing for a company of hermetic compressors, works about identification of eco-design application starting from the established indicators which allows to identify which that components should be firstly modified, function of their contribution to the sustainable development.

## References

1. Daly, H., - Steady State Economics - Island Press, 1991
2. United Nations, - Indicators of Sustainable Development - U.N. 1996
3. Projection Indicators of Sustainable Development SCOPE, Wiley & Sounds. 1997
4. Olsthoorn, X., Tyteca, D., Werhmeyer, W., Wagner, M. – Environmental Indicators for Business, A Review of the Literature and Standardisation Methods - Journal of Cleaner Production – November – 2000

5. Moldan, B., Billharz, S., and R. Matravers, - Sustainability Indicators: Report on the Porter, M., van Linde, C.. Green and competitive: ending the talement. Harvard Business Review (September-October), 120-134. 1995
6. Adriaanse A., - Environmental Policy Performance Indicators - SUK -1993.
7. Wackernagel, M., Onisto, L. - La Ecologica impronta. Quanta natura viene utilizzata e quanta è anchor disponibile? - Dossier of WWF – 1995.
8. Berkhout, F. et al. Measuring the Environmental Performance of Industry(MEPI) EC Environment and Climate Research Programme:Research Theme 4 - 2001
9. Wehrmeyer, W., Tyteca, D., Wagner, M. How many (and which) does Indicators are necessary to compares the Environmental Performance of Companies? The sectoral and statistical answerer. The 7th European Roundtable on Cleaner Production, 2-4 May 2001, Lund, Sweden
10. Ditz, D., & Ranganathan, J. Measuring Up: Toward Common Framework goes Tracking Corporate Environmental Performance. Washington: World Resources Institute. (1997)
11. Azzone, G., & Noci, G. Defining Environmental Performance Indicators: An Integrated Framework. Business Strategy and the Environment, Vol. 5, 69-80. (1996)
12. Wehrmeyer, W., & Tyteca, D. Measuring Environmental Performance goes Industry: From Legitimacy to Sustainability? The International Journal of Sustainable Development and World Ecology, Vol. 5, 111-124. . (1998)
13. Gee, D., & Moll, D. Information goes Sustainability: Eco-Efficiency Indicators. Copenhagen: European Environment Agency (EEA). (1998)
14. Wright, M., Allen, D., Clift, R., and Healthy, H. Measuring Corporate Environmental Performance: The ICI Environmental Burden System. Journal Industrial of Ecology, Vol. 1(4), 117-127. (1997)

# Concepts and definitions for product recovery

## Analysis and clarification of the terminology used in academia and industry

M. Lindahl, E. Sundin, J. Östlin and M. Björkman
*Division of Environmental Technology and Management, Division of Production Systems, Department of Mechanical Engineering, Linköping University, SE-581 83, Linköping, Sweden, Mattias.Lindahl@ikp.liu.se*

Abstract: This paper presents and clarifies the academic and industrial terminology used in the area of product recovery. It is concluded that there exist many different concepts and definitions in academia and industry, several of which are unclearly defined. Given this, a new way to define product recovery is presented through the use of a model. This model is based on actual industrial product recovery cases, existing academic product recovery concepts and definitions and product design theory. The presentation contains a holistic model that can be used for describing and analyzing different product recovery scenarios. In addition, several industry cases are presented as a verification of the model.

Keywords: Product Recovery, Remanufacturing and Refurbishment

# 1 Introduction

Product recovery has traditionally been viewed as an economically beneficial alternative to the ordering of new products. Product recovery is interpreted as a superior concept that involves concepts like reuse and recycling. The aim with product recovery is to retrieve a product's inherent value when the product no longer fulfills the user's desired needs. A product is in this context interpreted as an artifact produced to fulfill a user's desired needs. This also implies that material is interpreted as a product. During the last century, the industrialized world has put limited focus on product recovery. Instead, the main focus has been on the production of products from virgin materials (i.e. non-recycled). For several different reasons, the focus has now shifted to product recovery. For example,



*D. Brissaud et al. (eds.), Innovation in Life Cycle Engineering and Sustainable Development,* 123–138.

countries like China with a large population and huge material needs are driving up material prices. In parallel, society's awareness of the environmental problems of the present use of material and products has grown [1]. Today, much of the raw materials and products in society are used only once, and instead of product recovery, new products and materials are produced. The new production of products as well as the treatment of used and unwanted products causes a lot of environmental problems. As a result, the European Union (EU) has issued several directives to change the rate of material and product utilization in order to reduce its environmental impact [2]. Examples are Waste Electronic and Electrical Equipment (WEEE) [3], Restricting the use of Hazardous Substances (RoHS) [4], Integrated Product Policy (IPP) [5] and several producer responsibility directives, e.g. on vehicles [6]. For example, WEEE aims primarily to prevent the accumulation of waste containing electrical and electronic products, and at the same time promote reuse and material recycling of these kinds of products. According to this directive, the members of the European Union shall encourage design and manufacturing of electrical and electronic products that facilitates dismantling and recycling, especially reuse and recycling of components and materials from these products. [3] Furthermore, the EU has developed an Integrated Product Policy (IPP) that aims at reducing the usage of material and the environmental impact of waste [5]. Product recovery is breaking through into new industry sectors, e.g. because of increased material costs. Some industry sectors - for example the automotive industry - have extensive experience in this area when compared to other expanding sectors such as personal computers and cellular phones. The different industry sectors have developed their own, and not standardized expressions and terminology for their business within the product recovery area. They refer to themselves, for example, as "remanufactures", "rebuilders", "retreaders", "rechargers", "refurbishers", and "reconditioners" [7]. The vast number of terms within the product recovery area is a source of confusion within and between both industry and academia. Because of this, researchers and industrial practitioners could mean the same thing but name it differently, or alternatively use the same concepts but interpret them differently. The increasing number of companies entering the product recovery business, combined with the expanding research in this area, highlights the need for a holistic mapping of terminology in this area. Such a mapping, it is envisioned, could help clarify and describe the existing relationships between different academic and industrial concepts, and could facilitate both academic and industrial communication and collaboration.

## 2 Aim

The first aim of this paper is to identify and clarify the terminology for product recovery scenarios used by academia and industry. The second aim is to describe a holistic model for researchers to map different industrial product recovery systems in order to facilitate research in this area, e.g. communication and comparisons of different product recovery systems. The model describes different types of product recovery. The basis for this model is the existing experience found within both industry and academia, i.e. real industrial cases and existing concepts and definitions.

## 3 Methodology

The methodology for conducting this research was to study research within the area of product recovery, as well as the industry of product recovery itself. In regards to the theoretical study presented in this paper, a number of popular theories in academia were compared and analyzed. Furthermore, the industrial cases of Swedish product recovery presented here highlight the use of several concepts. These case studies were conducted through Internet surveys and telephone interviews. Finally, several industrial cases were utilized to help verify the authors' proposed model for product recovery, of which some, but not all, are presented in this paper.

## 4 Research definitions

Products can be recovered in many ways, and this recovery can be performed at different levels. For example, at a lower level, it could be the product materials that are recovered. In this case, it is often called "material recycling". Product recovery on a higher level, where product parts or modules are reused, is often called "remanufacturing", "reconditioning" or "refurbishment". A remanufactured product is often the term used to describe a worn-out/broken/used product that has been restored to its original specifications or has been modernized and upgraded to new specifications. Hence, remanufacturing not only promotes the multiple reuse of materials, but it also allows for the steady upgrading of product quality and functionality, and does this without the need to manufacture completely new products and scrap used ones [8]. It is important to avoid confusion in terminology, for example between reuse and recycling, as the environmental impact of reuse will usually be much less than recycling [9]. Recycling involves reprocessing products or components into basic material, which are subsequently used as input in new manufacturing. According to one observer, *"the confusion is arising because industry is keen to suppress the idea of re-use, whilst the environmental movement has failed to promote it"* [10]. In other words, industry is acting on an assumption that recycling and subsequent manufacturing will be more profitable than re-use, while campaigners have not been effective enough in promoting re-use. There exist many definitions for product recovery. The following definitions of reuse, remanufacturing, reconditioning, refurbishment, component cannibalization and material recycling illustrate the complexity of describing the product recovery area:

**Reuse is** ...the process of disassembling products to recover useable parts and assemblies for the purpose of utilizing them in newly manufactured products [11]...the additional use of a component, part or product after it has been removed from a clearly defined service cycle [12].

**Remanufacturing is** ...an industrial process in which worn-out products are restored to like-new condition. Through a series of industrial processes in a factory environment, a discarded product is completely disassembled. Usable parts are cleaned, refurbished, and put into inventory. Then the new product is reassembled from both old and, where necessary, new parts to produce a unit fully equivalent -and sometimes superior - in performance and expected lifetime to the original new product [13]... when a product is cleaned and repaired to return it to a 'like new' state; often requires completely disassembling the product [14]... the process of rebuilding a product. During this process the product is disassembled, defective components are replaced and the product is reassembled, tested and inspected to ensure it meets newly manufactured product standards [11]... an industrial process whereby products are restored to like-new condition. In contrast, a repaired or rebuilt product normally retains its identity, and only those parts that have failed or are badly worn are replaced or serviced [15]... the most economically sustainable form of reuse and recycling of manufactured goods, and it can be defined as the industrial process where worn out products, referred to as cores, are brought back to original specifications and condition [16]... the practice of disassembling, cleaning, refurbishing, replacing parts and reassembling a product in such a manner that the part is at least as good as, or better than, new. By remanufacturing a product, the product may be returned to service with a reasonably high degree of confidence that it will endure (at least) another full lifecycle [17]... an industrial process whereby products referred as cores are restored to useful life. During this process the core passes through a number of remanufacturing steps, e.g. inspection, disassembly, cleaning, part replacement/refurbishment, reassembly, and testing to ensure it meets the desired product standards [8].

**Reconditioning is** ...the process of restoring components to a functional and/or satisfactory state but not above original specification using such methods as resurfacing, repainting, sleeving, etc. [16].

**Refurbishment is** ...the process in which a product or component is cleaned and repaired in order to make a resell [18]... when a product is cleaned and repaired to return it to a 'like new' state [14].

**Component cannibalization is** ...the process in which a limited number of components are extracted from a product for recovery [18]... when parts or components are taken off of one item and used to repair or rebuild another unit of the same product [14].

**Material recycling is** ...the process by which materials otherwise destined for disposal are collected, processed, and remanufactured into new products. Composting is a form of recycling [19]...the process in which the structure of a product is destroyed in order to recapture its materials [18]...when a product is reduced to its basic elements, which are reused [14]...the series of activities, including collection, separation, processing, by which products or other materials are recovered from or otherwise diverted from the solid waste stream for use in the form of raw materials in the manufacture of new products other than fuel [12].

### *Summary of Research Definitions*

As previously stated in the introduction, one of the major problems for understanding and communication in the area of product recovery are the numerous definitions in use for the same concept. Furthermore, there are also many similar concepts describing more or less the same thing. Some of these different concepts are used by industry to define the quality level of the product being recovered. This is the case, for example, for the definitions of reconditioning and refurbishment. Reconditioning/refurbishing is often used when the product is only restored to its original specifications [20]. Remanufacturing, in any event, is becoming the generic term for the process of restoring discarded products to useful life [7]. If the rebuilding of the product is not extensive, i.e., if few parts are to be replaced, either of the terms *reconditioning* or *refurbishing* is commonly used by industry [8]. Another and related obstacle is that several of the existing definitions use general and unclearly defined concepts, making these concepts tricky to use due to the fact that different users may interpret them differently. An example of this is the confusion resulting from the ambiguity of the terms "product" and "components" (see Table 1).

**Table 1.** Relationship between several concepts used in the model and those used in academia and industry

| Model concepts | Alternative academic and industrial concepts |
|---|---|
| Function provider | Product, Core |
| Sub function provider | Product, Part, Spare part, Component, Core, Module |
| Manufacturing | Production |
| Recovery management | Closed-loop supply chain, Reversed logistics |
| Remanufacturing | Refurbishment, Reprocessing, Reconditioning |

## 5 Industrial cases

In the product recovery process, as we have seen, the condition of the returning cores can be in different states and thus suitable for different product recovery options. It is important to note that there is a gap between the terminology used in academia and that used in practice. To illustrate some of the problems associated with this gap, several of the industrial cases investigated for this paper are presented here.

### 5.1 Computer case

This case considers a process where faulty computers, monitors and other equipment undergo a recovery process. The condition of the computers is determined through an inspection, using specific software which can identify malfunctioning components that have to be replaced. Through the use of various software applications a reprocessing activity is created, whereby the memory of the computer is erased and then brought back into operation. The computer is then sold to a new customer, thereby entering another use phase. The company uses the term "refurbishment" to describe this process. Another company which primarily deals with warranty claims for OEMs and broken equipment uses a similar process. In this case, items are tested, and if possible, parts are replaced or repaired. The repaired items are then sent back to the same customer to be used again. This process is referred to as "reparation", and the product is repaired. There is, however, and important difference in these two cases: in the first, the computer is sold to a new customer; in the second, the computer remains with the original user. This implies that the remanufacturing company's relationship to the user is an important aspect for the terminology used to describe the product recovery process.

## 5.2 Engine case

Here we consider a process with two different outputs: one that meets or exceeds the level found in new production, and one that is in line with customer demands. In the first case, the item is completely investigated and all of the parts that not are in a "like new" condition are replaced. The items are then reassembled and tested, and the resulting item is referred to as "remanufactured". To create an option for the cost sensitive customer, another method is used. In this process, all of the parts are still inspected, but only those not meeting *customer demands* are replaced, meaning even parts that do not qualify as "like new" can be used once more. The resulting item from this process is referred to as "functional repaired". These processes are more or less the same, but the resulting product quality levels are different. This implies that the type of recovery process has no direct influence on the term used to describe these types of recovery processes.

## 5.3 Heat exchanger case

In another case of product recovery for plate heat exchangers, the plates are disassembled and thereafter cleaned and reprocessed. After reassembly, the heat exchangers are tested. According to the company, this process produces products which are essentially as good as new, although the terminology used by the company to describe these products is "refurbished". This shows that the quality of the product can be referred to with different terminology depending on the industry sector.

## 5.4 Toner cartridge case

A laser printer consists of many levels of sub assemblies and different parts; one of them is the toner cartridge. When the toner level in the printer runs low, a need for refilling arises, and the empty cartridge must be removed from the printer and sent for product recovery activities. First the cartridge is inspected to determine which of the parts can be replaced or repaired. Then the cartridge is reassembled, refilled and tested, exiting the process as a "good as new" product. The cartridge can continue its new use phase in the same printer or in a different one. This case illustrates how sub functions in a product that are considered to be maintained/repaired can undergo a recovery process. It also shows that, depending on which of the functions go through recovery, how different terminology is used.

## 6 Function providers

According to existing theory about product design, described by for example Roozenburg and Eekels [21], the presumptive user is interested in benefits the product may provide in fulfilling various kinds of needs. This could be, for example, a product's means to provide desired functions. A product is, according to Roozenburg and Eekels [21], a material system which is made by people for its properties. Because of these properties, it can fulfill one or more functions. By fulfilling functions a product satisfies needs, thereby giving users the possibility to realize one or more values. In some cases, the customer and the user are one in the same, but this is not a necessity. In accordance with the theory concerning product design, manufacturing companies tend to strive to fulfill users' needs by producing artifacts with a desired function. This is also in line with the terminology used to describe product recovery in this paper. Thus, instead of focusing on the recovery of "products" or "components" or "materials", the focus is put on the "functions" which manufacturers produce to satisfy their customers' needs. This implies that a product can be interpreted as a function provider (FP). Depending on the perspective and focus, a product or a component can be interpreted as a combination of products and components, in the same manner a FP can be seen as a combination of sub function providers (SFPs).

## 7 The generic life cycle of function providers

The mapping in Figure 1 presents some of the more used concepts relations to each other. Based on this, Figure 2 was developed to describe a generic model for describing different FP concepts life cycles. Before the model in Figure 2 is further described, however, some concepts used in the model are described.

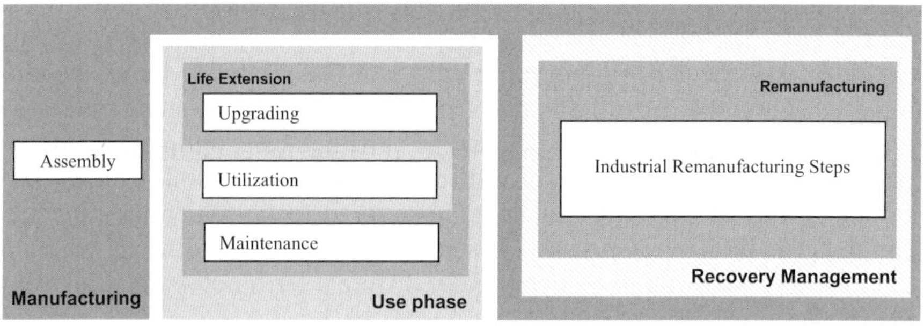


**Fig. 1.** Mapping of different concepts and their relationships

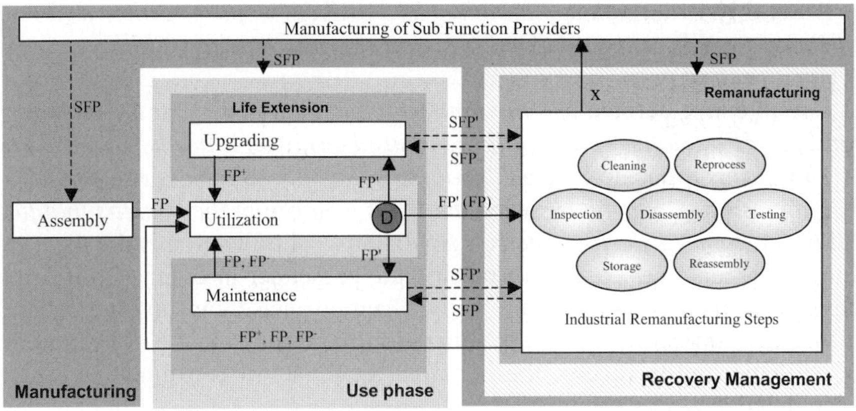


**Fig. 2.** Generic Life Cycle of Function Providers, with the abbreviations Function Provider (FP) and Sub Function Provider (SFP), where X represents one or more FP's or SFP's that are transformed into SFPs (One example of this is material recycling, where the X then represents the FP's or SFP's that are treated by "Sub Function Provider Manufacturing" e.g. at a refinery or recovery factory)

## 7.1 Definitions of concepts used in the model

All concepts used in the proposed model have been collected from the existing theory in order to avoid introducing yet more concepts that could cause more confusion. However, the different concepts' definitions have been transformed and translated in the respect that they are related to various kinds of FPs instead of, e.g. products, parts and components. Table 1 shows the relationship between some of the model's concepts and those used in academia and industry.

**Function provider (FP)** is defined as *"an artifact with a specific function that is provided by a combination of sub function providers, aimed to fulfill a customer's desired need"*. The FP's functions can be described and defined and must more or less fulfill and correspond to the *customer*'s need; otherwise, the FP will most likely be rejected.

**Sub function provider (SFP)** is defined as *"a subordinated FP that is used in a FP"*. An example of a SFP could be the toner cartridges for a laser printer. If the major focus is on the printer, then the printer becomes the FP and the cartridge the SFP. However, if the major focus is on the toner cartridge, then the toner cartridge is the FP and e.g. the toner material the SFP.

**Manufacturing** is defined as *"an organized process to manage, assemble and develop FPs"*.

**Assembly** is defined as *"an organized process to put together different SFPs into a FP, with a defined function that is aimed at fulfilling a customer's desired need"*.

**Use phase** is defined as *"the time from the FP delivery to the customer until the FP no longer can fulfill the desired customer's needs"*.

**Life extension** is defined as *"processes that extend a FP's use phase"*.

**Utilization** is *"the phase when the FP is generating the benefits that fulfill the customer's desired needs"*. An example of utilization is the time the car is in actual use, i.e. is transporting the passenger in order to fulfill the desired need of transportation. The utilization phase can involve several different *customers*. During the utilization there may exist a consumption of SFPs needed to receive the benefits. Examples of consumed SFPs are gasoline, detergents and toner.

**Maintenance** is defined as *"a process to preserve the FP's specific function"*. This may involve steps such as cleaning, inspection, disassembly, testing, storage, reassembly and repair. Examples include performing inspections to foresee and prevent malfunctions, or checking oil pressure and repairing malfunctioning SFPs, such as spark plugs and batteries.

**Upgrading** *is "a process when the FP's specific function is improved by adding, for example, better and more advanced SFPs"*. This may involve steps like cleaning, inspection, disassembly, testing, storage and reassembly. The result is an upgraded FP in comparison with the original. An example can be a computer (FP) that is upgraded with a new processor (SFP).

**Recovery Management** *"encompasses the management and industrial process that set out to retrieve the intrinsic value of a FP"*.

**Remanufacturing** is defined as *"an organized process to retrieve a FP's intrinsic value by combining and re-organizing different FPs into a new FP that corresponds towards a consumer's desired needs"*. This process can be divided into several process steps, which Sundin [8] refers to as the Generic Remanufacturing Process.

## 7.2 Principal description of the model

The *continuous lines* in the model shown in Figure 2 illustrate the flow of FPs, while the *dashed lines* illustrate the flow of various SFPs needed to accomplish the different processes during a product's lifecycle. Before using the model, there is a need to decide the focus for the description, i.e. the FP. Is it, for example, a forklift truck or a forklift truck's engine? In the first alternative, the forklift truck is the FP and the engine is the SFP, and it is this case which will be used as an example in the following description.

The forklift truck (FP) is assembled from many SFPs; the engine is but one example. During the utilization phase (which can involve many different owners as well as users) there is consumption of different SFPs, e.g. fuel, to receive the function. Later during utilization, either the FP's SFPs lose their ability to provide the specific function (e.g. the engine malfunctions) or customer needs are altered in such a manner that the FP no longer fulfills their desired needs[1]. An example of the later is when the lifting capacity is too small compared to the required capacity. The result is a degraded FP labeled FP'. The degraded FP can be treated in three different ways depending on the customer's decision taken at point D. The first two ways are life extension within the use phase, while the third is recovery management, implying that the FP leaves the present use phase.

In the first alternative, the customer decides to upgrade the FP to a $FP^+$. $FP^+$ represents a FP' that has been transformed into a new FP with, when compared to the original FP, higher functionality. The customer is in charge of the upgrading, and decides what to do with the FP. The upgrading process may involve several of the industrial remanufacturing process steps, and may imply that some SFP are replaced with new ones. It may also imply that certain SFP's (e.g. the engine) are sent away for remanufacturing at a remanufacturing company to upgrade the SFP (e.g. increase the power) and then sent back (note the dashed lines in Figure 2). The second alternative is that the customer decides to maintain the functionality[2] (FP) or want a lower level ($FP^-$) than the original functionality. This alternative is valid for ordinary performed maintenance. In accordance with upgrading, this may imply that some SFPs are replaced and that some are sent for remanufacturing and then sent back (see dashed lines in Figure 2). The third alterative is that the customer decides to reject the FP'. This alternatives differs from the two above in that the responsibility for the FP' is handed over to a manufacturer, usually a remanufacturing company. It is then up to the remanufacturing company too decide, depending on the status etc., what to do with the FP'. The results may be, for example, a $FP^+$, a FP or a $FP^-$ that are sent back for a new use phase, but it may also a X. Xs represents either one or many FP's or SFP's that are transformed to SFPs - SFPs that are used in other types of FPs. This implies e.g. that the fork lift trucks' engines are used for powering a pump station and that other SFPs are used in newer, more modern fork lift trucks. Important to note is that the resulting FP may be bought back by the original customer.

1 This is not valid for demo products.

2 Functionality is defined as "the combination of functions and their behavior that contribute to making the FP useful for an intended purpose."

## 8 Industry case examples

As an example of how the model works, some of the cases presented earlier are described using this model. The model has been evaluated with more than 20 industrial cases; however, only three of those cases are presented in this paper.

### 8.1 Computers case

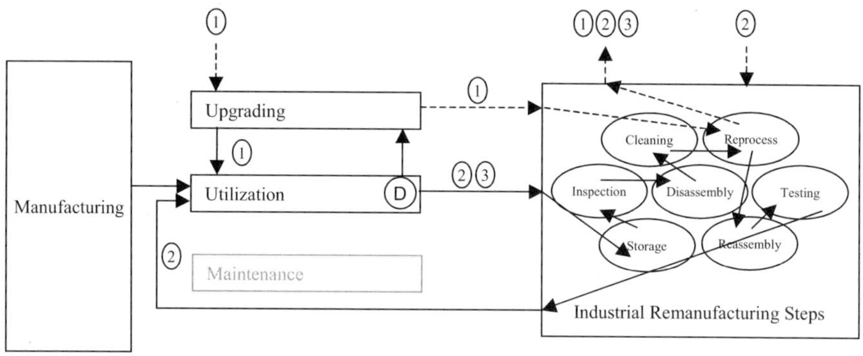


**Fig. 3.** Computer case (FP = A Computer)

Consider a computer (FP) not satisfying a customer's need; in this scenario, the function must be restored. There are several options: an upgrade can be made on the product, the product can be sold as it is to a different customer, the product can be resold or delivered to a remanufacturer for further use in the next use phase, or it can be sent to material recovery for scrapping. For the upgrading of a computer, sub functions (SFPs) such as the CPU are replaced with other SFPs, and sent for recovery for their material value (1). When the product leaves the use phase and enters the recovery process, various actions take place depending on the aim of the recovery option. Different quality levels exit the remanufacturing process and enter the next use phase (2). Some materials and cannibalized parts exit the process as SFPs, and can be used in other functions. The FP can also be material recovered (3). For an illustration of the options 1–3 applied on the model, see Figure 3.

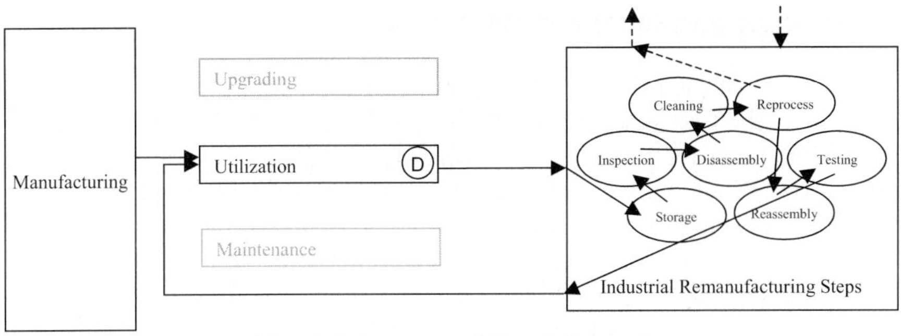


**Fig. 4.** Printer case (FP = A Printer)

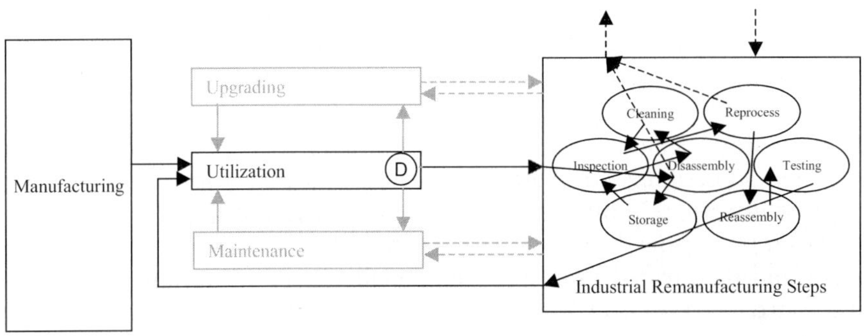


**Fig. 5.** Toner cartridge case (FP = A Toner Cartridge)

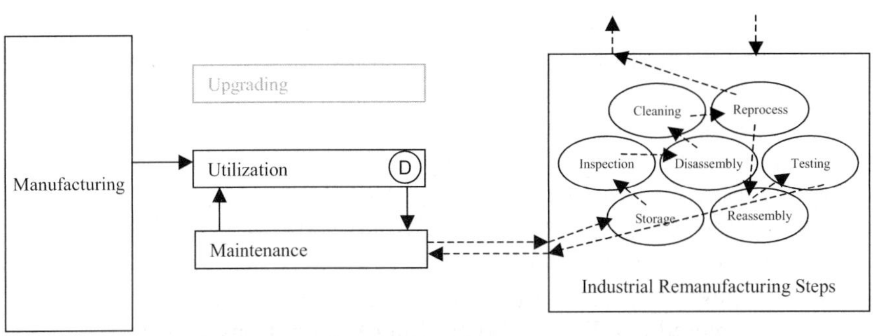


**Fig. 6.** The Single Use Camera Case (FP = A Single Use Camera)

### 8.2 Printer and toner cartridge case

The cases of the printer and the toner cartridge are illustrated in Figures 4 and 5. In the first case, the printer is considered as the FP, as seen in Figure 4. The toner (SFP) in the toner cartridge (SFP) is consumed during the utilization. When a toner cartridge runs empty, a replacement of the cartridge is needed to ensure the printer's (FP's) functionality. The printer usually newer change it is physical location during this process. A new toner cartridge (SFP) is asembled into the printer and the empty toner cartridge is sent for a remanufacturing process and undergoes different steps that are further described in the next paragraph. The remanufactured toner cartridge is then either returned to the same printer or a similar printer. In the second case, the toner cartridge is considered to be the FP, as shown in Figure 5. When the toner, in this case considered to be a SFP, is consumed, the toner cartridge loses its functionality and must be remanufactured. The toner cartridge is first sent to a remanufacturer, where it is stored until inspection of the cartridge. Next, an appropriate recovery option is chosen. Disassembly, cleaning and reprocessing of the items in the cartridge is carried out, some items are material recycled, and others are cannibalized to be used elsewhere. Following this, the items that exit the process are then replaced by various SFPs. In the next step, the cartridges are refilled with toner and then tested to ensure functionality before returning to the new use phase.

### 8.3 The Single-use camera case

The case of the single-use camera case is presented in Figure 6. Consider a camera (the FP) with which a customer has taken all available pictures on the camera film (SFP). The customer then returns the single-use camera to a photo shop, where the film is removed and the camera put in storage awaiting shipment back to the OEM. At the OEM, the single-use cameras follow the sequence of sorting (inspection of shape), disassembly, cleaning, inspection, repair, reassembly and inspection. In the reassembly step, new film is added. The reassembled single-use camera is then returned to a photo shop, and a new use phase begins.

## 9 Discussions and conclusions

This paper illustrates through a number of both academic and industrial examples the conflicts between several of the existing concepts and definitions used in area of product recovery. Furthermore, it is concluded that several of the concepts are inappropriately defined. One problem is the use within the definitions of general concepts where "product" and "component" are two such examples. These concepts are widespread, but the interpreted meaning differs depending on the interpreter's context and focus. For example, what may be considered for one person to be a product, e.g. a toner cartridge, might be viewed by another as a component in a printer. This may result in misunderstandings and confusion in the dialogue between researchers and in the analysis of different product recovery scenarios. Furthermore, some of the concepts, e.g. refurbishing, represents a degree of remanufacturing. At the same time, the definitions and borders between the different degree concepts are vague and unspecified.

One advantage with the proposed idea and concept of function providers is that the focus is on the function, the inherent value of the artifact. To describe the different used function providers facilitate the analysis and understanding of for example a product recovery system. The proposed concept also implies that there is no need to use an abundance of different concepts, e.g. product, part and component, to define the needed concepts, e.g. remanufacturing. In fact, the only needed concept is *function provider* and, as defined, the sub function provider is just a lower degree of function provider - but still a function provider. A major contribution of this paper is the holistic model with its related concepts and definitions. As described, this model can be used in order to describe and analyze different companies' products and product recovery concepts. The conclusion is that the function provider concept and the holistic model provide a more general and consistent way to describe product recovery scenarios. This, in relation to the existing terminology and models that have been found in academia and industry. Hopefully this holistic model will represent a starting point for continued work towards more harmonized and standardized terminology that would be beneficial for the product recovery area.

## Acknowlegements

This research has been conducted within the project "Sustainable systems and products for reuse and remanufacturing". This project is funded by VINNOVA, The Swedish Agency for Innovation Systems. We give a special thanks to Professor Yoshiki Shimomura and Doctor Tomohiko Sakao at Research into Artifacts, Center for Engineering (RACE). The University of Tokyo, Japan for fruitful discussion and co-operation.

## References

1. Ryding, S.O., 1995, Miljöanpassad produktutveckling, Sweden, Industrilitteratur AB.
2. Tukker, A., Haag, E., and Eder, P., 2000,Eco-design: European state of the art - Part I: Comparative Analysis and Conclusions - An ESTO project report. EC - Joint Research Centre Institute for Prospective Technological Studies: Brussels, Luxembourg. p. 60.
3. EU, Directive 2002/96/EC of the European Parliament and the Council of 27 January 2003 on waste electrical and electronic equipment, Journal of the European Union,.
4. EU, Directive 2002/95/EC of the European Parliament and or the Council of 27 January 2003 on the restriction of the use of certain hazardous substances in electrical and electronic equipment. Journal of the European Union.
5. EU, Integrated Product Policy - Building on Environmental Life-Cycle Thinking. 2003, Commission of the European Communities: Brussels. p. 30.
6. EU, Directive 2000/53/EC of the European Parliament and of the Council of 18 September 2000 on end-of life vehicles. Journal of the European Communities, p. 34-42.
7. Lund, R., 1996, The remanufacturing Industry: Hidden Giant. Boston, Massachusetts.
8. Sundin, E., 2004, Product and Process Design for Successful Remanufacturing, in Production Systems, Department of Mechanical Engineering., Linköping University.
9. Cooper, T., 1994, Beyond recycling: the longer life option. London, UK: New Economics Foundation.
10. Fairlie, S., 1992, Long distance, short life: Why big business favours recycling. The Ecologist. Vol. 22 (No. 6).
11. Seaver, W.B., 1994, Design Considerations for remanufacturability, recyclability and reusability and reusability of user interface modules. in Proceedings of IEEE International Symposium on Electronics and the Environment, San Francisco, CA, USA.
12. Keoleian, G., Menerey, D., 1993, Life Cycle design guidance manual - environmental requirements and the product system. United States Environmental Protection Agency.
13. Lund, R.T., 1983, Remanufacturing Technology Review. p.19-29.
14. Rogers, D., Tibben-Lembke R.S., 1998, Going backwards: Reverse logistics trend and practices.
15. APICS, APICS Dictionary. 9th Edition ed. 1998, Falls Church, VA, USA: APICS - The Educational Society for Resource Management. p. 81.
16. Amezquita, T. and Bras, B., 1996, Lean remanufacturing of an Automobile clutch. in Proceedings of First International Working Seminar on Reuse. Eindhoven.
17. Bras, B., Hammond, R., 1996, Design for Remanufacturing metrics. in Proceedings of First International Seminar on Reuse, Eindhoven.
18. REVLOG - The European Working Group on Reverse Logistics. 1999.
19. Kopicky, R., 1983, Reuse and Recycling-Reverse logistics opportunities. Council of Logistics Management.
20. Ijomah, W.L., Bennet, J.P., Pearce, J., 1999, Remanufacturing: evidence of environmentally conscious business practice in the UK. Proc. of IEEE International Symposium on Electronics and the Environment, Danvers, USA.
21. Roozenburg, N.F.M., Eekels, J., 1995, Product Design: Fundamentals and Methods, West Sussex, UK: John Wiley & Sons, Ltd.

# Remanufacturing of flat screen monitors

C. Franke, S. Kernbaum and G. Seliger
*Department Assembly Technology and Factory Management, Institute for Machine Tools and Factory Management, Technical University Berlin*

Abstract: The remarketing of used flat screen desktop monitors is evolving into a profitable market segment. Demand for used or non-functional monitors is growing especially in Eastern European nations and cannot be covered by End-of-Life (EoL) products, yet. Non-functional phase-out models are often sold to repair shops in Eastern European countries, which is often associated with improper EoL treatment of non-reusable components. To avoid negative environmental impacts caused by today's practice in demand markets with slack environmental regulations, remanufacturing must start at the returned product's origin. This paper discusses economical, technological and environmental aspects in remanufacturing of flat screen monitors. Legal recycling requirements for the monitor's components are described. Applying Method Time Measurement, a disassembly study is conducted for selected monitor models. Based on the results quality classes are defined for reusable components. Remanufacturing challenges in collection, testing, disassembly, and program planning are identified, and a prototypical hybrid disassembly system is presented.

Keywords: Remanufacturing, Flat Screen Monitors, Recycling, Disassembly, Reuse

## 1 Introduction

The ongoing shift from cathode ray tube (CRT) to liquid crystal (LC) displaying technology is evident, with the global market for all Flat Panel Displays (FPDs) surpassing US-$ 59,9 Bio. in 2004 [1]. The environmentally benign manufacturing, use, and End-of-Life (EoL) treatment of LCD products are a challenge for all manufacturers, users and recyclers dealing with this strongly growing technology (figure 1). In this paper, flat screen monitors, being a dominant market segment for the fast dissemination of LCDs serve as a representative product to access options for the environmentally benign treatment of used and returned products. By 2010, ninety percent of all desktop monitors are likely to use LCD technology. In the



*D. Brissaud et al. (eds.), Innovation in Life Cycle Engineering and Sustainable Development,* 139–152.

first step the technology is briefly presented to stress its economic significance and potential environmental impacts. Furthermore, legal recycling requirements and flat screen monitor specific product attributes are analyzed showing how monitors are being treated at their EoL today. A disassembly analysis of a representative model shows the effort required to conduct a non-destructive separation into components to enable the (partial) reuse of the product. Finally, challenges for a profitable remanufacturing are pointed out.

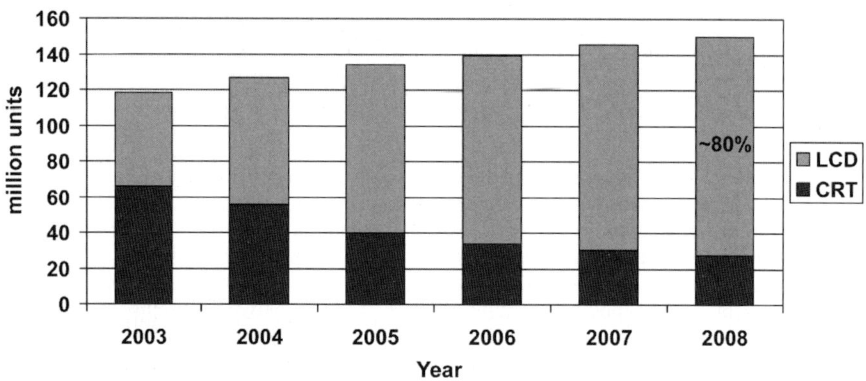


**Fig. 1.** Market share for CRT and LCD desktop monitors

## 2 Market, technology and environment

Today, the predominant technology for flat screen monitors are thin film transistor (TFT) displays, also known as active matrix (AM) LCD. Today, nearly all AMLCD production is located in Asia, including the manufacturing of glass substrates. Assembly of flat screen monitors is mainly concentrated in East Asia, also. Very few companies have located some low capacity assembly activities in Germany, e.g. the Korean manufacturer Hansol. The German company Merck AG has established its position as the worldwide leading supplier of liquid crystals, with a 60 % market share. An overview on the main brands for LCD monitors and the brands considered in this study is given in figure 2. Demand for flat screen monitors today is growing strongly. Used monitors are hardly available due to the low market penetration. In 2005 alone, about 137 Mio. desktop CRT and LCD monitors are expected to be sold. With 97 Mio. sold AMLCD monitors their market share will reach about 71 percent [3].

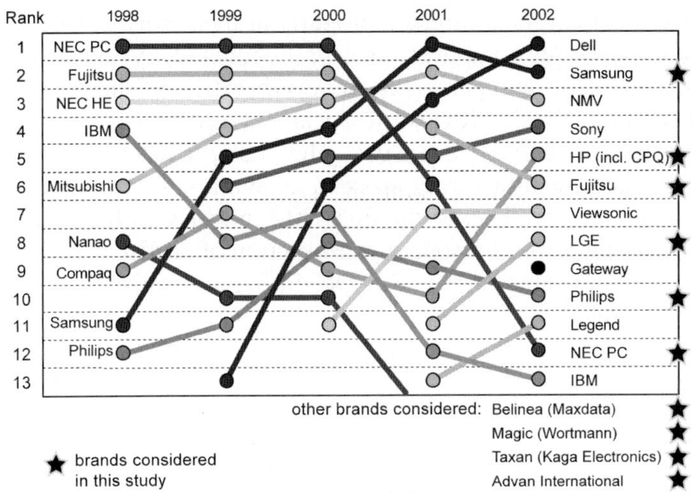


**Fig. 2.** LCD monitor brand ranking

To estimate the return volume of used monitors that are to be disposed in 2005 one can assume the average monitor's usage time being approximately 4 years, i.e., considering the sales volume of 2001. In that year, about 100 Mio. desktop monitors were sold, with the AMLCD monitor's market share being 16 percent only, i.e., 16 Mio. units. According to a US National Safety Council (NSC) Report [4] about 45% of computers and peripherals are being used for a second time, i.e. 55 percent, or 8,8 Mio AMLCD monitors will be disposed. In addition to that one needs to consider those monitors that were used more than once, i.e. those that will return in 2005 but were sold before 2001. According to NSC [4] the average period for second usage is 2,5 years, i.e., 45 percent of monitors sold between mid 1998 and mid 1999 are used until 2005. With 1,3 Mio. AMLCD monitors sold in 1998 and 4,5 Mio. in 1999 [5] one can estimate the number of units to be disposed in 2005 form this period with 1,3 Mio. [(0,5*1,3+0,5*4,5)*0,45], resulting in a total of 10,1 Mio. AMLCD desktop monitors to be disposed worldwide in 2005. Prices for used and non-functional 15"-18" monitors with control failures from private users – determined via the online market place eBay™ – usually range from € 40 to approximately € 110, as long as the LCD panel glass is not broken. Defective monitors from leasing returns collected by larger IT-remarketing companies from institutional users are frequently exported to Eastern Europe, where high demand meets cheap labor. An adequate EoL treatment of these monitors is at least very questionable. From an ecological point of view, the shift from CRT to AMLCD technology can be considered a step towards a sustainable development in the information and telecommunication sector, if monitors are treated properly at their EoL. A Life Cycle Assessment of desktop computer displays conducted by the University of Tennessee in 2001 compares the environmental and health impacts of AMLCD and CRT technologies [6]. As an example, AMLCD manufacturing is water and energy intensive. The manufacturing life stage, i.e., mainly the glass manufacturing process, accounts for over 75 % of the renewable resource use. Due to the 60%-70% lower weight as compared to CRT displays the natural resource consumption of AMLCD is significantly lower. Energy consumption of AMLDC during use is also 60%-70% lower than for CRT displays.

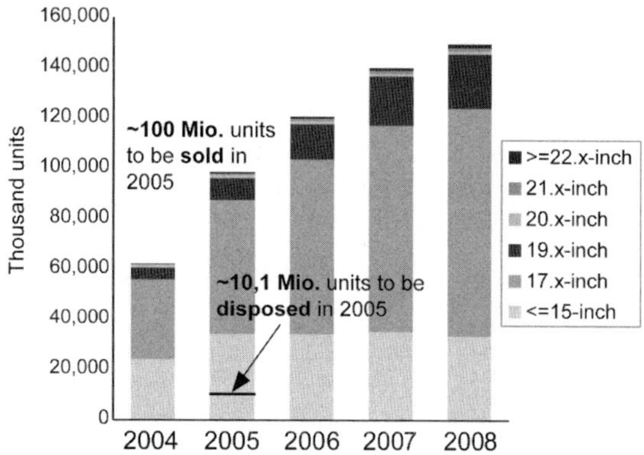


**Fig. 3.** Worldwide AMLCD desktop monitor sales, own calculation and [3]

## 3 Recycling

In Germany today, only a few hundred tons of LCDs from flat screen monitors, including LCDs from TV sets return for recycling. By 2012 the quantity of returned LCDs from these applications is expected to reach approximately 4.000 tons. In Article 7 (2) of the European Directive on Waste Electric and Electronic Equipment (WEEE), the reuse and recycling targets for LCDs (group 3) are defined with 65% of the product's weight [7]. The Annex II of the WEEE obliges disassembly of all LCDs with a surface greater than 100 square centimeters and all those back-lighted with gas discharge lamps, prior to material recycling. Figure 4 depicts the elements of a LCD module. The most relevant recycling facts for flat screen monitors are summarized below. Data obtained by a literature review is complemented by the analysis of a 4 year old 15" flat screen monitor of a popular brand.

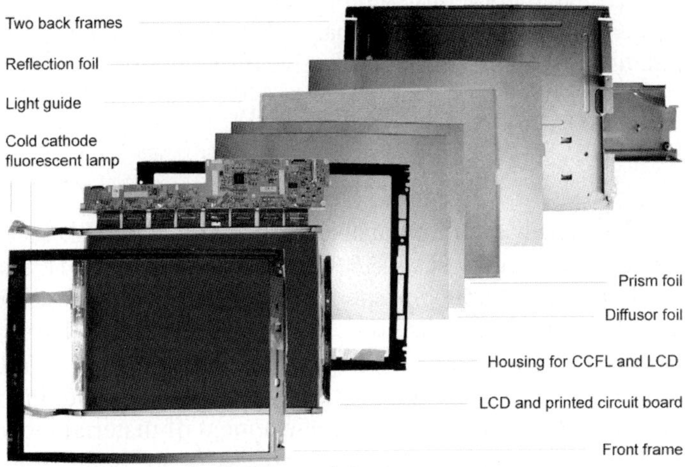


**Fig. 4.** Components of a LCD module

## 3.1 Liquid Crystal Display (LCD)

The LCD without printed circuit boards consists of approximately 87,2% glass, 12,7% foil and 0,1% liquid crystals [8]. According to a statement by the German Federal Environmental Agency concerning the ecotoxicology of liquid crystals in liquid crystal displays, LCDs do not require special disposal, due to the content of liquid crystals [9]. Available and economically profitable recycling processes use LCD's glass as surrogates, e.g., instead of quarry sand in steel smelting processes, and plastics, i.e. foils, as an energy source for the smelting process [8]. Recycling technologies for the recovery of the glass fraction for reuse in LCD manufacturing are still in an experimental state [10].

## 3.2 Cold Cathode Fluorescence Lamps (CCFLs)

CCFL contain small quantities of mercury and require special treatment. CCFLs must be disassembled from the LCD module. Following the prognosis of the IZT [10] between 290 kg and 480 kg mercury will have to be disposed for all 80 Mio. flat screen monitors that will be in use by the year 2010.

## 3.3 Printed Circuit Boards (PCBs)

Due to the mandatory disassembly and separate treatment of LCDs, printed circuit boards need to be separated from the LCD's glass and foil compound, usually by means of destructive disassembly.

## 3.4 Plastics

For the flat screen monitors analyzed, halogen free, i.e., bromide and chloride free plastics, e.g. PC + ABS -FR(40) or PC - FR(4), and acrylic glass were found. If separated properly these plastics can be material recycled for reuse in similar applications. Figure 5 depicts the relative distribution of material fractions for an analyzed 15" LCD module (1790 grams) and the flat screen monitor without stand (4070 grams) including the LCD module, as depicted in figure 5. The achievement of material recycling targets of 65% or higher can be looked upon as being realistic if material fractions are separated before recycling. With 78% of the monitor's weight being perfectly separable ferrous metal, and plastic components, targets can be achieved even neglecting the glass and copper fractions from LCDs, respectively PCBs, and wiring.

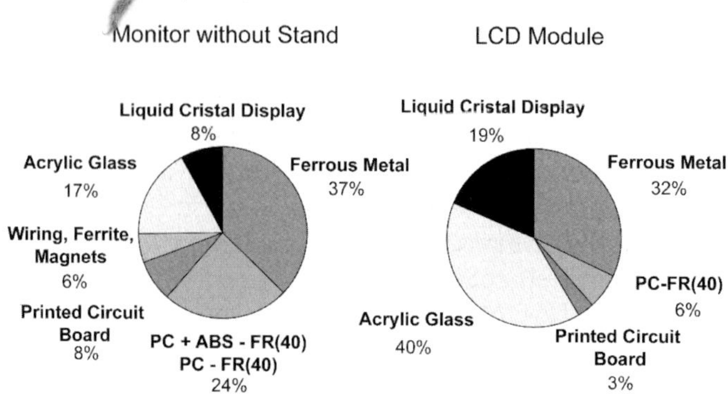


**Fig. 5.** Relative distribution of material fractions

Extending the usage phase of flat screen monitors beyond today average of four years is ecologically reasonable, due to the highly energy intensive production of LCDs and PCBs. Whether non-functional monitors can be prepared for another usage cycle economically depends on the costs for take back, identification, testing, disassembly, spare part procurement, and re-distribution. By means of a disassembly analysis components suitable for reuse, and the necessary disassembly effort are identified.

# 4 Disassembly analysis

The disassembly analysis was conducted for 17 flat screen monitors to module level. Non-destructive disassembly of the LCD module is usually economically unfeasible with the LCD being the module's primary value, and being hardly repairable. Destructive disassembly of the LCD including the separation of all components, i.e. frame, foils, housing, CCFLs, and PCBs requires less than one minute. The only common repair option for the LCD module is the replacement of the two CCFLs. Therefore in the following, CCFLs are defined as a module of the flat screen monitor.

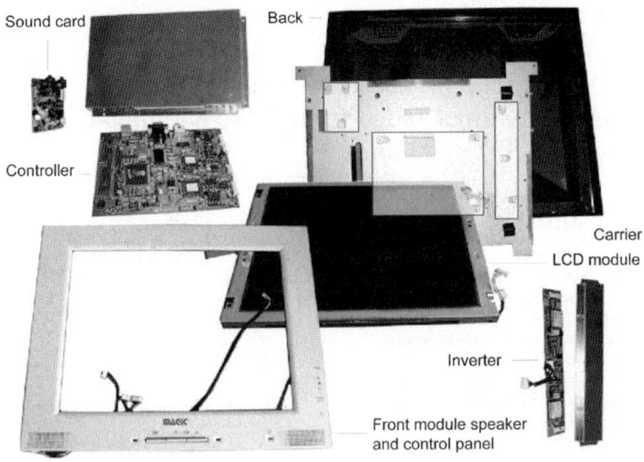


**Fig. 6.** Flat screen monitor without stand

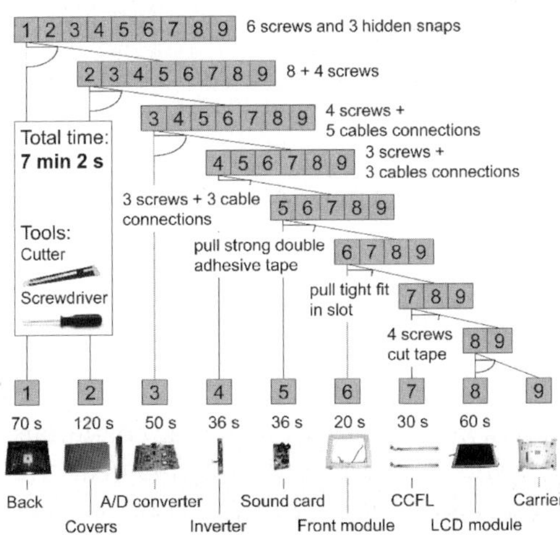


**Fig. 7.** Selected And-/Or-graph

The objective of the disassembly evaluation is the non-destructive separation of modules and components as depicted in figure 6. During disassembly, the flat screen is fixed at the back housing using a clamping device. Figure 7 depicts a selected And-/Or-graph for the complete disassembly. The total disassembly time was determined with approximately seven minutes. Partially destructive operations, e.g., the cutting or pulling of single and double sided adhesive tapes, are inevitable. To obtain comparable results regarding the disassembly effort required, a Method Time Measurement (MTM) analysis was conducted for all 17 monitor models, as presented figure 8. MTM analyzes and classifies method data into basic motions and establishes a relationship between the motions and the time required to perform them. It also serves as a data basis for the manual workplace design, helping to set Labor Standards. The MTM results show as wide range in disassembly times depending on monitor design. The disassembly times range from 3,6 to 8,7 minutes and the quantity of screws used from 33 to 108 units.

| | Operation | CODE | | | | TMU |
|---|---|---|---|---|---|---|
| ⋮ | | | | | | ⋮ |
| 310 | DISENGAGE monitor base remainder parts | D | NT | EH | | 4,0 |
| 320 | MOVE monitor to the work area to the container | M | C | 61 | F1,00 + 0,00 | 25,5 |
| 330 | REACH monitor | R | A | 30,5 | | 9,6 |
| 340 | GRASP monitor | G | 1A | | | 2,0 |
| 350 | MOVE monitor to the work area | M | B | 30,5 | | 20,6 |
| 360 | GRASP screw driver | G | 1A | | | 2,0 |
| 370 | SCREW PROCEDURE | | | | | 156,5 |
| 380 | SCREW PROCEDURE | | | | | 156,5 |
| 390 | SCREW PROCEDURE | | | | | 156,5 |
| 400 | SCREW PROCEDURE | | | | | 156,5 |
| 410 | DISENGAGE front module | D | T | EH | | 7,5 |
| 420 | MOVE front module to the container | M | C | 61 | F1,00 + 0,00 | 25,5 |
| 430 | REACH monitor | R | A | 30,5 | | 9,6 |
| 430 | GRASP monitor | G | 1A | | | 2,0 |
| 440 | MOVE monitor to the top | M | B | 30,5 | F1,06 + 2,20 | 16,4 |
| 450 | GRASP LCD module | G | 1A | | | 2,0 |
| 460 | MOVE LCD module to the work area | M | B | 30,5 | F1,00 + 0,00 | 13,4 |
| ⋮ | | | | | | ⋮ |
| | | | | | Total | 11712,5 |
| | | | | | 1TMU = 0,036 seconds | |

**Fig. 8.** Excerpt from MTM disassembly analysis

## 5 Reuse

Within the disassembly analysis, different conditions for monitor components were identified and are used to define quality classes, as depicted in Table 1. In the following, selected quality attributes are briefly described.

**Table 1.** Standard components of the flat screen monitor

| | Component | Quality Class C | Quality Class B | Quality Class A |
|---|---|---|---|---|
| 1 | Front housing module speakers and controls | Defective speakers or control | Scratched housing | Unscratched housing |
| | | Broken housing | | |
| 2 | Liquid Crystal Display Module | Glass substrate broken | Pixel defect category III or IV (ISO 13406-2) | Pixel defect category I or II (ISO 13406-2) |
| | | Beyond category IV (ISO 13406-2) | | |
| | | Defective row or column controller | Scratched polarizer | |
| | | Defective contacts, tape adhesive bonding (TAB) or data or gate driver | Aged color filter | |
| 3 | Cold Cathode Fluorescent Lamp (CCFL) | Broken | Weak emission | Bright |
| 4 | Covers | Bended | Not bended | |
| 5 | Controller | Defective integrated circuits | Functionality tested | Electronic fault diagnosis |
| | | Defective wiring | | |
| 6 | Inverter | Broken piezo element | | |
| | | Burned coil | | |
| 7 | Sound card | Defective | | |
| 8 | Carrier | Bended | Not bended | |
| 9 | Back housing | Broken | Scratched | Unscratched |

## 5.1 LCD module

The standard ISO 13406-2 defines the quality criteria for pixel error tolerance. Limits for constantly bright or dark as well as defective sub pixels are defined. The standard ISO 13406-2 provides customers and manufacturers a clearly defined and thus comparable warranty claim definition, in which the pixel fault classes are given. As an example for a 15" LCD panel a maximum of 4 defective sub pixels or 2 defective bright or black pixels are accepted for category II, i.e., quality class A. Up to 500 defective sub pixels or 50 bright respectively 150 dark pixels are considered quality class B. Inspection can be performed visually.

## 5.2 Controller, inverter and sound card

Functionality by itself does not guarantee the faultlessness of the electronic component. Intensive diagnoses under different environmental conditions using predetermined procedures are required when components are not only to be reused but re-assembled in new products. An approved method in remanufacturing is signature analysis, a diagnostic tool to examine critical performance parameters of subassemblies and components.

Signatures of used parts are compared to those of new parts. For example, signature analysis for an inverter might measure output current, frequency and voltage. Using signature analysis, it is possible to determine the remaining life and performance potential of used parts.

### 5.3 CCFL

Under the assumption of an average flat screen usage of 8 h per day, 250 working days, 4 years usage time, two CCFLs per flat screen, and today's common Mean Time Before Failure (MTBF) for CCFLs being 40.000 h, the probability for a backlight failure during the product's lifetime is 36%, as given in equation 1. Also, CCFLs can become less light intensive due to gas leakage during usage. Less light intensive CCFLs are considered quality class B.

$$R_{CCFL} = \left(1 - \frac{\text{Operation time}}{\text{MTBF}}\right) = \left(1 - \frac{8h * \frac{250 days}{year} * 4 years}{40.000h}\right) = 80\% \qquad (1)$$

$$P(backlight\ failure) = 1 - R(CCFL) * R(CCFL) = 1 - 0{,}8 * 0{,}8 = 36\%$$

### 5.4 Front and back housings

The determination of surface quality usually requires visual inspection. Cleaning before inspection can help to improve the rating process. The disassembly of control panels and speakers to be used as single spare parts can be economically under certain conditions.

## 6 Remanufacturing challenges

With a predicted 90 % penetration of the monitor market with flat screen technology by 2010, product returns will be sufficient to develop remanufacturing activities at large scale. The main challenge as in most remanufacturing activities is the handling of uncertainty regarding time, quality, quantity, and place of product returns that affects collection, testing, disassembly, reassembly, and the warehousing of spare parts likewise.

## 6.1 Collection

Today, IT equipment returns from leasing are frequently collected by IT remarketing companies. Due to the low market penetration of flat screen monitors, collection from private households is still uncommon. Future collection using today's municipal collection hubs is unlikely to guarantee an undamaged take back of flat screen monitors. To obtain access to product returns from private households, new take back models, e.g. charity recycling [11], are required. Due to the dimension and weight of monitors, as compared to mobile phones, collection using postal service is rather unlikely.

## 6.2 Testing

Electronic fault diagnosis of flat screen monitors and/or components requires the exact identification of components, and adequate diagnosis procedures. Signature analysis, see above, was identified as a suitable diagnostic method, requiring the identification of test parameters and allowed parameter values. The simple testing of the product's functionality is usually considered reasonable, as long as no security relevant functions are jeopardized. The fact that certain components, e.g., inverters, CCFLs and LCD modules are being supplied by a relatively small number of manufacturers can help in reducing the required effort for developing test procedures. An Original Equipment Manufacturer (OEM) wide component analysis for flat screen monitors is an essential requirement before larger remanufacturing activities should be planned.

## 6.3 Disassembly

Particularly important for cost efficient disassembly is the provision of a suitable disassembly plan under consideration of the acquired test results for each product. A certain order of disassembly steps is usually predefined by the products design. Automation of selected disassembly, i.e. mainly unscrewing processes, can be an option once higher quantities of used flat screen monitors are available. Based on the disassembly analysis of 17 monitor models basic restraints regarding the automation of disassembly operations were identified as follows:

- very strong snap fits in housings parts due to the strong release torques required,
- stands mounted to the monitor's inside,

- integrated video or power supply cables requiring manual disassembly of the back housing, and
- electronic components mounted on the top and the bottom of the carrier, necessitating the additional handling of the carrier.

The monitors selected for hybrid disassembly offer weak housing snap fits, externally mounted monitor stands, externally removable cables, and electronic components mounted to one side of the carrier. A prototypical hybrid disassembly system with one automated and one manual workplace connected by three square conveyor blocks is being realized, as shown in figure 9. A 4-axis Selective Compliant Articulated/Assembly Robot Arm (Scara) is applied for performing unscrewing and handling operations. A manual workplace, originally used for manual testing of mobile phones, is adapted for manual disassembly operations. The system can be extended to one more automated and five more manual workplaces to accommodate further disassembly, as well as testing and reassembly operations.

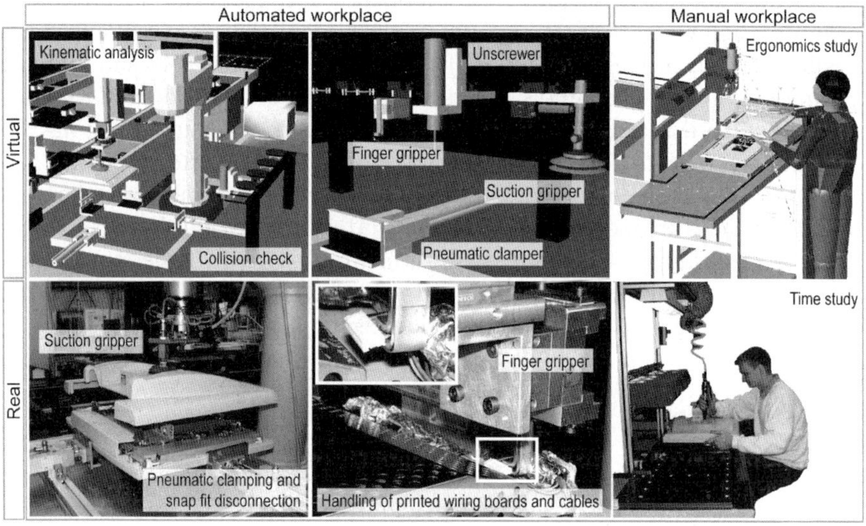


**Fig. 9.** Hybrid disassembly system for flat screen monitors

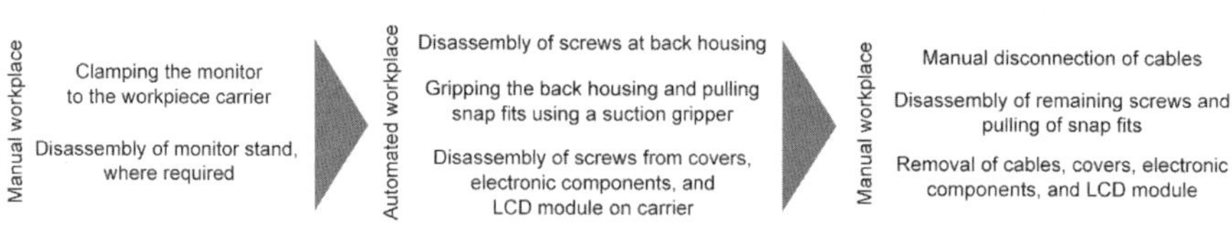


**Fig. 10.** Hybrid disassembly process

Operations currently being implemented are described in figure 10. At the manual workplace, the monitor is clamped to the work piece carrier. Small monitor stands can remain mounted to the back housing as long as they do not interfere with the accessibility of joining elements. The monitor is then transferred to the automated disassembly workplace where the screws securing the back housing are loosened and the back housing is removed using a suction gripper. Subsequently, all accessible screws from covers and printed circuit boards are loosened. The monitor is then transferred back to the manual workplace for the disconnection of cables, and remaining joining elements, as well as the removal of covers, electronic components, and the LCD module.

### 6.4 Program planning

Used products can generally be offered in different quality classes, e.g., as "non-functional", as "functionally tested" or "electronically diagnosed". To assign products to the most profitable reuse option, process requirements and costs must be known to derive profit margins for each reuse alternative. In order to plan the remanufacturing program resource capacities for testing, disassembly, repair and reassembly must be considered as well as availability of cores, spare parts and demand regarding models and quality classes [12]. The consideration of trade-offs between the effort expected in recovery operations and resale value is discussed in [13] with the key question being "How economically efficient is it to disassemble a unit and test a component given the opportunity to recycle it instead".

## 7 Summary and outlook

This paper described aspects from the economical, ecological and technological framework for profitable remanufacturing of flat screen monitors. A disassembly analysis for different monitor models showed strong differences regarding required disassembly time and force, variability of screws and snap fits, and moderate differences regarding the monitor's material composition. Test procedures and quality classes were identified as a fundamental requirement in the development of a reuse market. A hybrid disassembly system integrating selected automated and manual operations was introduced based on the identified potential for automation. Next steps will include the implementation of testing and reassembly processes into the presented system, as well as the specification of a generic test and remanufacturing process.

## References


1. DisplaySearch 2004. Available at www.displaysearch.com
2. Matsuno, S., FPD Market Overview and Outlook. DisplaySearch. 2002.
3. Alexander, R., Worldwide Monitor Market Tracker. iSupply, Stanford Resources. 3. quarter 2004.
4. National Safety Council, 1999, Electronic Product Recovery and Recycling Baseline Report: Recycling of Selected Electronic Products in the United States. Washington.
5. DisplaySearch Delivers Breakthrough Report on LCD Monitor Market, 1999. Available at www.displaysearch.com/press/1999/120999.htm
6. University of Tennessee, 2001, US-Environmental Protection Agency, Desktop Computer Displays - A Life Cycle Assessment, 2003. Available at www.epa.gov.
7. Directive 2002/96/EC of the European Parliament and of the council on waste electrical and electronic equipment (WEEE) of 27 January 2003.
8. Martin, R., Becker W., Simon-Hettich, B., 2004, Verwertungsverfahren für LC-Displays. In: Workshop Nachhaltigkeit als Chance und Herausforderung für Unternehmen im Flachdisplaymarkt. May 12,. Berlin.
9. Umweltbundesamt, 2000, Ecotoxicology of liquid crystals in liquid crystal displays.
10. Behrendt, S., Erdmann, L., 2003, Display-Märkte im Umbruch – Neuorientierungen für Umweltschutz-strategien. Roadmap des IZT, Berlin.
11. Seliger, G., Ba dere, B., Ciupek, M., Franke, C., 2003, Contribution to Efficient Cellular Phone Remanufacturing. CIRP Seminar on Life Cycle Engineering, Copenhagen.
12. Seliger, G., Franke, C., Ciupek, M., Ba dere, B., 2004, Process and Facility Planning for Mobile Phone Remanufacturing. Annals of the CIRP Vol. 53/1: 9-12.
13. Ferrer, G., 2001, On the widget remanufacturing operation. European Journal of Operational Research 135/2: 373-393.

# Improving product recovery decisions through product information

A.K. Parlikad, D.C. McFarlane and A.G. Kulkarni
*Institute for Manufacturing, Cambridge University Engineering Department, UK, aknp2@eng.cam.ac.uk*

Abstract: The advent of networked RFID-based automated identification approaches has resulted in the ability to enhance the quality of product information that is available to make decisions along the product lifecycle. One of the major impact areas of such a capability is in improving the effectiveness of decisions made during end-of-life product recovery. This paper investigates the link between product recovery decisions and information quality through observations made from a number of case studies. A Bayesian approach for quantitative evaluation of the impact of enhanced product information on product recovery option decisions is also presented.

Keywords: RFID, Product Recovery Decisions, Bayesian Decision Modelling

## 1 Introduction

Due to environmental legislations, continually increasing landfill costs, and realisation of cost benefits arising from effective product recovery management, manufacturers have started considering new technologies to manage returned and obsolete products in an efficient manner. Unlike conventional manufacturing and assembly processes, demanufacturing and disassembly operations are characterised by a high variety of products, uncertain product condition after usage, and a not so rigidly defined process goal [1]. As a result of such uncertainties associated with returned products, effective recovery of value from these products requires extensive information about the identity and the condition of the product when it is returned. However, product information shortage is widely blamed in the literature as one of the major obstacles for efficient recovery of value from returned products [2]. This problem becomes critical as government



*D. Brissaud et al. (eds.), Innovation in Life Cycle Engineering and Sustainable Development,* 153–172

regulations require manufacturers to share information that would facilitate product recovery with remanufacturers and recyclers. The concept of the so-called 'networked RFID' [3] developed by the Auto-ID Centre [4] makes it possible for a product to carry complete information associated with it throughout its lifecycle and ensure flow of this information between the various actors in the supply chain, for e.g., between the manufacturer and the recycler. To find out the exact requirements on the type and quality of information that is demanded by decisions made during product recovery, a case study exercise that covered nine remanufacturing and recycling facilities was undertaken by the authors. The case studies examined the various decisions involved while recovering end-of-life (or returned) products, their relationship and dependency on the type and quality of information associated with the product with an aim to evaluate the impact of RFID-based product identification technologies on the performance of product recovery operations. The ultimate objective of this research is to test and prove the hypothesis that the ready (or timely) availability of complete and accurate information associated with a product will lead to a significant increase in the effectiveness of product recovery decisions.The next section will use observations from the case study exercise as a basis for understanding how product recovery decisions are made and the information requirements for making these decisions, and identify the link between product recovery decision effectiveness and key information quality parameters. Having identified the information quality requirements imposed by product recovery decisions, in section 3 we will propose a networked RFID-based approach to manage and deliver enhanced product information. Acknowledging the fact that enhancing product information comes with a cost attached to it, in section 4 we propose an approach to quantify the impact of enhanced product information on product recovery decisions. Finally, section 5 provides some concluding remarks.

## 2 Case studies

### 2.1 Overview of companies

As discussed, this study involves nine remanufacturing/ recycling companies. In particular, the case studies concentrated on companies that were involved in the recovery of electric and electronic products. This was influenced by two major factors: (a) government regulations on waste electric and electronic equipment [5], and (b) the fact that electric and electronic equipment are most suitable for reuse and remanufacturing due to their long wear-out life and short usage period [6]. Seven out of the nine companies visited during the course of this case study dealt with brown goods, and two were white goods remanufacturers. The information presented in this paper has been collated from semi-structured interviews with company personnel ranging from top-level executives to factory floor managers, and also from observations made during on-site visits.

## 2.2 Product recovery operations

In this section, we will extract the common features found in the different companies and present a general picture of how product recovery is performed in the industry. Figure 1 shows the typical steps that are performed during product recovery.

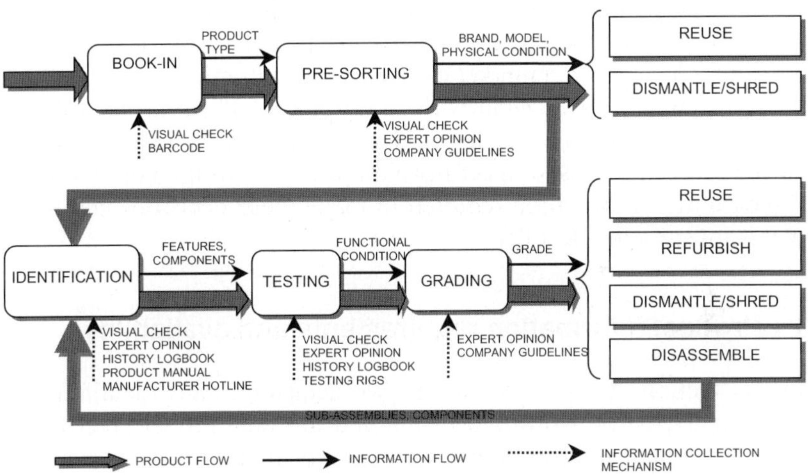


**Fig. 1.** Product recovery operations

**Book-in.** The first thing that is done when products come in through the remanufacturing shop floor is to book-in all the products into the company's inventory database. Essentially, this consists of noting the product type (for e.g., laptop, printer, mobile phone), and in some cases, the brand and model of the product as well.

**Pre-sorting.** After the products are booked-in, a preliminary sorting is performed on the products to filter out the products that do not apparently have much market value. This is the first decision point where there are three options available to the decision-maker: (1) *reuse* – if the product is perceived to be of a value less than the cost incurred for further inspection and testing, but can be sold "as seen" in the market at a lower price; (2) *recycle*, – if the product is 'evidently' bad or perceived to have no market demand; or (3) *inspect and test* – if the product is perceived to be valuable enough to warrant further inspection and testing before a recovery decision is made. There is an underlying assumption that due to regulatory requirements, no products are land-filled directly.

**Identification & Testing.** If a product is deemed to be valuable enough to warrant further inspection and testing, additional efforts are put in to gather the information required to choose the recovery option that would maximise the value recovered from the returned product. This involves inspecting the product to gather its technical specification as well testing it to ascertain its physical and functional condition.

**Grading.** After the product has undergone thorough inspection and testing, a final decision about the choice of recovery option is made at this point. Here, in addition to the options available at the pre-sorting stage, depending on the quality and condition of the product, it could also be (1) *reused* – if it is in good condition and there is a demand for it in the market, (2) *refurbished* – if a market can be found with minor modifications or parts replacement, or (3) *cannibalised for parts* – if the product does not have a market value as a whole, but has valuable components that can be retrieved and sold or used in refurbished products. In the next section, we will look at the information required to make these decisions and also see how this information is collected.

## 2.3 Product information requirements and availability

In this section, we will look at the two decision points identified previously, namely the pre-sorting decision and the final product recovery option decision, and try to understand the level of information availability under which these decisions are made. We will also look at the various mechanisms that are used to collect the information associated with the products.

**Information requirements for the pre-sorting decision.** The pre-sorting decision is mainly made on a minimum system requirement basis, (for e.g., computers above Pentium 550 MHz will be sent for inspection and testing) which is obtained from the guidelines provided by the sales department (which makes this decision on the basis of the market situation). It is assumed that anything below this specification is not worth testing as it would not be cost-effective to do so. However, at this stage, only an external visual check is conducted, and the only information about the product that is known is the type, manufacturer, brand, and the model, if any. Figure 1: Product recovery operationsOne should keep in mind that in some cases such as mass-customised products like computers, the components of the system could vary widely within the same model, or the product could have undergone changes during its use (for e.g. replacement/upgrade of parts), but that level of detail is not captured at this stage. Products are identified using various methods such as the manufacturer's barcode, manuals and specifications attached to the product, or in many cases, expert knowledge of the people handling the products (refer to Figure 1 and Figure 2). When the remanufacturer receives a product that he has not seen before, for e.g., a new product such as a digital photocopier, they are quarantined and expert opinion (sometimes from outside the company) is sought to identify them. This causes a costly delay in processing the products and the issue of invoices. The delay in identification of the product, or even worse, mis-identification of the product affects the marketability of the product. Evidently, the pre-sorting decision is made with limited/ inaccurate information, and therefore affects the effectiveness of the decision made. As an outcome, often low quality products are sent for inspection and testing, thereby decreasing the overall cost-efficiency of the process. It is also possible that many good products are sold at a price lower than its value, or even sent for recycling, thereby losing an opportunity to maximise profits.

**Information requirements for the final grading decision.** The products that make the 'first cut' go through detailed inspection and testing. To optimise the product recovery option decision, it is necessary to ascertain the complete identity and condition of the product. By "identity", we mean all the information that is required to completely describe the product (for e.g., technical specifications, components etc.). In companies where the primary objective is to recycle the product, it is required to know what the constituent materials are so that appropriate separation and purification techniques can be applied. Environmental regulations also stipulate that certain hazardous substances have to be separated first (for e.g., batteries) before the product can be shredded. This information can be collected by way of contacting the manufacturer, by consulting the product manual and the maintenance logbook (if available), or by detailed inspection of the product. Direct access to the manufacturer is limited to third-party remanufacturing companies, or in the case of a closed-loop supply chain where this information maybe readily available. Figure 2 shows the distribution of various product identification methods used by the companies that participated in the case studies. The maintenance logbook is an important source of information for products that undergo a lot of maintenance and parts replacement throughout their life. However, in most cases it was seen that the logbooks are not kept up-to-date, and hence the information obtained is inaccurate. For certain products such as computers, identification of the product and their components can be performed by running a program that scans the system for their components and features (for e.g., PC Check). Making a decision whether the product is to be re-used, refurbished, cannibalised for parts, or recycled requires more than just knowing the identity of the product. The residual value of the product depends on various quality parameters such as the age, functional condition (working, or not working), physical condition, functional age (as a measure of obsoleteness), remaining useful life, etc.

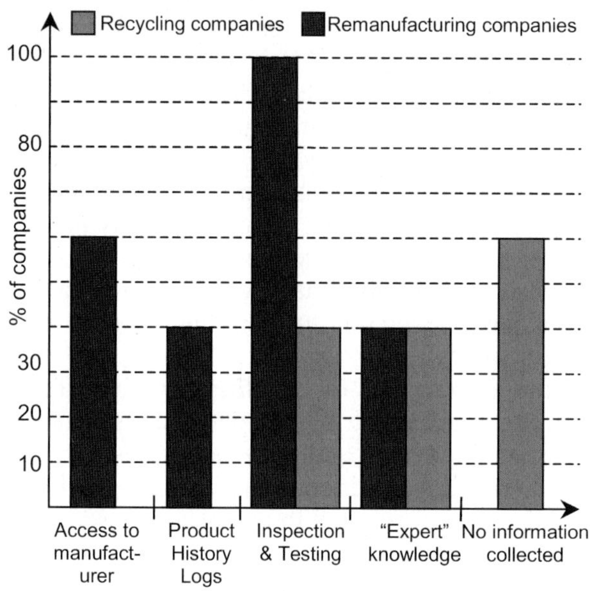


**Fig. 2.** Product identification method used

The value of these parameters, which define the current state of the product, are determined through testing or estimated from knowledge gathered through experience. In the case of computers, identification and preliminary testing can be performed at a single step, as the product-scanning programs mentioned earlier will also identify non-functioning components. Nevertheless, this program will not be able to identify the exact type of fault in a component, and hence further diagnosis will be required to be performed to ascertain that. If we examine Figure 2 closely, we can see that the companies that remanufacture or refurbish most of the incoming products are those that have more access to product information, such as access to manufacturer databases or history logbooks. Due to shortage of information available, rest of the companies end up recycling most of the products in spite of knowing that refurbishment or reuse will bring higher profits to their operations. Having understood the importance that the availability of information plays in product recovery operations, in the next section we will examine how the decisions made during product recovery are affected by the quality of information that is available to the decision-maker.

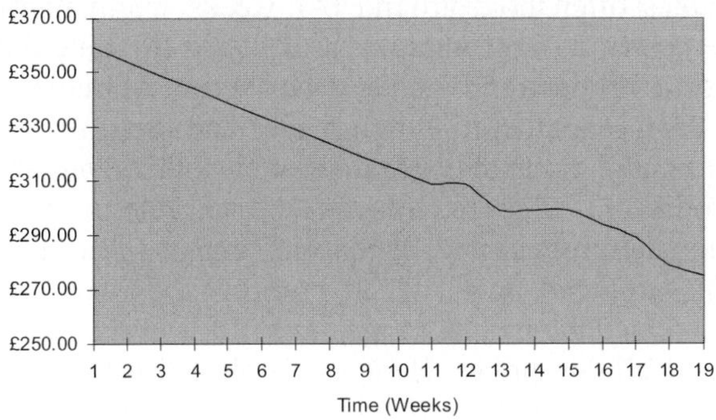


**Fig. 3.** Value depreciation of a 650MHz laptop

## 2.4 Linking product recovery decisions to information quality

In this section, we will identify the key information quality parameters that are critical for making effective product recovery decisions, so as to understand the requirements to be met by an information system that supports these decisions. From the above discussion, clearly the availability of complete product information plays a very crucial role in the effectiveness of decisions made during product recovery. The fact that each product is subjected to a different set of conditions throughout their lifecycle means that this information will be unique to every single product. By observing how product recovery decisions are made and how the information required to make these decisions is collected by the remanufacturers, it can be understood that it is not merely the "completeness" of information that is the critical issue here. The information required for making product recovery decisions "can" be collected by performing extensive inspection and testing. This being labour intensive and very expensive, in most cases is not economically justifiable due to the low-value nature of returned products, resulting in only those products that are evidently valuable on the outset being recovered efficiently. Hence, it is the absence of 'readily available' or 'timely' information that is the biggest hindrance to making effective decisions. Timely availability of information is also important due to the volatility in the market value of returned products. Figure 3 shows how the market value of a P3 650 MHz laptop deteriorates with time. The data used for this plot was obtained from one of the companies involved in the case study. This verifies the observation made in Blackburn *et al.* [7] that the value of volatile products like computers and laptops deteriorates at the rate of more than 1% per week and that the rate increases as the product nears the end of its lifecycle. Given the fact that remanufacturing companies often hold around 6-7 weeks worth of inventory, the products lose nearly 10% of their value between the time the remanufacturer receives the product and the time they sell it. Moreover, some of the contracts between manufacturing companies and their 3rd party operators) stipulate a maximum threshold period before which the remanufacturer has to sell the products. If the remanufacturer is not able to sell it within that stipulated time, the product will have to be bought from the manufacturer at a previously arranged price, which in most cases would be more than what the remanufacturer would be able to sell for in the secondary market due to rapidly declining prices. Hence, it is clear that the ability to collect complete information about the product is not just sufficient – it is necessary to be able to collect this information in a timely manner. Summarising, it is desirable that product information that supports product recovery should have the following qualities:

**Unique identification** – to enable individual information trails for each unique object throughout its lifecycle and across the whole supply chain.

**Ready availability (or timeliness)** – to ensure that information is readily available for decision-making and execution process with minimal need for manual inspection or testing.

**Completeness** – to ensure that all relevant information is available for optimising decisions.

**Accuracy** – to reduce or eliminate inaccurate representations of current and historical product information.

In the next section, we will look at how the concept of networked RFID supported can ensure the quality of information that is required by product recovery decisions.

## 3 A delivery mechanism for enhanced product information – Networked RFID

Radio-Frequency Identification (RFID) is a technology which allows remote interrogation of objects using radio waves to read data from RFID tags which are at some distance from an RFID reader. This has several advantages over manual scanning using optical barcodes, since many tagged items (or embedded sub-components of a composite product) could be simultaneously identified in an automated manner, very quickly and without the need for line-of-sight to each item. A recent breakthrough in enabling affordable widespread global deployment of RFID is the emergence of the so-called 'EPC Network' as a means of connecting a product tagged with an RFID to a network. After discussing the key technology components of the EPC network, we will discuss how RFID-based product identification technologies can improve product recovery decisions.

## 3.1 An overview of the EPC network

In this section we will provide a brief overview of the EPC network developed by the Auto-ID centre. For a detailed discussion of the EPC network and its applications in product lifecycle management, the reader is referred to Harrison *et al.* [3]. In order to achieve a networked RFID solution that satisfies the requirements for making effective product recovery decisions, the EPC Network consists of the following fundamental technology components, which work together to bring about the vision of being able to identify any object anywhere automatically and uniquely. These are:

- A Unique product identifier (*Electronic Product Code*);
- *Radio Frequency tags and readers* to ensure timely and automatic identification of product.
- A distributed product information database (The *EPC Information Service*) linked to an information lookup service (The *Object Name Service*) to ensure completeness and accuracy of product information.

**The Electronic Product Code (EPC).** The aim of the EPC [3] is to provide a unique identifier for each object. Designed from the outset for scalability and use with networked information systems, the EPC typically consists of three ranges of binary digits (bits) representing (refer to Figure 4): a) an EPC Manager (often the manufacturer company ID), b) an object class (usually the product line or SKU) and c) a unique serial number for each instance of a product.

21.203D2A9.16E8B8.719BA30C3

Version Manufacturer Product Type Serial Number

**Fig. 4.** Electronic Product Code

Thus, EPC introduces mass-serialisation, giving each instance of a product a unique identity and allowing information systems to store accurate, complete and relevant individual data (an individual life history) of each unique object.

**Radio frequency tags and readers.** Radio Frequency Identification (RFID) is a key technology enabling automatic reading of multiple items simultaneously, without requiring manual scanning of each individual item. The reader emits radio waves of a particular frequency. When the tags enter the range of a reader, their antenna absorb energy from the radio field, powering the microchip which stores the unique EPC identity code – and returning this information back to the reader via a modulation of the radio waves. Thus, RFID technology provides the ability for automated and efficient method for collecting product information in a timely and accurate manner.

**The Object Naming Service (ONS).** The Object Name Service (ONS) is used to convert an EPC into a number of internet addresses where further information about a given object may be found.

Recognising that the remanufacturer may require information about the product as it moves through its lifecycle, and that this information may be held by various parties along the supply chain, the ONS will provide the serial-level lookup for instances of a given product, pointing to the various other parties across the supply chain, which also hold information.

**The EPC Information Service (EPCIS).** The EPC Information Service [5] allows trading partners to access and exchange data through a standard interface, while interfacing the back-end to diverse databases and information systems from multiple vendors, without their partner needing to know the details or have direct access to the underlying systems. Such a distributed database tied with an ONS registry that provides pointers to all the EPC Information Services on the individual supply chain followed by each individual object in a more robust manner would help in collecting and retrieving complete lifecycle information associated with a product.

**The EPC Network.** These elements together form the core infrastructure of the EPC Network and provide the potential for automatic and unique identification of any tagged product. shows how information about the toaster is stored in a distributed manner in various EPCISs across its lifecycle. It also describes clearly how, a recycler could find the toaster's date of sale (possibly to find out how long it has been used). The RF reader detects the presence of the toaster when the toaster returns its EPC code in response to the reader's interrogation. The software application connected to the RF reader will then query the ONS to obtain the location of EPCISs that contains data linked to the EPC code of the toaster. It can then query the retailer's EPCIS to obtain the date of sale. Without a product-oriented information system similar to the one provided by the EPC network, it might be impossible for the recycler to obtain that kind of information.

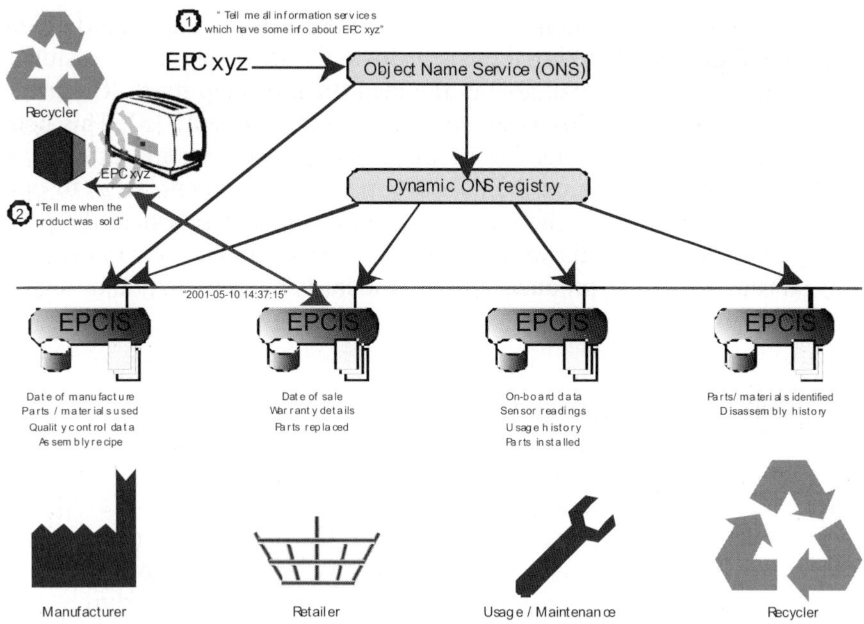


**Fig. 5.** Retrieving product information through the EPC Network

## 3.2 The role of networked RFID in product recovery

As seen in the previous section, tagging individual products using RFID will provide an automated means of capturing information about the product. We have seen that in many cases, the product identity is misinterpreted and wrong decisions are made. Apparently, automatic product identification enabled by networked RFID would eliminate these errors by providing complete and accurate information about the identity of the product, and by automating the booking-in process, thereby making it more efficient and less error-prone. Moreover, by enabling ready availability of product information through networked databases linked to the product, it is possible to gather complete identity information during book-in, thereby eliminating the need for further manual identification. As networked RFID would enable ready availability of complete lifecycle information about the product, the decisions made at this point are far better informed than how it is done currently. In addition, it is also possible to monitor critical performance parameters of the product (temperature, number of revolutions, etc.) throughout its life by attaching sensors that record crucial lifecycle data and by linking these sensors to the RFID tag, it is possible to make this information available at the identification step itself. The use of such devices has been discussed elsewhere in the literature [8]. This helps the decision-maker to filter the products going to the testing process, or make early assumptions on the cost of repair/refurbishment, so as to optimise the performance of the whole operation. In the case where the product is to be recycled, this would alert the recycler about potential hazardous substances in the product, and help identify the products that require special processing.

## 3.3 RFID Vs Traditional methods

As we saw in section 2.3, two methods are found to be used in the industry to gather information associated with returned products: (a) Manual (visual) inspection – where the products are visually inspected and identified; and (b) Barcodes – where the product can be identified by scanning a barcode attached to it and obtaining associated information from a database linked to the serial number represented by the barcode. Figure 6 presents a comparison between these two methods and RFID-based product identification. In the manual approach, very little information is collected during the book-in stage, and detailed inspection and testing is required to gather information about the product. When barcodes are used to identify the product, the serial number can be used to access the manufacturer's database and identify the product and its original specifications. However, information about any changes to the product after the point-of-sale has to be collected by disassembling the product and inspecting its internal parts as most barcodes identify only the product-type and do not identify products at the item-level. On the other hand, if the product and its major components are tagged with RFID, information about its current specification can be obtained without disassembling the product as RFID tags can be read without line-of-sight. As discussed earlier, with a networked RFID approach to lifecycle information management, information such as the age of the product (determined by its date of sale, as obtained from the retailer's EPCIS), and its usage rates (obtained either by using sensors to monitor crucial lifecycle data, or by connecting to maintenance EPCISs where history logs can be accessed) can also be obtained at the book-in stage itself, thereby eliminating the need for detailed inspection. Although such an approach would provide pointers to the product's residual life, the actual functional condition of the product will still have to be determined by testing the product.

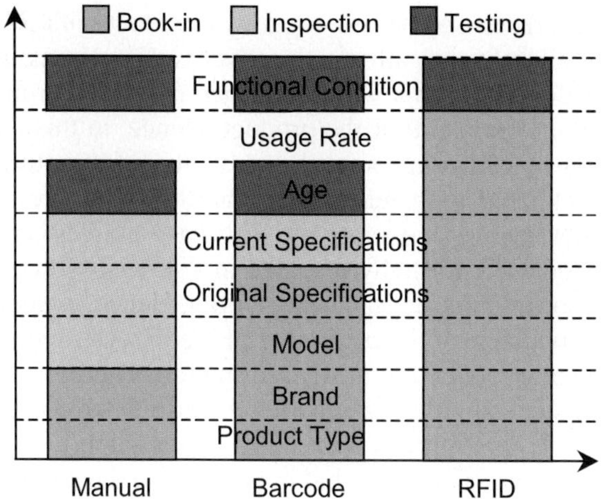


**Fig. 6.** Product data collection locations

From the above discussions, it is clear that RFID can bring two-fold benefits to product recovery operations: (1) process improvements – brought about by automating the product identification process, and (2) decision improvements – brought about by enhancing the quality of information available to the decision-maker. However, such enhanced information comes with a cost attached to it. In order to understand the cost-effectiveness of using such technologies, in the next section we will investigate means to quantify the impact of enhancing the quality of information available during product recovery.

# 4 Towards quantifying the impact of enhanced information on product recovery decisions

In this section, we will propose a novel method to model product recovery decisions as a means for quantitative evaluation of the relationship between product information quality and product recovery decision effectiveness.

## 4.1 Characteristics of product recovery decisions

The primary role of product information in product recovery decision situations is to reduce to the level of uncertainty that prevails in those situations (in fact, this holds true for every "information-decision" relationship). As discussed before, tagging a product with RFID would enable timely availability of information that is directly associated with a product, i.e., information that would enhance the decision-maker's knowledge about the *current state* of the product. Hence, in this analysis we are concerned with investigating how the reduction of structural and quality uncertainty associated with returned products would affect the performance of product recovery operations. The case studies showed that the identity of a product is normally revealed in a series of information gathering steps (pre-sorting, inspection, testing, etc.). Hence, even though limited information is available at the beginning of the process, recovery decisions are often made after collecting more information. Therefore, this is a situation where decisions are often made after assessing "new information" that becomes available through different information gathering methods. From the case studies it was also clear that in addition to the availability of product information, the ability to retrieve this information in a timely manner is critical to making effective product recovery decisions, as the net benefit from recovering the product decreases with the time taken to collect enough information to make the decision.

## 4.2 A Bayesian approach for analysing product recovery decisions

In this section, we propose the use of a Bayesian decision theory approach to analyse the effect of enhanced information on product recovery decisions.

**Decision-making under uncertainty.** A well-known rule for making decisions about taking a particular action given alternative uncertain outcomes is the 'principle of maximum expected utility' (MEU). The MEU principle dictates that we should take actions that maximise the value computed by summing together the value attributed to each possible outcome multiplied by the probability of that outcome [9]. In mathematical terms, if $H$ is a set of $n$ possible states of the product, $D$ a set of feasible product recovery decisions, then the utility of making a decision $D_j$ (for e.g., this could be the selling price, if the decision is to reuse the product) when the state of the product is $H_i$ is represented by a mapping $U_k$: $H_i \times D_j \rightarrow u(H_i, D_j)$. Then, the expected utility of an action/decision $D_j$, given some background information about the product $\xi$ is given by:

$$EU(D_j) = \sum_{i=1}^{n} p(H_i \mid \xi) u(H_i, D_j) \qquad (1)$$

The best decision, $D^*$, is the action with the greatest expected utility, given the probability (belief) distribution and the utility model, and is given by:

$$D^* = \arg\max_{D_j} \sum_{i=1}^{n} p(H_i \mid \xi) u(H_i, D_j) \tag{2}$$

The maximum expected utility is thus given by:

$$\max_{D_j} \sum_{i=1}^{n} p(H_i \mid \xi) u(H_i, D_j) \tag{3}$$

Equation (3) gives the maximum expected utility of making an immediate decision, given prior information. In the next section, we will see how the decision maker will evaluate her options if we give her the option to collect more information before making a decision.

**Computing the value of information.** Equation 3 gives the best decision to make under prior beliefs with prior information. If the decision-maker is often offered the ability to obtain more information (or evidences), say by conducting a test, and she would wish to know the value of the information provided by the test in order to decide whether it is worth going for the test or not. Such a situation occurs at the pre-sorting stage in the remanufacturing process where a preliminary evaluation is performed on the returned products and depending on the estimated value, a decision is taken as to whether further inspection and testing should be performed. In fact, each inspection/testing step in a remanufacturing process can be considered as an information collection process, and a decision about the recovery option for the product can be made after collecting one or more pieces of information. Now let us assume that the decision-maker has the option to perform a test $\tau$, which will reveal the value of an evidence $E_k$. We shall use $E_k^l$ to represent the value of this evidence, where $l$ indexes outcomes of the test or observation. We denote the values of $E_k$, by $E_k^1, \ldots, E_k^m$, where $m$ is the number of mutually exclusive values. We can compute the expected value of information of performing the test by conditioning the probability of different states of the product on different outcomes of the test, and determining the expected value of the best actions associated with the revised probability distributions $p(H \mid E_k^l, \xi)$ using Bayes theorem. The expected utility associated with each test outcome then is weighed by the probability of that outcome, $p(E_k^l \mid \xi)$. Therefore the expected utility of making a decision after conducting the test is given by

$$EU(\tau) = \sum_{l=1}^{m} p(E_k^l \mid \xi) \left[ \max_{D_j} \sum_{i=1}^{n} p(H_i \mid E_k^l, \xi) u(H_i, D_j \right. \tag{4}$$

Then, we calculate the difference between the expected utility of actions, dictated by the current state of information and the expected utility of making the decision to act after performing the test. Thus, the expected value of information (conducting the test) is:

$$EVI(\tau) = \sum_{l=1}^{m} p\left(E_k^l \mid \xi\right)\left[\max_{D_j} \sum_{i=1}^{n} p\left(H_i \mid E_k^l, \xi\right) u\left(H_i, D_j\right)\right] - \max_{D_j} \sum_{i=1}^{n} p\left(H_i \mid \xi\right) u\left(H_i, D_j\right) \tag{5}$$

In equation 5, we have not considered the cost incurred in conducting the test, or collecting information. It should be kept in mind that information can be collected using various methods, each with a different cost structure. Equation 5 will give the "gross value" of the information to the decision-maker, and enables her to decide the type of system that is best suited to collect the required information, given the benefit arising from it. This is particularly useful in this research, as we are comparing different information methods (RFID vs. traditional methods). Let us take a particular example, say a laptop with four different components of interest (1- hard disk, 2- processor, 3- CD drive, and 4- network card). Figure 7shows the results of sensitivity analysis conducted on the product recovery decision model for this example built using the Bayesian approach detailed above. The figure shows the relationship between the value of information and probability of the product being good $p(H)$.

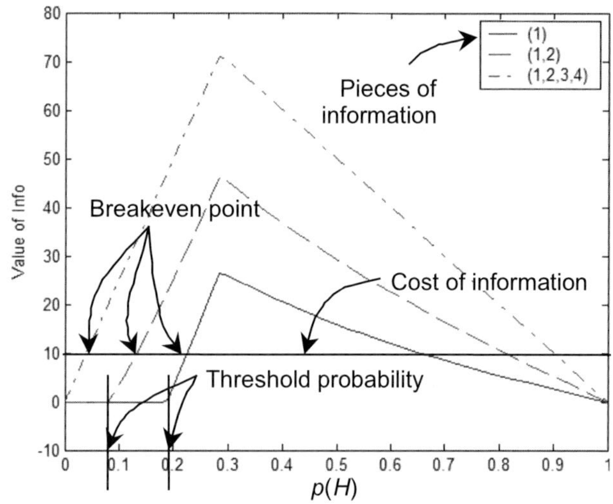


**Fig. 7.** The effect of information completeness

The different plots in the figure correspond to (1) – value of information associated with component 1, i.e., the size of the hard disk, (1, 2) – value of information associated with components 1 and 2, and so on. It was seen from the case studies that the probability of the product being good, i.e., the product quality varies with respect to the type of returns, the source of returns, as well as the type of business model used. From the figure, it can be seen that the value of information for products with very low as well as very high quality is low. In fact, below a threshold probability, information does not have any value at all. However, it can be seen that as increasing amounts of information becomes available, the value of information increases and the threshold probability decreases, and even for products with a low and high quality, there is some value for the information. The horizontal line shows the cost of collecting information. The region between the points where the value of information curve meets the cost of information line gives the range of probabilities where availability of product information brings benefits to product recovery decisions. Assuming that the cost of information collection remains the same, the figure shows that if increasing amounts of information is made available, more products will start seeing benefits due to better decisions.

**The effect of information timeliness in decision-making.** The utility of making any recovery decision decreases with time due to two reasons: (a) market value depreciation, which is a function of the product's value volatility, and (b) the costs involved in conducting tests to retrieve information about the product. We can adapt the concept of 'time-dependent utilities' presented in Horvitz [10] to model decision problems where the utilities vary with the time taken to make the decision. According to this concept, time-dependent actions can be represented by considering a continuum of decisions, each defined by initiating an action at a progressively later time, and by assessing the change in utility of the decision outcome as a function of time. In formal terms, if $H_i,D_j,t$ refers to an action $D_j$, taken at time $t$ when state $H_i$ is true, where $t$ is defined in terms of an initial time $t_0$, the time when the decision problem begins, the utility of that action at different times $t$ is given by u($H_i,D_j,t$). The losses with time can be modelled using, say, an exponential function as follows:

$$u(H_i, D_j, t) = u(H_i, D_j, t_0)e^{-\lambda t} - C_t \qquad (6)$$

where $\lambda$ is a parameter constant, which we call the *volatility* of the product value with respect to time, and $C_t$ is the cost of the tests conducted (for collecting information) till time $t$.

To understand the effect of information timeliness on product recovery decisions, we look at how the utility of making decisions evolve over a series of information gathering steps with time. Figure 8 provides a hypothesised picture of the dynamics involved in product recovery decision-making. As described before, product recovery decisions are made after performing a series of tests on the product in order to collect more information associated with it. Figure 8(a) show how the certainty about the product state would increase as we conduct more tests on the product. However, increasing the number of tests translates to a decrease in benefits due to the increasing cost of conducting further tests as well the decrease in product value over time (refer to section 2.4). Since the expected utility of any product recovery decision is a function of the certainty about the product state and the benefit of making the decision at that point of time, the decision-maker will have to stop collecting information and make his decision at a point where any further benefits arising from increase in certainty is offset by the decrease in the benefit of making the decision as time passes (as represented the peak on the "utility" curve). Now suppose the product has an RFID tag linked to networked information about its identity. This situation is represented in Figure 8(b) where an RFID enabled information system provides information that is normally obtained by conducting more than one test.

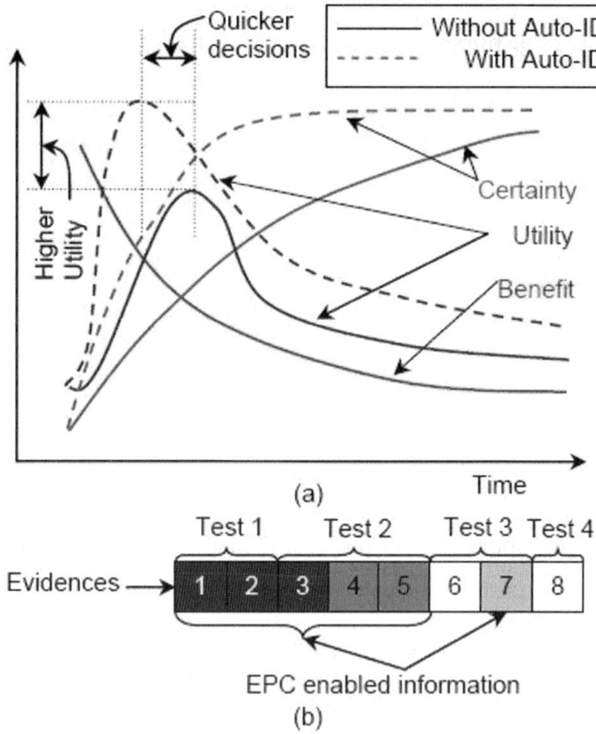


**Fig. 8.** The effect of information timeliness

The varying shades of grey tell us that while networked RFID can provide complete information about certain parameters, it would only be able to provide information that leads to increasing certainty about the values of other parameters. For e.g., suppose evidence 8 represents "wear" of a particular component. The networked RFID would potentially provide crucial lifecycle data such as usage rates and temperatures by attaching appropriate sensors to the tags, which would lead to a better estimation of the value of "wear". This would lead to a higher rate at which certainty about the product increases (as shown by the dotted lines in Figure 8(a), which is skewed to the left), resulting in an expected utility curve that is more skewed to the left than that without such enhanced information. The outcome of this would be that the decision will be made much quicker, and the expected utility of the decision would be higher due to higher certainty and higher utilities.

## 5 Conclusions

There is a marked lack of timely information availability for making product recovery decisions which hampers the efficiency of product recovery operations. The low margins and increasing volatility of returned products make timely information gathering a high priority. From the observations made during this study, as well as from case studies conducted by other researchers, it is clear that this is an area with a lot of potential for bringing improvements to. It was clear that the quality of product recovery decisions control the cost-effectiveness and viability of product recovery operations, and the quality of these decisions depend greatly on the quality of information made available to the decision-makers. Providing the ability to extract product information in a timely manner could bring two-fold benefits: (a) decision improvements – being able to make informed decisions in a timely manner that could lead to higher profits, and (b) process improvements – being able to facilitate automation of identification and sorting processes, thus improving the efficiency and cost-effectiveness of product recovery operations. Increasing the overall cost-effectiveness of operations could result in increased amounts of reuse of products and components in future. It cannot be disputed that final testing of the product cannot be dispensed of even if the products are embedded with identification tags that enable ready identification. Nevertheless, the availability of lifecycle usage data would greatly improve the quality of decisions made as in many cases it would decrease the rigorousness of testing required to be performed. The concept of networked product identity would enable lifecycle usage data to be collected using appropriate sensors and to be linked directly to the product. Although it can be seen on the outset that there are several benefits that can be brought about by such systems, it is important to realise that providing complete lifecycle information in a readily available format comes with a cost attached to it. In order to make a proper strategic decision about investment in product identification technologies, it is essential to quantify the impact of readily available product information on the effectiveness of product recovery decision-making and subsequently, the efficiency of product recovery operations as a whole.

## References

1. Zussman, E., Scholz-Reiter, B., Scharke, H., 1996, Modeling and Planning of Disassembly Processes, In Proc. IFIP WG5.3 International conference on Life-cycle modelling for innovative products and processes, Berlin, 221-232.
2. Thomas, V., Neckel, W., Wagner, S., 1999, Information Technology and product LC management, In Proc. Int. Symposium on Electronics and the Environment, 54-57.
3. Harrison, M., McFarlane, D.C., Parlikad, A.K., Wong, C.Y., 2004, Information management in the product lifecycle – the role of networked RFID, In 2nd IEEE International Conference on Industrial Informatics INDIN'04.
4. Auto-ID Labs Homepage (http://www.autoidlabs.org).
5. Directive 2002/96/EC of the European Parliament and of the Council of 27 January 2003 on waste electrical and electronic equipment (WEEE),13/02/2003, Official Journal of the European Union, L 037, 24-39.
6. Ishii, K., Rose, C.M., Mizuhara, K., Masui, K., 1999, Development of Product Embedded Disassembly Process Based on End-of-Life Strategies, In Proc. Environmentally Conscious Design & Inverse Manufacturing Symposium.
7. Blackburn, J.D., Guide, V.D.R., Souza, G.C., van Wassenhove, L. N., 2004, Reverse supply chains for commercial returns, California Management Review, 46/2, 6-22.
8. Klausner, M., Grimm, W. M., Hendrickson, C., 1998, Reuse of electric motors in consumer products: Design and analysis of an electronic data log., Journal of Industrial Ecology, 2/2, 89-102.
9. Marschak, J., Radner, R., 1972, Economic Theory of Teams, Yale University Press.
10. Horvitz, E., 1990, Computation and Action under Bounded Resources, PhD Thesis, Stanford University.

# Photocopier remanufacturing at Xerox UK

## A description of the process and consideration of future policy issues

A. King, J. Miemczyk and D. Bufton

*Mechanical Engineering, University of Bristol, UK, Andrew.King@bristol.ac.uk, Innovative Manufacturing Reserch Centre, School of Management, University of Bath, UK, Mitcheldean Remanufacturing Facility, Xerox UK*

Abstract: This paper explains and illustrates the process Xerox uses at its UK factory to return old photocopiers to an "as new" product. Through a series of industrial processes in a factory environment, a discarded product is completely disassembled. Usable parts are cleaned, refurbished, and put into inventory. Then the product is reassembled from old parts to produce a unit fully equivalent to the original new product. Although Xerox has made a profitable business from remanufacturing, there are many barriers to its wider development. This paper outlines some of these barriers to remanufacturing and suggests some ways in which they might be removed.

Keywords: Remanufacturing, Waste

## 1 The waste problem

Between 1980 and 1997, municipal waste in OECD countries increased by around 40%. Perhaps most alarmingly, it is predicted to grow by a further 40% by 2020 [1]. The main environmental pressure from this is that the current landfill sites are filling up, leading to pressure to use new sites with the loss of that land use for housing, leisure or agriculture. In England, of the 28.8 million tonnes of municipal waste produced in 2001/2, 77% of this was land filled [2]. A report in 2002 estimated that if use continued at the same rate there is only 6.5 years of remaining landfill space left in the UK [3]. The European Union has responded by introducing Extended Producer Responsibility legislation (EPR) which is defined by the OECD as "the principle that manufacturer and importers of products



*D. Brissaud et al. (eds.), Innovation in Life Cycle Engineering and Sustainable Development,* 173–186.

should bear a significant degree of responsibility for the environmental impacts of their products throughout the product life-cycle, including impacts from the selection of materials, the production process, and from the use and disposal of the products at the end of life cycle" [1]. EPR is the logical extension of the "polluter pays" principle. This rests on the argument that environmental impacts are substantially determined at the point of design where key choices are made – on materials, processing and finishing technology etc – that is, with the producer [4]. This principle has been most clearly implemented in the WEEE Directive on Waste Electrical & Electronic Equipment. From August 2004, EU OEM (Original Equipment Manufacturers) and importers are legally bound to take significant responsibility for the treatment and disposal of post-consumer products [5]. Electronic and electrical waste contains substances that may have a damaging impact on the environment when it is disposed of either in landfill sites or by incineration. In addition, the volume of electronic and electrical waste is growing by between 3 to 5 % a year – about three times faster than general waste [6]. Electrical and electronic waste is defined as all appliances run by electricity, which does not exceed 1,000 volts for AC and 1,500 volts for DC. Products operating at higher voltages are excluded because their high value makes them unlikely to arise in the waste stream. The objectives of the WEEE Directive include:

- Reducing the waste arising from end-of-life electrical and electronic equipment (EEE),
- Improving and maximising recycling, re-use and other forms of recovery of wastes from end-of-life electrical and electronic equipment, and
- Minimising the impact on the environment from their treatment and disposal.

A recent analysis on the likely effect of the WEEE Directive in the UK concludes that it may only reduce Municipal Solid Waste (MSW) by 0.1% because high levels of WEEE waste is already recycled [7]. But this overlooks the fact that end-of-life waste also generates commercial waste during the production phase for the electrical product. One rule-of-thumb cited is that every tonne of consumer waste has also generated 5 tonnes of manufacturing waste and 20 tonnes of resource extraction waste [8]. Thus, if the WEEE Directive were to lead to an increase in product repair and remanufacture, this would in turn reduce both consumer, manufacturing and resource extraction waste. However, regardless of the quantified difference in waste, an undisputed effect of the Directive will be to transfer the cost of waste management from the general taxpayer to the individual OEMs, and therefore consumers.

## 2 The waste solution

In order to achieve a step change in practice, designers need to consider the entire "lifecycle" of a product from raw material extraction, through manufacturing, product use and final disposal. Thus, a key concept to true sustainability is identified as "closed loop design" where disposal streams are diverted to become new raw material/manufacturing streams [9] as shown in figure 1. There are only two possible long-term fates for waste materials: reuse (closed loop) or dissipative loss (open loop). This is a straightforward implication of the law of conservation of mass. Thus, a relatively simple proxy for "sustainability" in environmental terms is the ratio of recycled/reused material to the total supply of virgin and recycled/reused material [10].

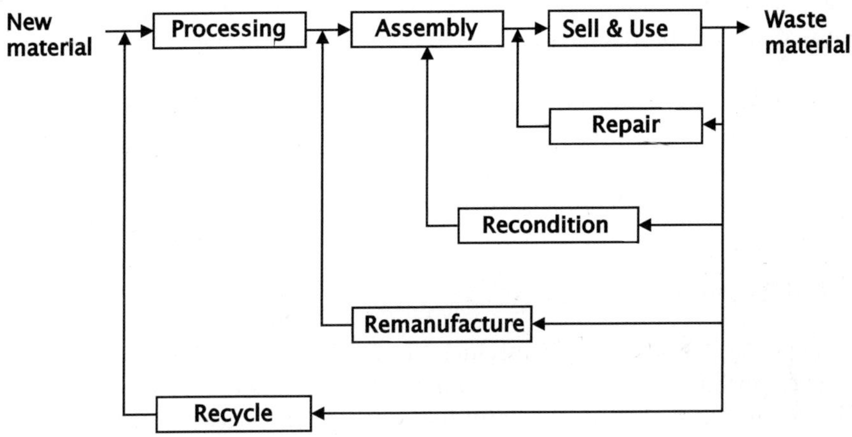


**Fig. 1.** Closing the loop through repair, remanufacturing or recycling

### 2.1 Repairing

A logical approach to closing the loop on product use is simply to repair and extend the product's life. Repairing is simply the correction of specified faults in a product. Generally, the quality of repaired products, when repaired and sold is inferior to those of remanufactured and reconditioned alternatives. When repaired products have warranties, they are less than those of newly manufactured equivalents and may not cover the whole product but only the replaced component.

## 2.2 Reconditioning

Reconditioning involves the rebuilding of major components that have failed or that are on the point of failure. This practice is well established and has created what is called a "green goods" market where original "white goods" products such as fridges and washing machines are reconditioned after a single life and returned for sale as "green goods". Often such products are either sold directly to economically disadvantaged or indirectly through local government social service providers.

## 2.3 Remanufacturing

Remanufacturing is the only process where used products are brought at least to Original Equipment Manufacturer (OEM) performance specification and, at the same time, are given warranties that are equal to those of equivalent new products [11]. This is because remanufacturing requires the total dismantling of the product and the restoration and replacement of its components. Remanufacturing is particularly applicable to complex electro-mechanical and mechanical products which have cores that, when recovered, will have value added to them which is high relative both to their market value and to their original cost [12].

## 2.4 Recycling

Recycling is the series of activities by which discarded materials are collected, sorted, processed, and used in the production of new products. However, although it is currently the most mature waste avoidance strategy, with established rates as high as 80% for certain products, many designers are reluctant to use recycled materials because of uncertain quality or supply standards [13]. In addition, whilst the materials recycled reduce virgin material use, they do still require additional energy to be used to reform them into manufactured products because the embodied energy used in production is lost during the recycling process [14].

## 2.5 Which solution is best?

Stahel states that smaller return loops are more profitable [15]. Thus, repairing or remanufacturing products ought to be more common (if it is more profitable) than rccycling. And yet, the reality is the opposite: recycling is far more common than repair or remanufacture. This has been due to a lack of product lifetime liability. Until the recent introduction of extended producer responsibility legislation (such as the WEEE Directive) a manufacturer had no liability (outside of a short warranty period) for the product sold. Therefore, as recycling is essentially disconnected from individual manufacturers (the material is mixed with other material and processed remotely), this has been the dominant return loop. With the introduction of legislation such as the WEEE Directive, manufacturers are now liable for their products through and beyond their end-of-use life. Whilst a strategy to ensure recycling would meet their obligations, many manufacturers see that this is an additional cost with little or no financial benefit (scrap values are often less than recycling costs). Thus, interest in repairing, reconditioning or remanufacturing products is increasing because the potential profits from these smaller loops will be the most "value adding" way to meet extended producer responsibility. They also have the added societal benefits of providing gainful employment for low to medium skilled labour because much of their tasks are simple to learn. Effective end-of life strategies are however dictated by product characteristics and will therefore vary from product to product [16].

## 3 Remanufacturing at Xerox

### 3.1 Remanufacturing facilities

Xerox has a long history of 'unintentional environmental improvement' as a result of leasing its copiers. By retaining ownership, they have been able to work towards their environmental goal of making "waste-free products in waste-free plants to help customers attain waste-free workplaces". In 1987, Xerox started a new programme called "asset recovery" and created a new, wholly owned subsidiary next to its manufacturing plant in The Netherlands. Its aim was two-fold: firstly, to remove old copying machines from the waste stream and, secondly, to process these machines for resale. This was called the Asset Recovery Operation (ARO). In 1989 5% of scrapped machines were remanufactured; by 1997 this had risen to 75% of the 80,000 copiers returned. At the beginning of 1993 landfill accounted for 41% of manufacturing waste but by 1995 this was only 21%. To encourage return, an incentive scheme was introduced in The Netherlands, and although the remanufactured copiers compete with new Xerox machines, the company claimed to have saved $65million by 1996 [17]. By 2001, Xerox had remanufacturing facilities in the USA, the

UK, The Netherlands, Australia, Mexico, Brazil and Japan [18]. These facilities not only run at a profit, they also enhance Xerox's environmental image and reputation all over the world. In Europe, Xerox remanufacturing facilities are located in two factories: Mitcheldean (UK) and Venray (NL) respectively. At Xerox Venray remanufactured parts are put onto the assembly line for reuse in brand new copiers. This is a process more complicate than the one at Mitcheldean because it integrates both manufacturing and remanufacturing. At Mitcheldean end-of-life photocopiers are either cannibalised for parts and then scrapped or remanufactured as complete products for resale.

## 3.2 Operations management

Xerox's remanufacturing operation is coordinated at the European level through the European Fulfilment Operation. Through this organisation Xerox brings back parts for servicing and if they are classified as "end of life" they can be brought back in to an Asset Recovery Centre. Of the 39,000 photocopiers that were processed in 2002, 25% were refurbished and resold, 25% were cannibalised for parts and 50% sent as scrap (recycling or disposal). As well as the Xerox facilities, two additional companies are involved in the remanufacturing logistics and recovery process:

- TNT (a logistics provider) has a contract with Xerox to collect photocopiers from customer sites and deliver them to an appropriate destination.
- Covertronic is part of the AGR Group based in the Ruhr Germany and specialises in the disassembly and re-sale of parts and scrap.

TNT carry out a simple machine filtering operation at the consolidation centres based on machine condition using a database maintained by the Xerox Operating Companies (the companies that manage the customer contracts, sales or leasing). Therefore TNT is able to identify which machines ought to be sent straight to Covertronic for material recycling and which go to the Asset Recycling Centre (ARC) at Mitcheldean for remanufacturing. TNT also carries out a simple disablement operation to prevent the machines accidentally being placed back into the market before remanufacturing. Co-located at Mitcheldean, Covertronic employs eight people to deal with all Xerox photocopiers under the management of a single person. As the operation is essentially one of disassembly for scrap, the skill level is relatively low with temporary workers being trained within a couple of days. One of the key aspects of the relationship between Xerox and Covertronic is the transparent operation. In order to share all the profits from the re-sale of parts and scrap with Xerox, Covertronic operate an open book system to show both the mass balance of materials recycled (for

WEEE legislation compliance) and revenue information on the sale of materials and costs of disposal. A mass balance log is kept for each of the machines at each stage of disassembly. The cost effectiveness of this operation depends entirely on the metal and plastic revenues, with the material breakdown as follows:

- Metal (as light iron is 90% of the output)
- Printed Circuit Boards (PCBs)
- Cables
- Low grade electrical, power assemblies, hard disks
- Plastic of three types, ABS, PC ABS and ABS fire retardant (for ABS plastic 98% is recycled)
- Miscellaneous (such as aluminium).

The logistics process is shown in figure 2 [19]. TNT receive approximately 100 machines a day in the UK from customer sites and deliver them through four Xerox dedicated consolidation centres to Xerox Mitcheldean.

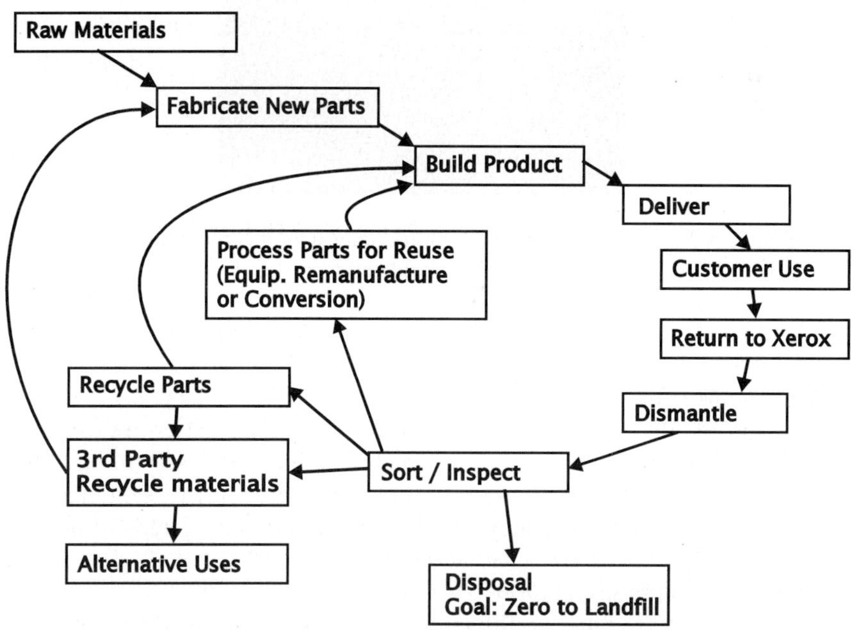


**Fig. 2.** Operations Management for Xerox Remanufacturing

## 3.3 Process management

The Asset Recycling Centre at Mitcheldean employs 58 people to remanufacture two Xerox photocopier models: the Hodaka model and the Silverstone model. In both cases the entire photocopier is remanufactured. Figure 3 shows the Hodaka 220 DC model.

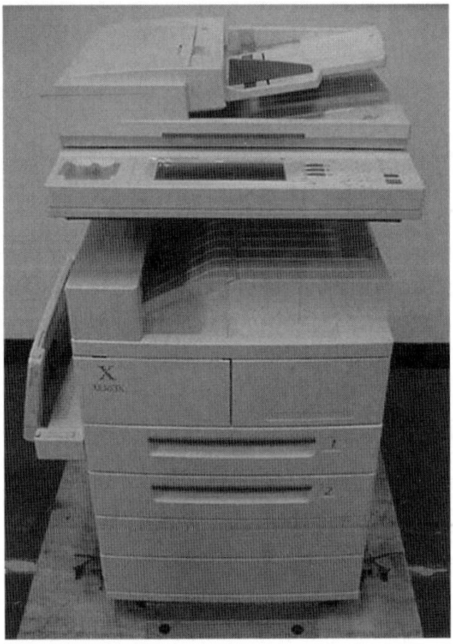

**Fig. 3.** A Xerox Hodaka 220DC Photocopier which is remanufactured at Mitcheldean

There are 7 stages for remanufacturing of entire photocopiers (illustrated in figure 5), which are described as follows:

*Stage 1: Machine Receipt*

The end-of-life photocopiers delivered by TNT Logistics are known as "carcasses" to signify their state as "dead machines". These will have been sorted or tested by their condition before they arrive in Mitcheldean plant. After the carcasses arrive, they are unpacked and segregated by machine modules.

*Stage 2: Parts Disassembly*

The returned Hodaka or Silverstone carcasses are stripped into three main frames: DADF/IIT, IOT/TTM and HCF/CAB/FIN. These main frames are then moved to the small parts stripping process. These parts such as fuser, duplex, Bypass etc are disassembled into subassemblies and then move to stage 3 for clean and paint. Disassembly is regarded as a non-value-adding stage.

*Stage 3: Parts Clean & Paint*

There are three methods to clean subassemblies. They are: hand clean operation, CO2 clean operation and ultrasonic clean operation. Different subassemblies are progressed into different clean operations. For example, Duplex parts and Stripped DADF subassemblies are cleaned by hand operation; IOT parts by a CO2 cleaning operation; and DADF panels are subjected to a thorough ultrasonic cleaning operation. The hand clean operation is shown in figure 4.

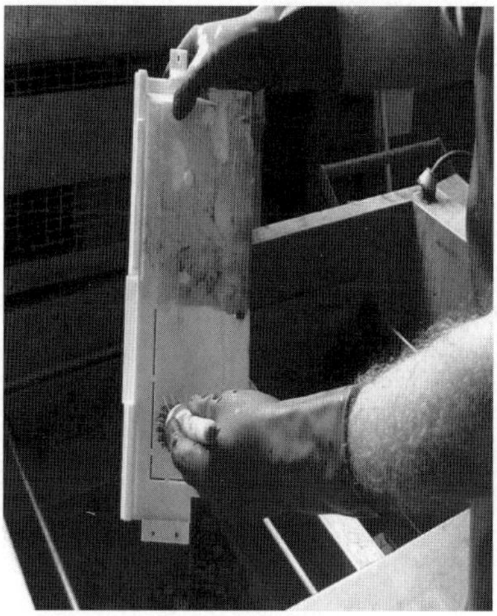

**Fig. 4.** The Hand clean stage to remove toner dust

*Stage 4: Parts Rebuild*

In this stage, the cleaned and painted subassemblies from the previous stage are collected and sorted for rebuild. Brand new parts and materials are used together with some reprocessed parts to build the remanufactured machines. For example, brand new panels replace all the old panel sets. The electronic parts are reprogrammed or upgraded. Then the completed re-skinned components of these machines are remanufactured.

*Stage 5: System Integration to Customer Orders (SITCO)*

The remanufactured components are then configured to individual customer requirements. Any options ordered for a particular photocopier are fitted at this stage. For example, some parts like finishers are integrated into the completed product according to customer requirements.

*Stage 6: Final Quality Assurance (FQA)*

Following the SITCO process, the completed remanufactured machine is subjected to quality and electrical safety tests. Final Quality Assurance

(FQA) is a critical stage of the whole process. During this stage, the necessary controls will be provided such that only the finished products that satisfy all customer requirements are released to distribution. The Quality Plan for each model machine is applied and a sheet detailing various checks is completed. Defective machines are recorded and reported to Quality manager as necessary.

*Stage 7: Packing & Shipping*

When the product is fully configured and has been tested, it is packed and shipped either directly to the customer or via the European Logistics Centre (ELC) in Venray.

Some parts for remanufacturing are obtained from cannibalised copiers or field returns. The components are only remanufactured if it is feasible, based on the following conditions:

- The profit gained from remanufacturing parts;
- The material value added parts;
- The amount of demands for the parts;
- The frequency of replacement for the parts;
- The condition of returned parts.

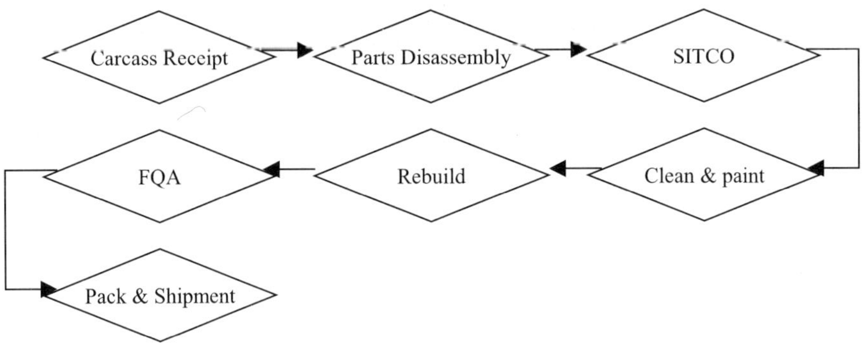


**Fig. 5.** The 7 remanufacturing stages at Mitcheldean

The components recovered from cannibalised machines are sold to service engineers who service Xerox copiers for 40 % of the price of a new part. This price is higher than the cost of asset recovery and thus Xerox continues to make a profit through the asset stripping and recovery process. The print and toner cartridges (which are regarded as Customer Frequently Replaceable Units - CFRUs) are remanufactured by other Xerox facilities all over the world.

# 4 Future policy issues

## 4.1 The WEEE Directive

Whilst Xerox's remanufacturing programme puts them ahead in terms of already have a "return loop" established, there are still three policy issues that need to be addressed in response to the WEEE Directive. Firstly, the UK Government may choose to require all manufacturers to support (via a tax) a central 3rd party vendor to collect and recycle electrical waste. Whilst this would provide a robust accounting mechanism to measure the amount of diverted waste, it would remove the incentive for individual manufacturers to remanufacture. Unless an "opt out" was possible, Xerox may need to end its remanufacturing activity. Secondly, (and subject to the first) the remanufacturing activity at present is limited to the photocopiers originally sold under a rental agreement. Whilst this accounts for approximately 80% of EU sales, new policies and procedures would need to be developed to include the remaining 20% currently sold outright to a customer. This is a company policy issue. Thirdly, Xerox anticipates that (subject to the first point) the introduction of the WEEE legislation will significantly alter the scrap materials market because there will be a dramatic increase in supply, yet without planned increases in demand (that is, policy initiatives to promote greater use of recycled materials). This is a governmental/EU policy issue and may require future legislation or fiscal measures such as reducing value added tax on recycled materials to enforce a % use of recycled materials in new products in order to fully close the reuse loop.

## 4.2 The ROHS Directive

The Reduction of Hazardous Substances (ROHS) Directive is a sister EU legislation to the WEEE Directive. It is aimed at banning certain hazardous substances (Lead, Cadmium, Mercury, Hexavalent Chromium & Flame retardants such as PBB & PBDE). One curious anomaly with the legislation is the concession to allow these substances to be used to repair existing products but not to allow them to be used in a remanufactured product because it is resold. Together with another manufacturer called Oce, Xerox has campaigned to the EU for remanufacturing to be included in this concession.

### 4.3 The 1968 UK Trade description act

The 1968 UK Trade Description Act states that it is illegal to sell as entirely "new" any product (or part of a product) that has previously been sold [20]. This legislation was famously tested when the Ford Motor Co. was deemed to have contravened the Act by repairing a car that was damaged on route to a dealer before going to a customer. Whilst this legislation is intended to protect the consumer from sub-standard goods, it also precludes an "entirely new" product from including a remanufactured part – even if the quality of that part is assured to the same level as a virgin product, and provides the same warranty. The authors question whether, in the current world context where the need to reuse waste materials new policy measures are needed to adjust this Act to accommodate the proper and regulated reuse of components.

### 4.4 Module remanufacturing

The rate of technological changes in the photocopier market is continually increasing, resulting in shorter product cycles and rapidly changing product designs. Thus, Xerox has released many new models of digital photocopiers to the market. In response to these changes, Xerox is making a policy shift from a focus on remanufacturing entire photocopiers to remanufacturing component parts and subassemblies. The fullest use of these remanufactured items will, however, need to be moderated by the implications of the 1968 UK Trade Description Act.

### 4.5 Labour flexibility

From an Operations Management perspective, one of the hardest aspects of a remanufacturing operation is the supply of labour needed to meet demand. The intrinsic reason for this is that, although predictions can be made, the actual rate of photocopier return is not stable. Many photocopiers return to Xerox due to changes in business demographics. If a company relocates, is bought-out or ceases trading, new contracts may be made thus releasing an additional supply of old photocopiers at the Mitcheldean site. Whilst a second worker shift is sometimes possible, peak demands can only be met through increased recruitment and training. Whilst this is a good thing for the local economy, it is equally a difficulty for Xerox's image in the community when demand drops and staff are no longer needed. A possible policy solution to this is discussed in the next section.

## 4.6 Sector-wide facilities

Due to the relatively low production volumes, remanufacturing is not presently as cost efficient as current manufacturing processes. Although the free "raw materials" from the returned carcasses makes the activity profitable.Well-documented evidence has shown that traditional "virgin manufacturing" in Western countries has only been sustainable where production efficiency has been increased through higher production volumes [21-22]. However, whilst the trend has been towards higher production volumes, there has also been a trend towards greater customisation [23]. Although at present a few companies operate individual remanufacturing facilities, one way to increase both the financial and environmental efficiencies would be the establishment of sector-wide remanufacturing facilities. Such larger facilities would have much higher production volumes leading both to efficiency gains and more stable supply and demand. In addition, the higher volumes would justify the introduction of leading edge technologies that minimise the environmental burden. Xerox is currently considering the financial and environmental feasibility of such a venture.

# 5 Conclusions

In Europe, Xerox remanufacturing facilities are located in two factories: Mitcheldean (UK) and Venray (NL) respectively. At Mitcheldean end-of-life photocopiers are either cannibalised for parts and then scrapped or remanufactured as complete products for resale. As well as the Xerox facilities, two additional companies are involved in the remanufacturing logistics and recovery process. Although the free "raw materials" from the returned carcasses makes the activity profitable, the present activity faces a number of policy issues including supply instability, the wider implications of environmental legislation on material and parts markets and the longer-term need to become more efficient.

# References

1. OECD, 2001, Environmental Outlook Report 2001, http://www.oecd.org/ data-oecd/51/6/2088589.pdf (visited 10th October, 2003).
2. DEFRA, 2003, Municipal Waste Management Survey 2001/02, Summary www.defra.gov.uk/environment/ statistics/wastats/mwb0102/wbsummary.htm (visited 10th October 2003).

3. Biffa, 2002, Future Perfect, An analysis of Britain's waste production and disposal account, with implications for industry and government of the next twenty years, ISBN 0-952-3922-3-2, p44, www.biffa.co.uk.
4. Gertsakis et al., 2000, Industrial Ecology and extended producer responsibility, A handbook of Industrial Ecology, Elgar E, Cheltenham, pp 521-529.
5. ICER, 2000, UK Status Report on Waste from Electrical and Electronic Equipment, Industry Council for E&E Equipment Recycling (ICER), London.
6. PriceWaterhouseCoopers, 2002, Environmental Life Cycle Assessment and Financial Life Cycle Analysis of the WEEE Directive and its implications for the UK. A study produced for the UK Department of Trade and Industry, p42.
7. Meadows et al., 1992, Beyond the Limits: global collapse, or a sustainable future? London, Earthscan.
8. King A. M. et al., 2005, Design for End-of-life: Repair, Recondition, Remanufacture or Recycle? ASME DFM Conference, 2004, Salt Lake City 2004.
9. Ayres & Simonis, 1993, Industrial Metabolism, University Press, Tokyo.
10. Ijomah, 2002, A model-based definition of the generic remanufacturing business process Doctoral thesis, University of Plymouth, UK.
11. Lund, 1985, Remanufacturing: The experience of the US and implications for developing countries. UNDP Project Management Report No 2, pp 24-34.
12. Chick & Micklethwaite, 2002, Obstacles to UK Architects and Designers Specifying Recycled Products and Materials, Design History Society Conference, Aberystwyth.
13. Jacobs, 1991, The Green economy: environment, sustainable development and the politics of the future London: Pluto Press, pp 110-116.
14. Stahel, 1994, The Utilization-focused Service Economy: Resource Efficiency and Product Life Extension, in The Greening of Industrial Ecosystems, National Academy Press, pp 178-190.
15. Rose, Ishii, & Masui, 1998, "How Product Characteristics Determine End-of-Life Strategies." IEEE Symp. for Electronics and the Environment Conf.
16. Ayres R. et al., 1997, Eco-Efficiency, Asset Recovery and Remanufacturing, European Management Journal, Vol. 15, No. 5, pp 557-574.
17. Kerr & Ryan, 2001, Eco-efficiency gains from remanufacturing. A case study of photocopier remanufacturing at Fuji Xerox Australia, Journal of Cleaner Production, Vol. 9, pp 75-81.
18. Xerox 2003, Environment, Health & Safety Progress Report 2003, Xerox. www.xerox.com.
19. Regina v Ford Motor Co Ltd, 1974 All England Law Reports, All ER 1974 Volume 3.
20. Kanter, 1995 World Class: Winning Locally in the Global Economy Simon & Schuster Publishers, New York, pp 60-70.
21. Roth A. et al., 1995 Manufacturing Strategy: Process & Content, Chapman & Hall, pp 133-165.
22. Erens F. & Verhulst K. "Architectures for product families", Computers in Industry, Vol. 33, pp 165-178.
23. Pine J. 1993, Mass customisation: the new frontier in business competition, Harvard Business School Press, Boston.

# Dynamic process and operation planning for hybrid disassembly

H.-J. Kim
*Korea Institute of Science and Technology (KIST) Europe Forschungsgesellschaft mbH, Saarbrücken*

Abstract: The European Waste Electrical and Electronic Equipment (WEEE) directive and the End of Life Vehicle (ELV) directive will result in an increasing demand for cost efficient disassembly systems at a high capacity. But major differences between the disassembly plan from the planning system and real disassembly situations cause difficulties for execution in the workshop in line with what was planned originally. The objective of this work is to reflect availabilities of disassembly tools in the real disassembly cell as well as to provide a new concept for generating a set of operation sequences by using disassembly process planning. Algorithms developed for dynamic process and operation planning takes into account the disassembly system conditions and the actual disassembly steps for the specific product.

Keywords: Disassembly, Process Planning, Operation Planning, Cell Control

## 1 Introduction

Integrating the product and material cycles has emerged as a new paradigm for industry in the 21st century. Disassembly plays a key role in a life cycle economy since it enables the efficient recovery of resources [1]. A partly automated disassembly system should adapt to the large variety of products and the different degrees of devaluation. The amounts of product to be disassembled can also be varied frequently [2]. Besides the difficulties of industrialized disassembly, the interfacing tool between the process planning system and the shop floor controller is desirable. In this study, at first, a system will be described for automatic verification of a disassembly processes, generated by a disassembly process planning system. Alternative processes are substituted for processes not available at the time. The results of the verification will be used to connect the planning system to a



*D. Brissaud et al. (eds.), Innovation in Life Cycle Engineering and Sustainable Development,* 187–198.

cell control system. This verification module focuses on the products at the end of the product life cycle stage. The second objective of this research is to develop an "operation sequencing tool," which can satisfy the demands mentioned above: the seamless integration of the process planning system with the shop floor controller.

# 2 State of the art

## 2.1 Availability of process

An examination of available tools and processes has to be executed, in order to verify a disassembly process plan. Within the verification of a disassembly process, the technical feasibility of proposed tools for each disassembly step should be guaranteed. Availability is defined in different sources such as DIN 40041 [3], VDI 4003/4 [4], VDI 4004 [5] or VDI 3423 [6]. Basically, availability is described as the probability that to a certain time under given conditions, no relevant failure causes non-operability of a unit. Therefore, availability can be interpreted as the ratio of service time to the entire application time. The availability of a technical system is separated into:

- Theoretical availability,
- Technical availability,
- Practical availability

Theoretical availability is determined using a simulation of a production system. Several statistical methods can be implemented within the model in order to obtain a realistic image of the system. Technical availability of tools within a system could be controlled via intelligent tool control, which is executed either centralised by a cell Programmable Logic Controller (PLC) or distributed by co-ordination of the handling devices. In both cases, the mechanisms are the same. A tool used by an automated handling device is equipped with an active or passive coding element. In order to define the practical availability, all system states have to be evaluated according to usage, idle or service time. This could be managed via a Plant Data Acquisition System. Such a system is based on a real time database. Real time is required to capture all incoming data according to their generation time. Time differences between several time stamps are used to define tool or system availability, reliability or the utilisation ratio.

## 2.2 Technical feasibility

Nowadays, process and tool selection for disassembly depends strongly on the empirical know-how and even qualitative selection performed by the process planner rather than rational and systematic planning. For an automated or hybrid disassembly system, the selection of disassembly processes and tools should be quantitative and systemised. The advantages of such a process planning are as follows:

- An optimised selection of disassembly processes is possible.
- Under a given framework for process selection, a rapid adaptation is possible on new unknown products.
- If appropriate processes, devices and tools are not available, alternative processes could be generated by the dynamic process planning.

In this study, the technical feasibility of the disassembly process is declared as an applicable range of usage for a specific process and their required tools. In practice, the technical feasibility is limited by boundaries of product- and process-related characteristics.

## 2.3 Operation planning

Normally, a process planning system does not specify detail operations in a process, such as, "take the tool", "deposit tool" and "execute unscrewing process". But a cell controller needs the specific information to coordinate automation devices, that is, robots or transport systems. It has been a difficult job for the cell controller operator to develop a control program with the input of process plans since the disassembly system should accommodate diverse product variants, unknown product conditions and production volumes. The activity, which provides the specific functions in disassembly processes, is defined in this work as "operation sequencing".

# 3 Analysis of disassembly processes

## 3.1 Structuring of disassembly processes

In this chapter several possibilities for disassembly process execution are examined. Disassembly processes are examined with respect to disassembly objects, devices and their corresponding connections. First, possible disassembly processes and already existing disassembly devices are analyzed by the test disassembly. In addition the analyzed processes and the devices are structured regarding the process-related criteria. From this analysis, generalized disassembly functions are derived, which must be used in the disassembly operation planning. The general disassembly functions consist of detecting, handling, separation and special functions (see Fig. 1). These can be divided into several sub functions, i.e. Detecting, Handling [7], Separating [8] and Special operation. All elements of disassembly process are defined for the systematic structuring of all available disassembly processes (Fig. 1). The disassembly order is declared as a sequence of disassembly steps with the aim of completing disassembly of a product. The disassembly sequence is a task-driven subset of this sequence, for example, the disassembly of a motor. The disassembly step is defined as a unit in these disassembly sequences. The step consists of a particular subassembly or a component reference and a process reference. The disassembly process again is a combination of disassembly methods, disassembly devices as well as tools. The disassembly method can be categorized into four fundamental types of disassembly methods: detecting, handling, separating and special operation.

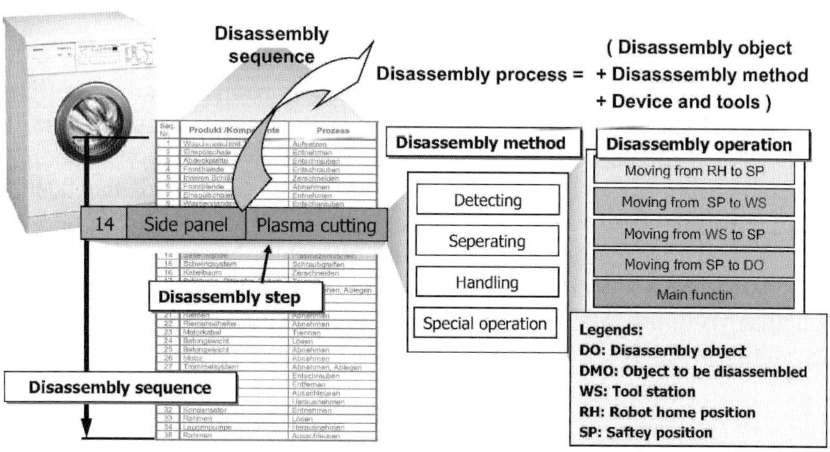


**Fig. 1.** Structuring of disassembly process

In the context of the disassembly operation, an individual disassembly step is described in detail. This activity in the manufacturing technology often also described with the terms, 'sequence in detail' or 'fine plan' [9]. Disassembly step is divided into several disassembly operations. When automatic devices are used for the disassembly, an operational planning is necessary. Therefore, the operation plan plays a role of a link between the process plan (disassembly sequence) and the robotic and the PLC control.

### 3.2 Process specific & general main and secondary functions

Based on the definitions and the structuring, generalized and process-specific main and secondary functions are determined. The generalized main and secondary functions are also grouped into the mentioned disassembly methods: detecting, handling, separating and special operation. These generalized functions build a unified disassembly step and are independent of disassembly objects, tools and processes. Those particular functions can be applied to the specific disassembly objects, special disassembly processes, devices as well as tools. Besides the displayed main functions, additional functions might be required.

## 4 Concept

### 4.1 Dynamic process planning

In order to improve the efficiency of the disassembly shop floor, the methodology for selection of alternative disassembly methods and devices will be addressed in this chapter. If a disassembly process or tool, addressed in the disassembly plan, is not available on the shop floor, alternative available processes should be suggested. In this chapter, a procedure is presented and discussed for selection of alternative disassembly processes for occasional abnormal situations on the shop floor. Figure 2 describes a systematic methodology for dynamic process planning within a hybrid disassembly system by generation and selection of alternative disassembly processes. As a result of the planning, a updated process plan can be generated. The data, i.e. components of the disassembly object, disassembly method, disassembly devices as well as workstation in the disassembly system, serve as an input for dynamic process planning. Components to be disassembled are linked to appropriate disassembly processes including tools and their parameters, provided by the product and process database. Once the database is provided, the availability tests are executed. If the disassembly tool is available in a disassembly step, the disassembly step will be handed to the next disassembly step, if not, a dynamic process planning takes place. The result is an alternative disassembly step including a technologically suitable disassembly process and related utilities.

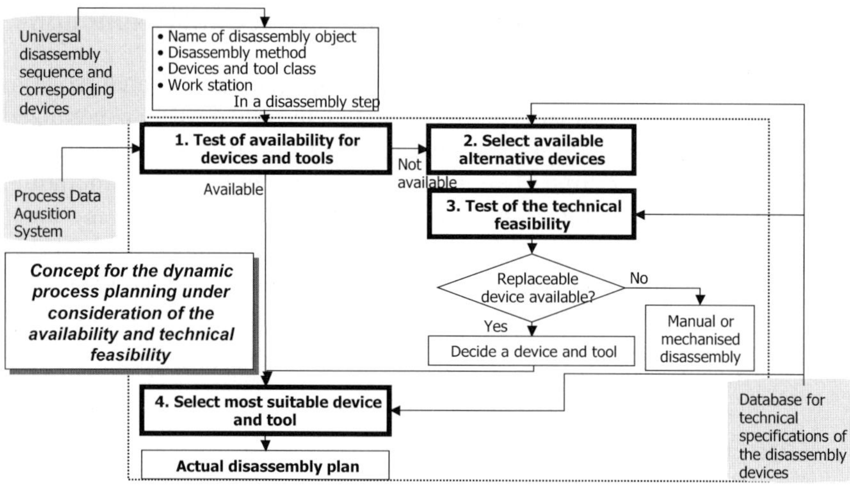


**Fig. 2.** A dynamic process planning procedure for a hybrid disassembly system

## 4.2 Operation planning

Each disassembly operation sequence is dependent on the assigned disassembly method in the disassembly step. In this chapter a concept is represented for generating the disassembly operation sequence. The universal disassembly sequence is entered in the form of structured files by the disassembly planning system. The components to be disassembled and the appropriate separation processes and devices are determined by these data. The product and process related parameters are chosen by the product and by the process database. Based on this information, requested secondary functions are specified for the selected separation process, device and tool. These secondary functions are defined through process-specified disassembly functions considering the temporal sequence. Afterwards disassembly operation sequence is designated with the degree of automation and work station. From these, a product-specified disassembly operation sequences are generated. Figure 3 shows a procedure concept for generating the disassembly operation sequence.

In this chapter, a procedure is declared to generate the operation plans for each disassembly steps. The method deals with the disassembly plan with the information disassembly objects, the disassembly functions, the available devices and the work stations (Fig. 3). First, each disassembly step in this method determines whether that should be carried out with the manual or the automated process. The establishment can be carried out through the scanning of the attributes of the device types. If the steps are supposed to be executed manually, the steps are informed without any generation of the operations to the disassembly cell. However, in case of automated disassembly step, the steps are determined whether they are supposed to be accomplished with one or two handling devices. In order to generate operation sequences, characteristic parameters of the disassembly object, the disassembly functions, the devices and the workstations should be identified by the product and process database.

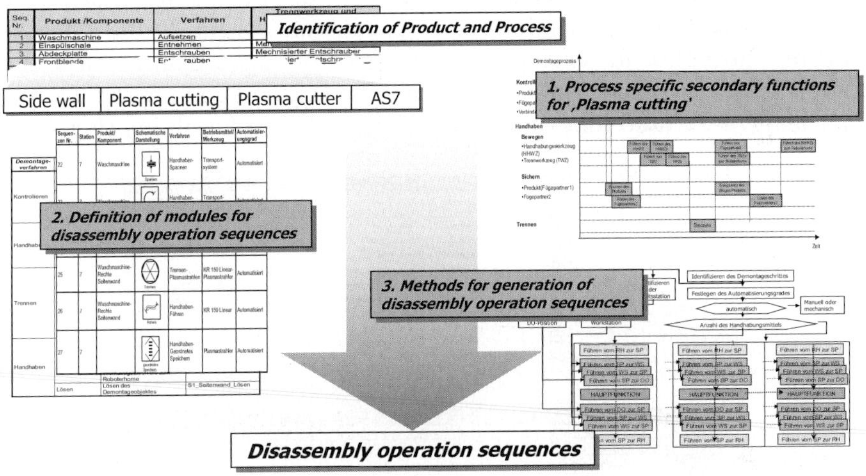


**Fig. 3.** Approach to the generation of disassembly operation sequences

Then, the position of the tool station and that of the disassembly object can be calculated subsequently with this information. The position and geometry calculation has not been observed in this work. The changeable positions are informed from the database to the robot controller whether they are variable by product variants or process changes. With this information, the robot controller knows whether the position calculation will be required or not. If it is required, the relevant information for the position calculation is delivered to the robot controller. A disassembly process needs only one handling device for separation if it does not need a disassembled part to handle. But the process needs two handling devices when the disassembled part should be handled. However, if it is carried out with two handling devices, two operation sequences should be generated. The operation sequences for two handling devices should be determined the order between the sequences. These sequences are used for the process specific main and secondary functions. In addition, the special operations should be inserted into the operation sequences when it is necessary. The information can be obtained from the process database.

# 5 Realization and application in a pilot disassembly system

## 5.1 Dynamic process planning

A software technical system can be generated. The system consists of a database system and dynamic process generation system. In the database system, Attributes, for example, the name of the disassembly object and the name of the separation processes will be selected and appropriate parameters will be used for examination of the technical feasibility (Fig. 4).

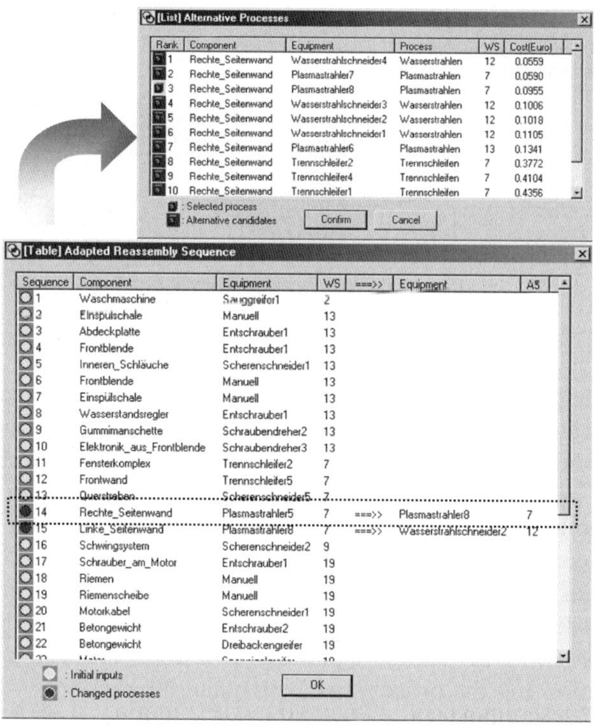


**Fig. 4.** Test of availability and generation of alternative disassembly processes

As an example, a side panel of a washing machine has to be disassembled by plasma cutter. As a first step, disassembly object-related parameters, i.e. material, thickness of the wall and surface of the wall are imported, then process-related parameters will also be included from the database. In the dynamic process generation system, for example, a separating process should be determined first to execute the disassembly of the side panel of the washing machine. Parameters for the component to be disassembled and a disassembly process (separation process) are called from the product and process database. Product-related parameters of the side panel are:

- Material: steel,
- Thickness of the panel: 1.5 mm,
- Dimension: 90 mm x 60 mm and
- Bending of the panel surface.

The concept of searching for alternative processes consists of two parts. One part is the generation of alternative processes based on material and geometric information for the disassembly object and the joining element used. The other part is the generation of alternative processes using the joining method, applied to the disassembly object. Finally, the most suitable process and respective tool are selected by evaluation. The evaluation module considers not only technical but also organisational factors since estimated process time, investment and management costs play a key role in economic evaluation. All technically suitable processes are compared to the original planned process and valued in terms of efficiency. In this way, the optimum process is selected.

## 5.2 Operation planning

The developed algorithm has been developed as a PC based system. The output of the processing step is the product-specified disassembly operation sequence, which is the pre-stage of the actual control sequence. The operation sequence can be verified by the graphic dynamic simulation tool, for example, the eM Workplace. The generated output is used as an input for the cell control system. Modular and parametric control program units can be connected with these operation modules. The operation planning system will contribute as a subsystem for the disassembly control system. The attributes of the disassembly plan, that is, components to be disassembled, disassembly methods, devices, tools and work stations are used as inputs. With these inputs and with the data base inquiry, a set of modularized operation sequences are generated. From the database, the inquiry product and the process information can be specified. If necessary, specified geometrical restrictions are input manually by operation planners. The operation plans correspond themselves one-to-one with the modularized control programs. Whole plans, and the individual operation plan for the disassembly step, can be generated by the automatic mode. Figure 5 represents the software for the disassembly operational planning.

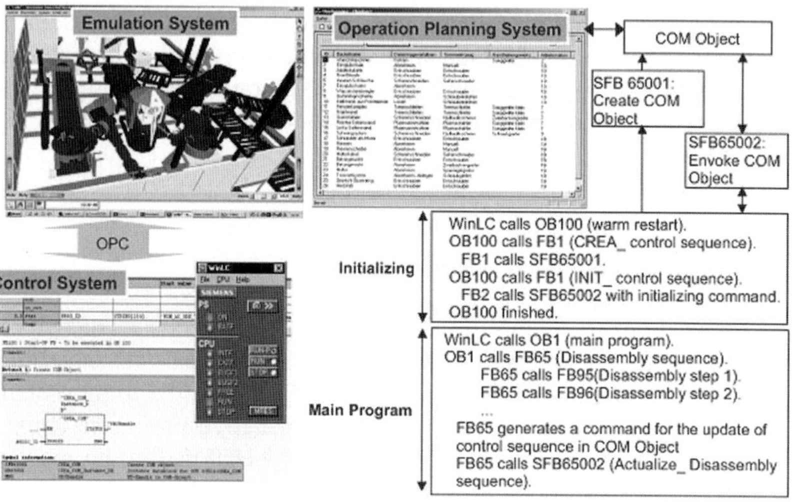


**Fig. 5.** System realization and verification by a 3D-simulation

## 5.3 Verification by a 3D-simulation

Generated disassembly operation sequence can be verified with the available 3D graphic-dynamic simulation system. In this work, the simulator named eM Workplace was used by the company Tecnomatix. With the system module Sequence Of operation (SOP) of the eM Workplace can be represented diverse operations of the disassembly processes with several robots, in temporal successions and dependence. The operability and validity of operations can be examined by comparing the respective results of the operational planning system with those of the system module of the SOP. The Figure 5 represents the system realization and verification by the 3D-simulation as an example of an operation sequence.

## 5.4 Integration into a disassembly control system

The generated operation sequences will be used as an input to the control sequence generation system. The complete control sequence will be built by the integration of control processing application. They are selected as a set of specific process-dependent and the general tool independent control processing application. The entire control program or parts of the program, which have to be exchanged, will be downloaded to the disassembly control system, which is a PLC. The disassembly control system is configured using the SCADA control structure as seen in Figure 6. The communication and control system consists of a product and process database system, which is used for the system operation sequence generation, the material flow simulation system. The master computer containing the PLC. Database -, Simulation- and the operation sequence generation system are connected via Ethernet. The PLC is linked via Ethernet and Profibus DP and serves also as a gateway between two levels of automation. For high flexibility of the PLC system, a PC based Software PLC is used. These interfaces are written in C++ code. Different communication processors, e.g. Siemens CP 5611® are used for connecting automation devices via Profibus DP. Some Input/Output Devices are connected with various Actuators and Sensors via the ASI Bus system. Modular control programs are generated and rearranged by using the data exchange concept between the operation planning application, which are programmed in C++. The actual control programs are programmed in PLC code.

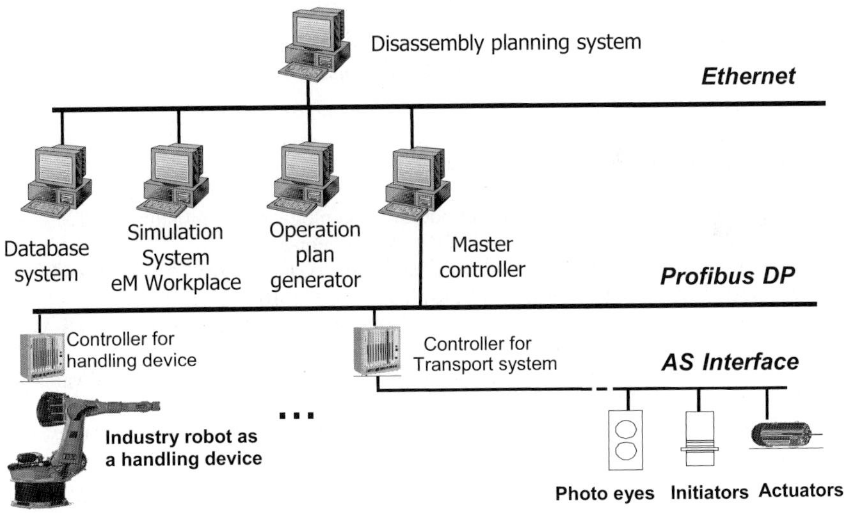


**Fig. 6.** Structure of the disassembly control system

## 6 Summary

The economic recovery of the valuable resources in the closed product and material cycle counts as the new model of production in the 21st century. In this sense, the disassembly is considered as the important preliminary stage for the high-quality recycling. A disassembly system should be able to adapt itself to the high variety of disassembly objects, the uncertain product conditions after usage and the uncertain product amounts to be disassembled. In this paper, a dynamic process planning system is introduced in order to fit a disassembly process plan to the actual conditions of a disassembly system. And a system for generating a set of disassembly operation sequences for the partly automated disassembly has been presented. The dynamic process planning system receives availability information for requested devices and tools. If a device or tool is not available, the system generates available alternatives with a database supported procedure for generation of alternative processes and a test procedure for checking technical feasibility. The operation planning system generates with the help of product and process-related parameters the required main and secondary functions for the selected separation processes as well as devices and tools. In the generation temporal sequences are also considered among the handling devices by the operation planning system. The developed procedure has been realized by a PC based system. The result has been verified by using a commercially available graphic dynamic simulation tool. The system can be used as a link between the process plan (disassembly sequence) and the robotic/PLC control system to result in the integrated planning and control for the disassembly plant.

## References


1. Seliger, G., 2003, Proceedings Colloquium e-ecological Manufacturing, Technische Universität Berlin, Universitat der Künste, Verlag uni-edition.
2. Seliger, G., Zussman, E. 1999,Product Remanufacturing. Kapitel 54, In: Handbook of Industrial Robotics, John Wiley & Sons, Inc. S. 1037 – 1043.
3. DIN 4004/1, 1986, Influence Characteristics for Reliability, Survey. VDI-Verlag, Düsseldorf,.
4. VDI 4003/4, 1985, General Requirements for a Safety Program, Class A. Availability. VDI-Verlag, Düsseldorf.
5. VDI 4004, 1986, Influence Characteristics for Reliability, Influence Characteristics for Availability. VDI-Verlag, Düsseldorf.
6. VDI 3423, 1991, Investigation of Efficiency for Machines and Facilities. VDI-Verlag, Düsseldorf.
7. N.N., "VDI 2860, 1990, Assembly and Handling Technique, Handling Functions, Handling Facilities, Notions, Definitions, Symbols." *VDI*, -05
8. N.N., "DIN 8580 E, 1985, Manufacturing Methods. Definition, Division (Draft)." Berlin: *Beuth Verlag*.
9. Eversheim, W., 2002, Organisation in the production technology, Ed. Springer Verlag.
10. Cuiper, R., 2000, Integrated computer supported Planning and Control of automated Assembly processes. München: Techn. Univ., Diss.
11. Buchholz, A., Ciupek, M., Kim, H.-J., Seliger, G. , 2003, Product Driven Control Sequence Generation for Disassembly. Proceedings of CIRP 2nd International Conference on Reconfigurable Manufacturing. Ann Arbor, 20-21.
12. Seliger, G., Kim H.-J., Keil, T., 2002, Generation of Control Sequences for a Pilot-Disassembly System, in Environmentally Conscious Manufacturing II, Proceedings of SPIE Vol. 4569, pp.81-92.

# Clean technologies for recycling

## A case study on automotive batteries in Brazil

H.V. de Medina

*Chercheur au CETEM -Centre de Technologie Minéral, en stage post doctorat au CREIDD – Centre d'Etudes Interdisciplinaires sur le Développement Durable – de l'UTT -Université de Technologie de Troyes, bénéficiant d'une bourse CAPES du Ministère de l'Education du Brésil.*

Abstract: During the last decade product design were enlarged to cope with environmental requirements from materials selection to final assembly and disassembly for recycling. DFE, DFR, DFD (Design for Environment, Design for recycling and Design for Disassembly) are new designing approaches that incorporate the environment issues from the beginning of a product development. But recycling does not intrinsically prevent environmental impacts. For instance, regarding toxic materials such as heavy metals, lead, mercury, chrome and cadmium, new and clean technologies and sound recycling processes management are required avoiding extra environmental impacts. This is exactly the case or lead car batteries that claims for cleaner recycling processes and for technological innovation. In this way new technological trails leading to more efficient batteries, are suppose to have a great impact on recycling process in medium-term and even short-term. Car batteries are recycled all over the world at different rates such at 90 % in European Union and at 50 % in less developed countries. In Brazil we do not have official data on batteries recycling but a case study made at CETEM by V. Trouche(1) under my supervision got to an estimated rate on an interval from 65 to 80 %. The main propose of this article is to present some results of this case study, highlighting the best environmental practices and giving additionally some insights on environmental and economical aspects of car batteries recycling in Brazil. This case study was conducted not only for its importance for the Brazilian market, but also for clean technologies and sound recycling process management required in order to avoid extra environmental impacts from lead car batteries discharge.

Keywords: Batteries Recycling; Clean Technologies; Lead Recycling Technologies; new Batteries Generation.



*D. Brissaud et al. (eds.), Innovation in Life Cycle Engineering and Sustainable Development,* 199–208.

## 1 Introduction

During the last decade product design were enlarged to cope with environmental requirements from materials selection to final assembly and disassembly for recycling. DFE, DFR, DFD (Design for Environment, Design for recycling and Design for Disassembly) are new designing approaches that incorporate the environment issues from the very beginning of a product development, during the design phase. The idea is to design products with the environment in mind, avoiding rather than controlling environmental impacts of those products during theirs role life, from cradle to grave, or from mining extraction to materials production to end of life products recycling to materials recovery. In Europe, since October 2000 according to European Directive on ELV's recycling [2], car makers are supposed to be responsible for their vehicles from cradle to grave. That means that they have to close automotive materials life cycle loop. In fact the strategies adopted by North American and European companies regarding environmental requirements are similar and the way automobiles are being conceived and manufactured from now on is changing the label of this former essentially polluter product into an almost 100 % recyclable one and maybe soon available in zero emission version. But recycling does not intrinsically prevent environmental impacts. For instance, regarding toxic materials such as heavy metals, lead, mercury, chrome and cadmium, new and clean technologies and sound recycling processes management are required avoiding extra environmental impacts. In this scenario the importance of lead batteries recycling is strongly linked to the environmental requirements in both senses claiming for cleaner recycling processes and for technological innovation, such as rechargeable batteries or long lasting batteries for electric or hybrid vehicles. In short new technological trails leading to more efficient batteries, are suppose to have a great impact on recycling process in medium-term and even short-term. For ELP in general the major contribution to a sustainable recycling is required from automobile industry mainly from the electric and electronics car components that are the largest materials intensive manufacturing sector. Besides that the economic viability of ELV recycling depends on disassembling techniques and materials separation processes. Best disassembling practices and soundest methods on separation economically feasible are nowadays the most important challenger for engineers and designers. For automobile industry, according to ACEA European Automobile Manufacturers Association [3] the factors that determine the recyclability of single materials and components include the purity of the recovered products, the market for the recovered products, the monetary value of the material, the cost of collection and transport, the cost of sorting, the cost of transformation into reusable material and the cost of disposing of any residual material. This is exactly what DFR (Design for Recycling) practices are looking for: to provide an added value recycling condition to new products at the end of their life. DFR is a new concept of the design activity that incorporates materials recyclability from the very beginning of the product creation. So complex products like automobile has to be designed to be assembled and disassembled. In fact, nowadays materials and products are being redefined and designed according to the consumers' expectations, reaching to improve the engine performance as well as to rend new products technologically innovative and environmentally friendly. In this context materials selection is one of the key elements of DFR some times called eco-design. And the life cycle engineering is the technical expression of waste prevention approach. "Since the rise in awareness of global ecological problems, the emphasis in pollution-control policy for industry has shifted away from end-of-pipe treatments and controls to waste prevention at the source and life-cycle engineering of products." [4] This article aims at presenting an overview of the main aspects of lead recycling from car batteries illustrated by a six months case study conducted, in 2003, under my supervision, by Vincent Gilles Trouche a former scholarship that developed at CETEM his Projet de Fin d'Etudes (PFE) en Génie Productique, to INSA de Lyon. Alternatives clean technology by hydrometallurgical processes pointed out as well as the economic and environmental aspects that represent the strengths and the weakness for future development of the lead recycling market in Brazil.

## 2 Lead recycling from lead-acid batteries

The importance of lead recycling is strictly linked to the environmental requirements in both senses claiming for cleaner recycling processes and for technological innovation, such as rechargeable batteries, long lasting batteries for electric or hybrid vehicles. In short it demands new technological trails leading too much more efficient batteries. In between nickel, zinc oxide, titanium nitride, and lithium are in perspective for future developments for vehicle uses. In this context the design of new batteries and the materials selection criteria taking into account materials recyclability are suppose to have a great impact on recycling process in medium-term and even short-term. It means that the development of a new generation of batteries is catching up the principles of life-cycle engineering. Batteries are a good example of practical R&D of the so-called eco-materials. "In the R&D of ecomaterials, the end application of the material is obviously of great importance. In the field of consumer materials, where great quantities are consumed daily, efforts focus mainly on finding substitutes for harmful substances to reduce the environmental burden during and after use, examples being materials for batteries, packing, soldering, painting" [4]. For the time being the lead remains the major substance for car batteries as well as for industrial uses. The largest use of lead in U. S. for instance, around 80%, goes to automotive and industrial lead-acid batteries, consisting of a polypropylene case, lead lugs, electrodes, plastics spacers between them, lead oxide paste and sulphuric acid (5). Scrap automotive batteries (lead-acid batteries) are the major source of secondary lead in Brazil as all over the world. Since 1996 Brazil have no primary lead extraction and, nevertheless, it has one of the biggest domestic's apparent consumption increased in the last two years as showed at Table 1and 2.

**Table 1.** Recent Primary and Secondary Production of Lead in Brazil

| MINERALS | SPECIFICATIONS | 1995 | 1998 | 2001 |
|---|---|---|---|---|
| Lead | Primary Metal | 13 958 | - | - |
| | Secondary Metal | 50 000 | 48 000 | 52 000 |

Source: DNPM, Mineral Summary 1996, 2000, 2002 Note: (-) Null

**Table 2.** Apparent Consumption of Metallic Lead Products in Brazil

| | 1997 | 1998 | 1999 | 2000 | 2001 |
|---|---|---|---|---|---|
| Secondary lead production | 53 | 48 | 52 | 52 | 52 |
| Import | 60,7 | 60 | 56 | 70,7 | 73,4 |
| Consumption | 113,7 | 108 | 108 | 122,7 | 125,4 |

Source: Trouche V. 2003 (estimate based on DNPM data)

ULAB (usage lead acid batteries) recycling has important environmental concerns such as the risk of water, soil and air contamination. ULABs are a hazardous waste for recycling proposes as well as final disposal. Additionally to lead, automotive batteries have also acids and other metals and minerals such as arsenic, antimony, tin, calcium - in suspension and solution. The lead (oxide and metallic) represents around 70% in car batteries' weight and that why it's so important to have sound technologies and processes to recovery it from ULAB. The recycling industry for batteries to cope with environmental regulations have to deal with complex tasks such as handling with toxic materials from collecting to transportation up to dismantling and separation of a number of materials, avoiding extra contamination of soil, air or water throughout all these phases. In addition it is also a complex market that depends directly on many actors such as scrap dealers, brokers, dismantlers and smelters and indirectly has to deal with governmental and environmental agencies and so on and so forth.... Nevertheless, car batteries are recycled all over the world at different rates such at 95% in United States, 90% in European Union and at 50% in less developed countries. In Brazil we do not have official data but Trouche's study estimated this rate on an interval from 65 to 80%, varying according to the region.

## 2.1 Technical aspects of lead recycling

Technically lead from batteries can be recovered by pyrometallurgical refining or by hydrometallurgical process. Pyrometallurgical refining (on blast furnace and rotary furnace) is traditionally the most used process worldwide for both primary and secondary lead productions. Wernick and Themelis (1998) resumed the pyrometallurgical ordinary recycling process "At a typical recycling plant, batteries are crushed or sliced and separated into streams: lead materials (about 60% lead, 15% $PbO_2$ and 12% $PbO_4$) polypropylene scrap, and sulphuric acid. The lead–containing materials are smelted to produce lead bullion and a molten silicate solution containing all of the lead oxides. This slag by-product is smelted and reduced with carbonaceous material and fluxing agents in a lead blast furnace, similar to those used in the primary smelting of lead oxides. The low-lead slag in such furnaces is environmentally inert and is disposed in industrial landfills". Nevertheless monitoring and preventing the escape of dust and fume is crucial. In Brazil a recent study (6) conducted by Chemical Institute of UFRJ (Federal University of Rio de Janeiro) and at the FIOCRUZ (Oswaldo Cruz Foundation) provided indicators of environmental pollution in areas adjacent to a source of stationary lead emission based on dust and air contamination. This study measured lead concentration in both the outdoor air and the household dust from houses located around a lead-acid battery repair shop. As a result over 50% of the air samples exceeded the standard limit (1.5 ug $Pb.m^{-3}$) (for more details see http://www.scielo.br). This situation occurred some years after a dozen of small smelters have been shut down by public environmental authorities, from 1995 and 1999, in the same region. Actually in the recent years, pyrometallurgical process has been largely improved to cope with the environmental requirements as well as to face the challenger of hydrometallurgy alternative that avoids gas emissions. Wernick and Themelis also pointed out some of these innovations: "in another lead recycling process, used extensively in Europe; lead components are smelted at 1000° with coke breeze and sodium carbonate flux and iron filings, used to fix sulphur by forming an iron sulphide matte, in a short fuel-fired rotary furnace. It is reported that use of bulk oxygen and other innovations can increase the production capacity of such furnaces by 40%. The ISASMELT furnace process used in Australia and South Korea; introduces lead materials with lump coal into a vertical reactor containing a slag bath. Air and oxygen are injected in this reactor that similar to the top blown BOF furnace used for steelmaking. During smelting, metallic lead containing less than0.01 antimony settles in the bottom and is periodically tapped out of the furnace. When the upper slag layer in the reactor reaches a certain depth the slag is reduced from 40–60% lead and 5–6% antimony, to 2–4% lead and less then 1% antimony by continuous injection in the presence of coal. Most of the treated slag is then tapped out the reactor and the cycle repeats. Thus, the two-reactor process used in U.S. (reverbatory and blast furnace) is replaced by a two stage process in a single reactor" [5]. Hydrometallurgical methods are the newest and the soundest technology developed but its economic and industrial viability has to be proved. Hydrometallurgy is a chemical metal processing technology to dissolve a metal from its concentrate by using water, oxygen and other substances on a pressurised vessel. Usually there are a further series of chemical processes, involving a number of separation and purification steps, which result in the production of a high purity metal. Compared to the pyrometallurgy hydrometallurgy process is technically more efficient (recovers a higher percentage of the metal concentrated) more environmentally friendly and uses much less energy. On the other hand economically speaking its industrial scale is over 10 times smaller than the conventional pyrometallurgical smelters plants. Nevertheless Engitec Technologies presents its CX Compact process that is able to treat 5t/h batteries as and advantage compared to their CX traditional system. The first steps on this via were made by US Bureau of Mines by the end of the 70's and the first results were published in 1981. At industrial level the best initiatives were provided by Engitec Technologies in 1992 (CX-EW and next the CX compact). The CX plants in operation are considered as state of art units for environmentally friendly design, for the quality of the recovered products and for the quality of design and manufacture of the equipment. As described by Engitec the CX systems plants were environmentally friendly designed, for the quality of the recovered products and for the quality of design and manufacture of the equipment. For instance in all units operating around the world the process water is continuously recycled. The main advantages of the CX Compact according to Engitec are:

- Minimizing the requirements for civil works and costs (pre-assembled)
- It can be easily transported and expanded (even at a later stage)
- Long life are guaranteed by stainless steel equipment
- As well as high performance and low maintenance costs

(for further information see details at http://www.engitec.com or email info@engitec.com). In Brazil, CETEM carried out a 5 five-year research on recycling domestic batteries and automotive batteries. The aim of the first project was to determine the level of contamination of final disposal of domestic batteries on proper conditions. The second one aimed to reduce the environmental impacts of the pyrometallurgical process by associating a hydrometallurgical phase for lead paste desulphurisation from car batteries [7]. These research at CETEM were encourage by CONAMA's Resolutions on batteries final disposal and recycling, by the end of the ninths (228/97, 235/98, 257/99) inspired on the 1994 Basel Convention (8) the ban of exports of hazardous waste (including ULAB) from developed to developing countries. The photos, which we got from one of the greatest Tires and Batteries Collection and Transportation Company - Mazola Pneus, show how Brazil is nowadays packing, stocking and transporting ULAB according to Basel Convention recommendations. (see more at http://mazolapneus.com.br).

Source : Mazola, São Paulo, Brazil.

## 3 Environmental and economical aspects of lead recycling market in Brazil

European legislation proposal is seeking for a closed loop system, resulting in the collection of all batteries in Europe to cope with Basel Convention. Neverthless, the quantity of batteries sold in relation to batteries recollected varies, depending on lifespan of automotive batteries, from 3 (in Brazil) to 7 years (in Canada and West Europe), the percentage of cars exported and the evolution of the lead price. In Brazil, the more restrict environmental regulations and the continuous falling down of international metallic lead prices threw the lead-recycling sector into an ever-present crisis. The Figure 1 shows the international lead price evolution in dollar per tonnage form 1997 to 2003.

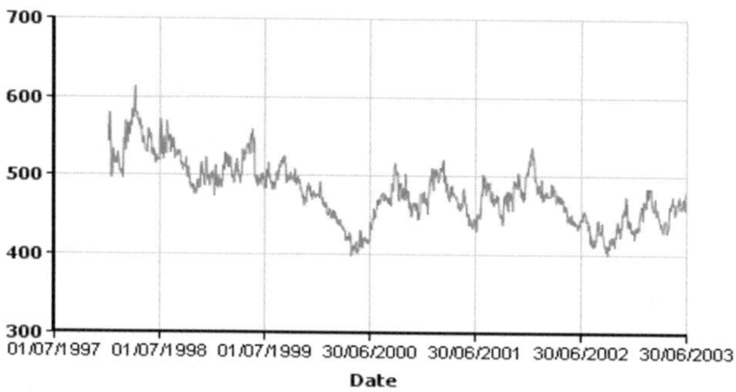


Source: London Metal Exchange. *www.lme.co.uk*

**Fig. 1.** International Metallic Lead Price 1997-2003

Some of them were shut down after 1995 by Environmental Federal Authorities under the pressure of the NGOs. Other got into tough financial and economic difficulties as Cobrac a primary lead producer, shut down in 1995 due to the depletion of mineral reserves. Some of the most importants in this situation are: **FAE** an independent one that had 2 industrial plants (São Paulo and Rio Grande do Sul) with an annual capacity of 24.000 tons, **and Saturnia - Microlite** a vertically integrated Co with a capacity of 18000 tons per year. Among the remainders there are: **Tonolli,** an independent recycler in São Paulo State with a capacity of 36.000 t. /year that produced around 12.000 in the year of 2001; **Moura,** the biggest integrated national company, located in Pernambuco State, with a capacity of 22.000 t. per year; **Tamara Metais,** an independent company for metals recycling located at The State of Parana producing 12.000 tons per year; and **Sulina de Metais** located at the State of Rio Grande do Sul with annual production of 11.000 tons [9]. Nowadays the Brazilian Lead Recycling Market is passing through a risk situation in both senses economic and environmental. There are a number of informal lead smelters and even some big companies, which are operating under risky situations [10]. Recent Greenpeace investigations, for example, revealed that Moura, one of the largest manufactures of car batteries in Brazil, is still importing ULAB. For automotive batteries the national scenario also includes some important threats concerning the sound management practices requirements for recycling activities in Brazil such as:

- The lack of a National Programme for collection vehicles batteries, as we have for the aluminium cans.
- The large number of irregular final disposals at 90% of the Brazilian municipalities, mainly at the Northeast, North and Central regions.
- The environmental legislation is not enforced by an effective system at local level;
- The out-dated recycling technology based on pyrometallurgy processes.

In a short run perspective the over capacity of Brazilian Batteries Producers and Recyclers associated at the depletion of the mineral reserves can be taken as opportunities to a enhancing the development of lead recycling sector. From a medium or long-term perspective the following factors can be considered as the main challenges to a national policy for a sound recycling management:

- To improve the collection system organisation at local and national level
- To minimise waste transportation since it represents a high cost due to Brazilian geographic dimension
- To promote materials, product and process development towards cleaner recycling technologies
- To extend the polluter-pays principal to product manufactures concerning end of life products responsibility.

## 4 Final remarks

Broadly speaking DFE and DFR practices is surely a great opportunity for less developed countries' to get into the Research, Development and Designing of materials and manufactured products, which up to now have concerned only the headquarters bureau employees. Because environmental impacts is a global concern that have no national frontiers and that calls for partnership and sharing different knowledge. Summing up: innovative solutions for clean products demands team work. It can be also a means to drive all actors to global commitment towards the sustainability of mining, materials and industrial production. Working in network as partners they can share risks and profits and also get more innovative solutions to improve the recyclability of ELP on technical, economic and environmental bases. Briefly we can say by our experience on this case study that life recycling for sustainability is a great challenge that means not only extending natural resources lifetime, reducing costs and wastes, and enforcing energy conservation; but also demands great effort on protection of human health and environment cares, clean materials recovery techniques development, well-organized collection systems establishing, and large markets for recycled materials assurance.

## Acknowledgment

My best acknowledges go to: CETEM researchers Roberto Trindade, Ramon Veras and Roberto Villas Boas for their support and advice, Vincent Trouche for the extensive data collecting and analysis on the subject and, CNPq for supporting financially my participation at the CIRP-2005, where this article was first presented.

## References

1. Trouche, V.G. 2003, "Le recyclage de Batterie Automobile au Brésil" INSA de Lyon, France., 82 p.
2. IPPC -European Union Integrated Pollution Prevention and Control - Integrated Pollution Prevention and Control. Reference document on Best Available Techniques in the non-ferrous metal industries. EU December 2001.
3. European Automobile Manufacturers Association (ACEA) (http://www.acea.be)
4. Halada, K. and Yamamoto, R., 2001, "The Current Status of Research and Development on Ecomaterials around the World", MRS Bulletin, 871-879.
5. Wernick, I. and Themelis, N. J., 1998, " Recycling Metals for the Environment", Annual Review Energy and Environment, vol 23, pp 465-497, copyright by Annual Reviews (http://www.AnnualReviews.org).
6. Quitério, S. L., Silva, C.R. S. da, Vaitsman, D. S., et al. 2001, "Use of dust and air as indicators of environmental pollution in areas adjacent to a source of stationary lead emission", Cadernos de Saúde Pública.
7. CETEM RT 58, Araujo R.V.V., Soares P.S.M., Trindade R.E. 2001, "Estudo da desulfurisacão da pasta de bateria automotiva visando a reciclagem de chumbo." Rélatorio Tecnico do Centro de Technologia Mineral.
8. Basel Convention Technical Working Group, Draft Guidelines on the Environmentally Sound Management of Lead-acid Battery Waste, report elaborated for the Basel Convention Secretariat. April 2001.
9. Stone, H., 1999, Effects of Amendments to the Basel Convention on battery recycling", journal of Power Sources, 78, pp. 215-255.
10. Marchetto, I., 2000 "A reciclagem do chumbo no Brasil", Seminário de Reciclagem de Metais Não-Ferrosos, ABM Associação Brasileira de Metalurgia, October – São Paulo.

# Identifying availability contribution of lifecycle-adapted services

J. Fleischer and D. Nesges
*Institute of Production Science, University of Karlsruhe Germany, Fleischer@wbk.uka.de*

Abstract: Assuring constantly high machine availability is a central challenge for production. Increasingly, manufacturers get involved in respective tasks by taking over warranties on specified availability targets. The compliance with a specified availability target can be accomplished by a suitable configuration of additional equipment elements in line with product-accompanying services. However, varying boundary conditions across the machine life require an adaptation of these additional measures, especially of the services. The article presents a method for planning a lifecycle-adapted service configuration in order to achieve availability targets at lowest cost. It focuses on the measurement of availability contribution.

Keywords: Lifecycle, Maintenance, Availability

## 1 Introduction

In many production settings, the manufacturer of the production system has more knowledge of its operating behaviour than the customer. Thus, the operator can often achieve higher system efficiency by making use of the manufacturer's expertise. A central task in this regard is availability assurance. As the constructing manufacturer is anyway responsible for many technical failures, his complete liability in terms of availability assurance can be reasonable [1]. On account of this, customers require that manufacturers assume risks in connection with machine operation (figure 1). These risks range from extended warranties over full technical responsibility to market risks, the latter of which are not observed here. In a typical agreement, manufacturer and customer agree upon a specified availability target, where the included aspects of availability and its minimum percentage are clearly defined [2].



*D. Brissaud et al. (eds.), Innovation in Life Cycle Engineering and Sustainable Development,* 209–222.

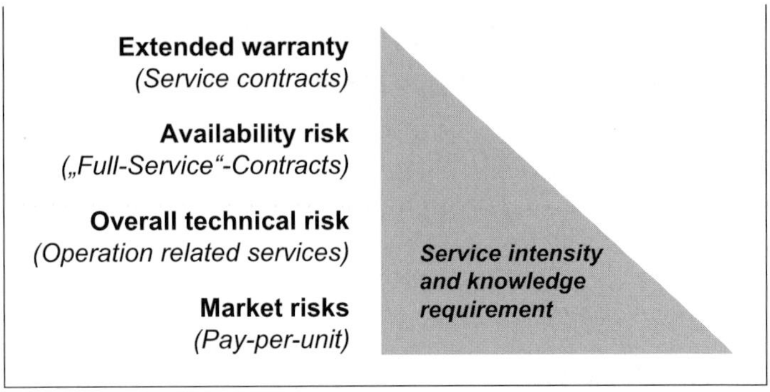


**Fig. 1.** Degrees of risk assumption

The more a manufacturer assumes such availability warranties, the more he has to provide the equipment with alternative elements and additional field services. Recognizing the fact that machine configuration is for the most part accomplished when availability agreements are negotiated, the manufacturer has predominantly to rely on cost-effective services to reach the availability target over the lifecycle.

## 2 Problem description

Even if machine manufacturers have profound information on component reliability, they are mostly unaware of the attainment cost for required availability targets. This is because of their limited experiences concerning the actual effects of their service measures on availability. A comprehensive method to produce satisfactory availability data is missing, notably one that is manageable for medium-sized machine manufacturers. Previous scientific approaches have explored the impact of technical availability on lifecycle costs and established some optimization algorithms [3–5]. However, organizational and logistic aspects of availability are not sufficiently included in existing methodologies. Meanwhile, researchers have detected the relevance of administrative and logistic performance on availability, but the viewpoint remains on machine design aspects [6–8]. In view of the present deficiencies, an integrated approach for the modelling of relations between equipment elements and accompanying services, achieved availabilities and the resulting cost has been elaborated. Before introducing a solution approach, some elements of the problem are discussed.

## 2.1 Operational availability

The applicable measurement category regarding availability warranties is the operational availability. Three factors determine operational availability: reliability, maintainability and maintenance supportability (figure 2). Compared to technical availability, it comprises the aspects of maintenance supportability, taking into account all administrative, organizational and logistic delays [9]. It refers therefore to all relevant time units that can be directly influenced by the maintenance organization. Since many product-accompanying services affect just these time units, service benefits can be expressed in improvement of operational availability. In the practical case that availability assurance is collaboratively carried out by manufacturer and operator, the included aspects of operational availability should be limited according to the manufacturer's responsibility.

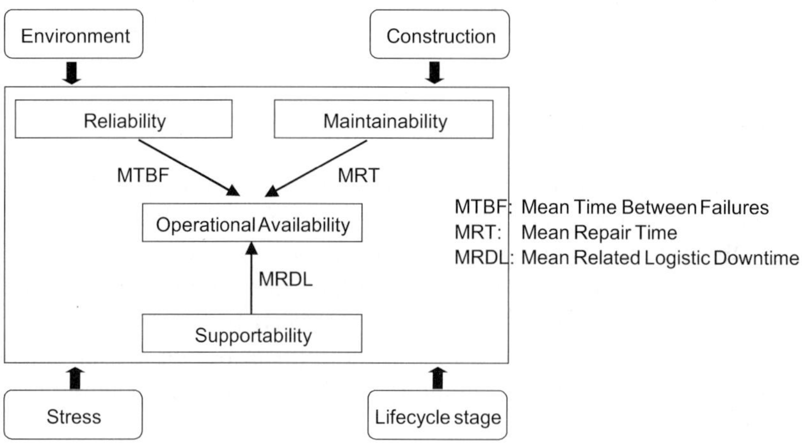


**Fig. 2.** Determinants of operational availability

## 2.2 Enrichment of standard equipment

Standard production machines can be enriched by alternative and supplementary technical components as well as product-accompanying services. Regarding the first group, the following effects on operational availability of the superordinate system are recognized: durability, fault-tolerance, maintainability, and diagnosis capability.

Extended durability of mechanical components with different dimensions or material properties directly affect reliability. Coherences in this context are amply demonstrated in literature [10–11]. Further options of the machine manufacturer include subassemblies and auxiliary equipment from different suppliers, which have proven to be less susceptible to failures. Fault-tolerance can be enhanced by means to avoid false alarms or by redundant items. If redundancy provides the possibility to repair the system on line, it also refers to maintainability. Finally, monitoring and diagnostic devices affect maintenance supportability.

Concerning product accompanying services, we observe the following effects on availability: preventive maintenance, reduction of administrative delays, logistical support, and qualification of operating personnel. While preventive maintenance tasks conserve reliability, services aiming at reducing administrative and logistical downtimes affect the maintenance supportability. Finally, after-sales services concerning qualification like extended training or provision of operating personnel may have positive impacts on several influencing variables of operational availability. Many product-accompanying services are directly related to additional equipment. This is true in particular for teleservice. Methods to investigate availability effects must take interdependencies carefully into account.

## 2.3 Variations over lifecycle

Many boundary conditions of the operational availability alter within the machine lifecycle. Calculation must account for the following aspects:

- Abrasion, fatigue, ageing and similar factors affect the intrinsic failure behaviour of technical items. The failure rate of most mechanically stressed items grows for this reason over time.
- Perfective maintenance measures can eliminate design and production weaknesses. This leads to reliability growth due to higher durability or better interaction of items.
- Increasing expert knowledge of the operating personnel increases maintenance supportability. Moreover, maintenance strategies and spare part provisioning benefit from increasing experiences.
- Collaboration of the manufacturer's field service with the responsible operating personnel advances as they become acquainted with each other. For this reason waiting times for spare parts, auxiliary means etc. shorten. These aspects, again, can be classed into reliability, maintainability and supportability. As for system reliability, the characteristic bathtub curve over the lifecycle is applicable. When it comes to maintainability and supportability, an early lifecycle stage can be specified, too. At this stage, expertise for the maintenance of the respective assets still has to be built up (figure 3).

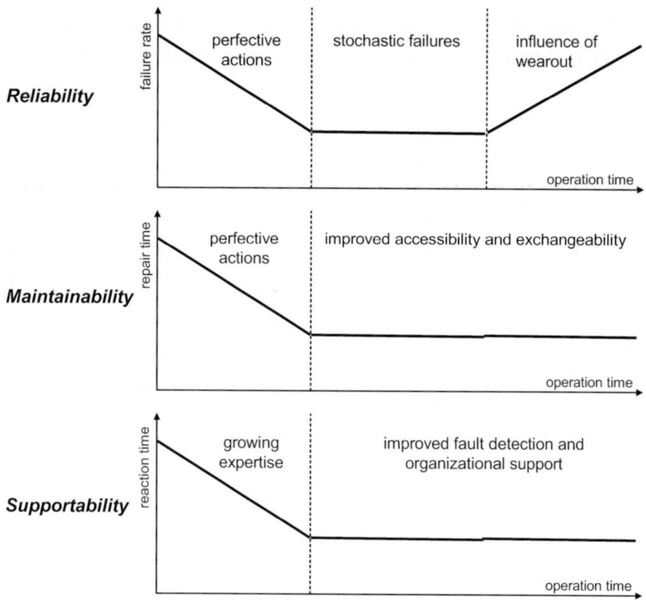


**Fig. 3.** Variation of availability factors over lifecycle

## 2.4 Scope of the approach

The presented approach integrates the modelling of the relations between equipment elements and accompanying services, attained operational availabilities and the resulting cost. It takes into account system reliability and maintainability as well as organizational aspects of maintenance supportability. The attainment cost of an availability target, respectively the resulting availability for a given budget for availability assurance is to be detected for characteristic lifecycle stages (figure 4). The methodology is designed for the use during the planning process of availability assurance, where measures and responsibilities are collaboratively defined between the operator and the manufacturer. At this stage, the standard production machine is finally designed. This implies that constructive changes are excluded from consideration despite their general effects on availability factors and lifecycle cost [12]. Forms of enrichment of standard machines with supplementary components and services as presented above are included in the approach. Beyond, modifications of the required functional level due to changes in users needs and technology development can be part of maintenance and service tasks [13]. However, machinery with improved functional specification or performance definition is excluded from the presented approach as it imparts additional performance and flexibility potentially counter-balancing availability losses [14].

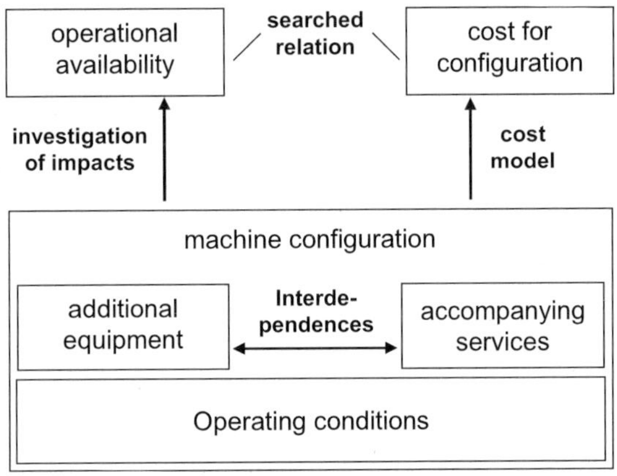


**Fig. 4.** Searched relation

## 3 Measuring availability contributions

The approach to identifying operational availability encompasses four major procedures: reliability model, contributions of equipment elements, contributions of services, and cost structure and selection.

### 3.1 Reliability model

Initial point for availability calculation is a reliability model of the technical structure of the system, which integrates the properties of the items in connection with technical downtimes. It represents relevant subcomponents in an adequate tree structure and allows for the residual items by combination on different levels. If desired, the user can confine the reliability tree to subsets of special interests. The underlying data for the reliability analysis have to be brought together and assigned to the respective items diligently, because they determine the outcome of the calculation to a great extent. In practice, the methodology relies on field data originating from operators, spare part orders, technical test data as well as expert ratings. Along with reliability data, the repair cost and the downtime statistic of each machine item are also recorded. Secondary failures and cause-consequence chains are reproduced by accordant information with each item (figure 5). In view of dissimilar stress loads and ambient conditions, correction factors are introduced. They come into operation, if operating conditions and resulting reliability deviation of an item can be properly identified. Reliability models have to integrate lifecycle variations because of ageing, wear and other effects [15]. Therefore the method takes into account early failures and increasing failure rates in the wear-out zone. Four basic failure patterns over lifetime are distinguished:

1. Items with primarily stochastic failures,
2. items with increasing failure rate,
3. items preventively maintained at fixed intervals, and
4. items predictively maintained on condition base.

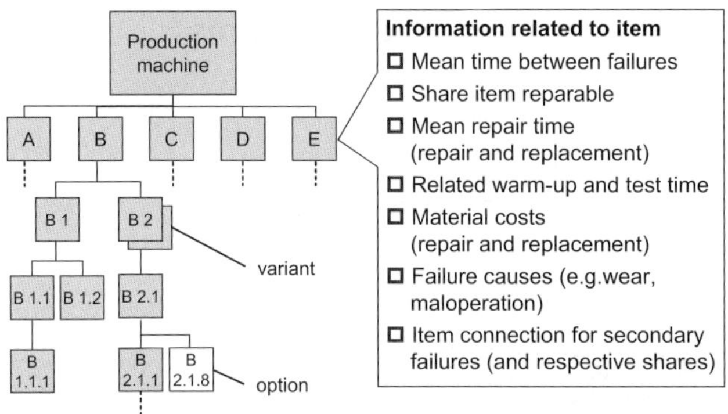


**Fig. 5.** Reliability tree

The basic failure patterns are divided into typical usage phases (figure 6). Reasonably, the first phase should encompass the ramp-up time, in which early failures occur and production processes are not sufficiently robust yet. The second phase covers the rest of the warranty period, where failure rates are mostly stable. The third phase covers the period, where ageing induces increasing failure rates for many items. It closes with end of life of the production system or with the planning horizon for availability assurance. At this stage, additional services can sustain the specified availability target that would probably be missed otherwise. Within each usage phase, we assume a constant failure rate. This is expedient as far as availability targets apply to the usage phases and as service configuration is supposed to be unaltered within each phase. Therefore, variant failure rates are converted into means for each usage phase. For the patterns (1) and (2), all technical failures go into calculation. Items conforming to (3) and (4) are only relevant in respect of sudden failures, because preventive maintenance is assumed to be carried out outside occupied time.

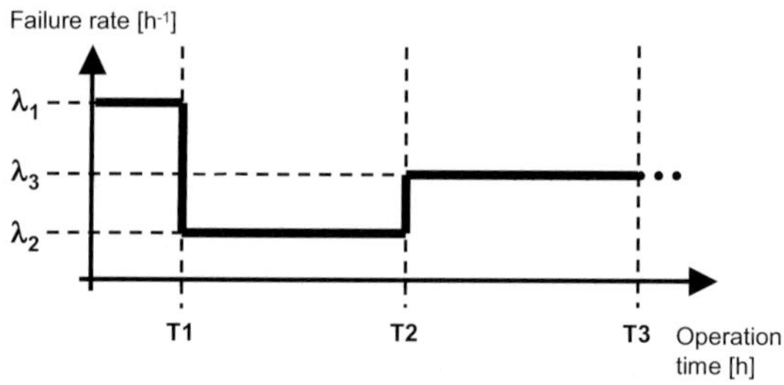


**Fig. 6.** Failure rate in different lifecycle stages

## 3.2 Contributions of equipment elements

| | | | |
|---|---|---|---|
| Mean time to repair MTTR | Predominantly depending on customer | Time to failure notification | MAD 1 |
| | | Travel time | MAD 2 |
| | | Waiting time for customer personnel | MAD 3 |
| | | Waiting time for auxiliary means and resources | MAD 4 |
| | Predominantly depending on service | Waiting time for free capacity | MLD 1 |
| | | Time until technician ready to work | MLD 2 |
| | | Time for diagnosis | MLD 3 |
| | | Waiting time for spare parts | MLD 4 |
| | Predominantly depending on technical item | Mere repair time | MRT |
| | | Warm -up time | MTD |

MAD = Mean administrative downtime
MLD = Mean logistic downtime
MRT = Mean repair time
MTD = Mean technical downtime

**Fig. 7.** Classification of characteristic time units

Availability improvement is brought about by alteration of specified time units making part of the availability definition $A = MTBF/MTBF{+}MTTR$ (figure 7). These time units are supposed to be chosen in order to cover all typical field service processes. The responsibility for these fractions of technical downtime must be clearly allocated to the liability of the manufacturer, the customer or a third party. An improvement of a time unit due to a supplementary feature or service is referred to as 'availability contribution' here. The identification of the availability contribution of alternative machine configurations initiates with the calculation of technical availability inherent in the physical system. As the general reliability model specified above encompasses all equipment variants and options, the first step to calculating technical availability is the reduction of the general model to the actual configuration. Thus, the availability contribution of alternative equipment elements is immediately identified by comparison of different machine configurations. Nevertheless, some additional elements have features, of which the availability contribution refers to the superordinate system. For example, hardware modules for teleservice may affect downtime due to diagnosis for the complete production machine. In such a case, the contribution of the supplementary equipment element is scrutinized together with the product-accompanying services. This is anyway expedient because of many interdependencies.

## 3.3 Contributions of services

The principle prerequisites for the quantitative evaluation of an accompanying service or service combinations are the standardization of the service quality and consistently used time units, against which the availability enhancement can be measured. For example, scope and intensity of training measures or the composition of spare part packages have to be defined with clear specifications. By this means, obtainable field data can be better compared. The method follows the principle of service catalogues allowing for different service types. To each service type a limited number of possible values is assigned (figure 7). In this connection, an industrial trend towards a modular composition of defined service packages can be detected [16]. The smaller the number of defined service and service packages, the more significant conclusions can be made concerning their effects on availability.

| | | | |
|---|---|---|---|
| **Spare parts package** | None | Small | Large |
| **Training** | Instruction M | Instruction M+E | |
| **On-site support** | None | 8h/d | 12h/d |
| **Telemetrics** | Std. Teleservice | Ext. Teleservice | Control Room |
| **Reaction time** | 24h | 12h | 8h |
| ... | | | |

**Fig. 8.** Catalogue of service qualities

There are two principal options to identify the availability contribution of product-accompanying services. Firstly, if only few relevant services are considered and many field data on hand, we can use appropriate regression analyses, which approaches the impacts of the service measures. This is only practicable, if the measures are stochastically independent. The second case where only fragmentary information on hand prevails in industrial practice. The following steps are conducted for each specified time unit:

**Structuring of the service programme.** As stated above, only a standardized service programme can be properly analyzed because of the many influence factors. To this end, the standard service catalogue of the machine manufacturer should be the basis of the considered service programme. It has to be discussed with the competent technicians, if other potential services shall be included. On the other hand, services can be excluded from consideration where appropriate, in particular if they do not appear to have observable availability contributions.

**Interdependencies between services.** In this step, interdependencies between the services in respect of their availability contributions are determined. It has to be detected, if potential availability contributions can just be added together or if services have different effects in combination. This is the case, for instance, if one service dominates the other in respect to a certain time unit. For example, the service 'on-site-personnel' dominates the function 'teleservice' in terms of diagnosis.

**Determination of service combinations.** The service combinations to be necessarily considered together directly derive from the previous step. For this purpose, all service combinations, where the potential availability contribution of the single services can not just be added together are registered. However, if combinations are not supposed to be provided together or do not make sense, they are excluded from further consideration.

**Evaluation of time units without services.** Identification of the mean durations and occurrence probabilities of the specified time units initially takes place for the non-service case. A distinction of cases is done in respect of repair execution by field service, by the maintenance department of the operator, and by local service providers (figure 9). According to the individual task sharing, the applicable time units can be selected and weighted for further calculation.

**Evaluation of time units with application of services.** The identification of the alterations of mean durations and occurrence probabilities of the time units is now executed for each service and combination thereof. In practice, impacts on availability must be gathered individually by scrutinizing service records, by analytical calculation of single effects and by expert ratings. This quantification task can turn out to be extensive. However, the accuracy of the calculation largely depends on it. Finally, the availability contributions are identified and can be used for optimization purposes.

| Time unit \ Operation by | Field Service (manufacturer) | | Maintenance (operator) | | Local service providers | |
|---|---|---|---|---|---|---|
| Failure notification | 0:25 | 100% | 2:30 | 100% | 2:50 | 100% |
| Diagnostics | 2:00 | 100% | 2:10 | 100% | 2:20 | 100% |
| Waiting for spare parts | 8:00 | 15% | 8:00 | 30% | 11:00 | 30% |
| Travel | 5:00 | 100% | 0:00 | - | 1:00 | 100% |
| ... | | | | | | |

| Mean duration | Occurrence probability (where applicable) |
|---|---|

**Fig. 9.** Time units dependent on maintenance agent

## 3.4 Cost structure and visualization

Parallel to the analysis of availability contribution of equipment elements and product-accompanying services, the associated costs are structured. They are broken down in activity-based and stand-by cost and distinguished in cases, e.g. dependent on the maintenance agent. Again, interdependencies in the provision of different services are detected and quantified. The quantification, however, can be restricted to the above service combinations. Furthermore, a calculation of failure cost or, alternatively, contract penalties has to be executed. In view of the long time span for the planning of availability assurance, risk analyses about technical, organizational and financial hazards should be executed, too. As basic operations for the application of the method these tasks are not illustrated further here. Facing the complexity of calculations, the methodology has been transferred into a software prototype. Impacts of alterations in equipment elements or services are immediately calculated and visualized in terms of availability contribution and cost. The user has to execute three guided data input steps: entry of machine configuration, entry of service configuration, and entry of operating and contractual conditions. Subsequently, the programme forecasts the availability for each usage phase individually. In this context, sensitivity analyses regarding production settings, variations of the typical usage phase, contractual regulations etc. can be executed without difficulty.

| Phase I | Phase II | Phase III |
|---|---|---|
| | Basic configuration | Current selection |
| Techn. availability | 98,7 | 99,1 |
| Oper. availability | 94,7 | 96,3 |
| Annual costs [€]: | | |
| *Equipment components* | - | *2300* |
| Service measures | - | 2800 |
| Personnel* | 7400 | 3600 |
| Material* | 5400 | 4800 |
| Failures (contractual) | 4800 | - |
| **Total** | **17600** | **13500** |

*in connection with repair

**Fig. 10.** Visualization (schematic)

## 4 Combination of services

With the identification of operational availabilities and costs, the basis for decision upon the services to be provided together with the delivered machines is completed. The following decision making is often subject to further corporate objectives, so that strict optimization programmes are rather academic. Nevertheless, some critical issues concerning service combination are discussed here.

**How to combine services.** The catalogue specifying the possible qualities of services and service combinations can be used again in the selection phase. It is deposited in the software assistant, acting as a morphologic box of possible combinations. In many practical cases, the number of possible service combinations is manageable, so that all solutions can be enumerated. If not, search algorithms have to be employed, which solve the combinatorial problem for both fixed availability targets and budgets for availability assurance.

**Adaptation to lifecycle requirements.** It is obvious that many boundary conditions may change during the planning horizon respectively the contract period. Among other things, the utilization ration may drift or new technologies emerge that leave former technical limitations behind. Since the equipment configuration is usually static along the lifecycle, the significance of services, which can be applied short-time is high. Combinations of services should be allocated to each usage phase according to the prevailing requirements. However, some scale effects have to be considered, which have influence on the optimal selection. In particular services generating primarily stand-by costs, such as the provision of adequate spare part packages, have high decreases in cost if applied continuously through all usage phases. These issues have to be discussed with the customer in such a way that service provision goes in line with the operating and maintenance strategies over the lifecycle. Result of the considerations are effective and customer-oriented lifecycle business models [17].

## 5 Summary and outlook

The presented approach assists the machine manufacturer in planning availability assurance against the background of contractual availability agreements with the operator. Central element of this approach is the detection of the impacts of services on operational availability. By this means, manufacturers can point up the extent, to which lifecycle-adapted measures contribute to the attainment of availability targets. In first industrial applications, a mean deviation of the availability forecast of 8 % could be verified. However, even if the method is applied for standard machines with an ample population, two challenges persist: First, as both machine breakdowns and service measures are often unique, experiences and field data can turn out to be little significant. Second, usable operating data records and downtime statistics for the application of the method can be scarce. For these reasons, sound procedures for data collection and expert questioning are fundamental. Since the scope of the approach is limited to single production machines, it must be combined with supplementary availability analyses considering material flow [18] and breakdown interdependencies [19], if interlinked systems are evaluated. Moreover, additional sensitivity analyses can be carried out, e.g. regarding different maintenance strategies [20]. For most of these tasks, the necessary simulations can be based on the presented approach for availability calculation. The next extensions of the methodology will encompass the impacts of capabilities of the operating organization as first experiences indicate a high relevance. Besides, the inclusion of quality capability simulation [21] is envisaged for future work.

## References

1. Spath, D., Sternemann, K.-H., 2001, The operating models of tomorrow require new control concepts today, IFIP TC5 / WG5.2 & WG5.3, 494-505.
2. Grob, M., 2004, Neue Zusammenarbeitsmodelle zwischen Werkzeugmaschinenherstellern und OEMs am Beispiel der Motorenfertigung, Life-Cycle-Performance Conference 2004, Institute of Production Science, Karlsruhe.
3. McNichols, R.J., Messer, G.H., Jr., 1971, A Cost-Based Availability Allocation Algorithm, IEEE transactions on reliability R-20/3:178-182.
4. Moore, L.T., Ntuen, C.A., 1984, RAMCOST – A Simulation Algorithm for Availability-Based Life Cycle Cost Analysis, Proc. of IE Conf. 536-540.
5. Werden, R.B., 1984, A practical method of optimizing system life cycle costs vs. availability, Microelectronics and Reliability, 17:1-7.
6. Gershenson, J., Ishii, K., 1993, Life-Cycle Serviceability Design, in: Kusiak, (ed.): Concurrent Engineering: Theory and Practice, John Wiley and Sons.
7. Goffin, K., 2000, Design for Supportability: Essential Component of New Product Development, Research-Technology Management, 43/2:40-47.
8. Markeset, T., Kumar, U., 2003, Design and development of product support and maintenance concepts for industrial systems, Journal of Quality in Maintenance Engineering, 9/4:376-392.
9. Verein Deutscher Ingenieure, 1986, VDI-Richtlinie 4004, Blatt 4, Zuverlässigkeitskenngrößen – Verfügbarkeitskenngrößen, Beuth, Berlin.
10. Schlottmann, D., Schnegas, H., 2002, Auslegung von Konstruktionselementen: Sicherheit, Lebensdauer und Zuverlässigkeit im Maschinenbau, Springer.
11. Birolini, A., 2004, Reliability Engineering, Theory and Practice, Springer.
12. Weule, H., Trender, L., 1999, Entwicklungsbegleitende Kostenkalkulation – prozessorientierte Bestimmung von Produktionszykluskosten unter Beachtung von Zielkosten, Konstruktion, 51/9:35-40.
13. Takata, S., et al., 2004, Maintenance: Changing Role in Life Cycle Management, Annals of the CIRP, 53/2:1-13.
14. Shewchuk, J. P., Moodie, C. L., 1998, Definition and Classification of Manufacturing Flexibility Types and Measures, IJ. of Flexible Manufacturing Systems, 10:325-349.
15. Gadhi, K., Simeu-Abazi, Z., 2001, The modeling of a manufacturing system taking account of the equipment ageing and wear, in: Emerging Technologies and Factory Automation, Proceedings, 1:503-507.
16. Schuh, G., et al., 2004, Erfolg mit intelligenten Produkt-Dienstleistungs-Kombinationen, Werkstattstechnik, 94:301-312.
17. Meier, H., 2004, Lifecycle-based Service Design for Innovative Business Models, Annals of the CIRP, 53/1:393-396.
18. Wiendahl, H.-P., Hegenscheidt, M., 2002, Reliability and Availability of Complex Production Systems, Advanced Manufacturing Systems and Technology, CISM Courses and Lectures, 437:225-232.
19. Weule, H., Fleischer, J., et al., 2003, DiHyPro – a diagnostic tool for electro-hydraulic machines, Production Engineering, 10/2:97-100.
20. Kimura, F., Hata, T., 2003, Life Cycle Design and Management based on simulation of Service Quality Variation, CIRP seminar on LCE, 2003.
21. Fleischer, J., et al., 2003, Quality simulation for Fast Ramp Up, 36th CIRP International Seminar on Manufacturing Systems, 149-153.

# Product Development for Sustainability

# Designing products that are never discarded

P. Zwolinski and D. Brissaud
*3S Laboratory, University of Grenoble, France, Peggy.Zwolinski@hmg.inpg.fr*

Abstract: With new environmental directives, manufacturers become responsible for the disposal of their products at the end of their life. These new constraints imply new costs to take into account in the product definition. So, for manufacturers, designing products, which are never discarded, is not a utopia, but one more objective to reach during the design process of the products. In fact, to control end of life costs, manufacturers have to develop new end of life strategies to increase the economic interest. But, end of life considerations are not sufficient, environmental aspects have also to be respected during the product life cycle. These new objectives imply major changes during the design of the products: A real use of the concept of product life cycle, in order to build product life cycle scenarios that correspond to economic, environmental and society needs, this in a sustainable logic of development. The construction of expertise specific to the various scenarios which can appear, in order to help products designers to integrate the new life cycle constraints related to the new types of consumption, to new supply chains, to new technologies... In this article, we will present researches that we currently carry out in order to help this concept of "products, which are never discarded" to emerge.

Keywords: Product Life Cycle, End of Life, Remanufacturing, Disassembly

## 1 Introduction

With new environmental directives, manufacturers become responsible for the disposal of their products in end of life. With these new rules, the concept of property of the product has to be reconsidered, and the manufacturers have to develop new strategies to limit product end of life costs [1]. They can adopt two main approaches:

- The manufacturer pays an external firm to recover and recycle its products. In this case, the recycling will certainly be limited to the material recovery, because of the diversity of the products treated by the external recycling firms.



*D. Brissaud et al. (eds.), Innovation in Life Cycle Engineering and Sustainable Development,* 225–244.

- The manufacturer recovers the products by himself to try to make a benefit while keeping more added value on his product than this obtained by the simple price of the recycled materials.

The last option has a great interest due to its scenario that is the most likely to go towards a real environmental benefit. Indeed, the classical scenario that recovers material for a new transformation encountered technological and economical obstacles that can impact environmental issues. For the technological aspect, the question is about the use of a secondary material: the recycling and the use of the new material depend on physical and chemical characteristics of the recovered material that are not always controlled. From the economical point of view, it is noticed that the recycling profitability is too low. Moreover, the uncertainty on the original raw material prices and the competition introduce risks for investment in a revalorization processes (as shredder for example). So, designing products, which are never discarded, is not a utopia, but one more objective to reach during the design process of the products. A product that is never discarded can be defined as a product designed to be reused in end of life, that is to say, a product that be can put back in the production loop. Thus, it keeps more added values than if there is only a material recycling process and its life cycle is expanded because a second life cycle (at least) is defined for it (or a part of it). But, end of life considerations are not sufficient to validate the benefits of this second life of the product. Many others life cycle considerations have to be taken into account. For example: more recent products can be more environmentally friendly in their use phase than those to be reused; the technology continuously evolves, so the product can be obsolete; the market prices can have decreased that makes the product less profitable… So, to help the scenario of "products, which are never discarded" to emerge, the issues are:

- To develop knowledge on "life cycle" in order to build more precise product life cycle scenarios [2]. Expertise on product/process integration has been developed but is now limited because it doesn't really integrate all the life cycle aspects. The main and urgent challenge is to build a knowledge base made up of life cycle scenarios that integrates functional, technical, financial, environmental and legal requirements within the design process.
- To construct specific design methods and tools linked to the various scenarios that can appear. This is necessary in order to help the product designers to integrate new constraints related to the new types of consumption, to new supply chains, to new technologies... Indeed, if designers have to really integrate all the life cycle constraints during the design process, they will be confronted to new design situations calling upon new competences, knowledge and tools. The current solution that leads initially the industrialist to gather the actors involved in the same

project is not possible. Indeed, designers' teams have increased over the last few years due to the integration of experts of different disciplines and cannot continue their growth without becoming less efficient. Moreover, actors cannot increase their competencies and skills without any effect on their own expertise.

So, the challenge is to manage knowledge and teams of designers in this new design approach that considers all aspects of the product life cycle. To understand what to take into account to design product, which are never discarded, section 2 describes the concept of "life cycle" from different points of view, to point out what is needed for proposing new life cycle scenarios of products. Section 3 focuses on the remanufacturing scenario, to show how life cycle data are used to define the design project goals and to guide designers in this new way of thinking. Section 4 illustrates how life cycle data can be used to design the product characteristics early during the design project. In this particular case, the disassembly characteristics are examined. To conclude, new research areas to help the concept of "products, which are never discarded" to emerge are explained.

## 2 Different points of view on product life cycle concept

For the design process, it exists at least two different kinds of tools that are called life cycles:

- The first one is about the life of the product in the broad sense of the word, with a point of view more focused on marketing and strategic aspects. The life cycle is then decomposed in five phases: Introduction, expansion, maturity, saturation and decline phase.
- The second one is about the product life from its design to its recycling. This tool is associated to the functional analysis of the product, an approach that should be in integrated in the mind of all the people in charge of research and development or involved within the design process.

With these two kinds of tools, it appears different levels to think about product life cycle and those have to be managed together to increase the competitiveness of a product under development. Three levels of analysis have been identified for a firm:

- The firm level, based on the marketing concept of the product life cycle
- The team level, based on the functional analysis concept of the product life cycle
- The expert level, based on data management and expert's knowledge.

After the presentation of these three levels, we will focus on the needs linked to the use of the product life cycle concept during the design process and explain what kind of information should be relevant for the design phase of "never discarded" products.

## 2.1 LC as a marketing concept: enterprise level

The product life cycle concept is certainly one of the best-known if not most important concepts in marketing. Product life cycle is an almost inexhaustible concept because it touches on nearly every facet of marketing and drives many elements of corporate strategy, finance and production [3]. The products life cycle analysis is a mean to identify in which stage of the life cycle a product of the company is situated. This is a mean to establish priorities to chose product design strategies to realize a continuous growth of the turnover. Usually, there are 5 phases to describe the product life cycle from a marketing point of view, and each phase produce different turnovers and profits. These phases are represented on a curve called S-shaped curve. An example of this representation is given figure 1.

1. Introduction phase: During this period, the product introduces the market. Benefits are negative depending on the high sales costs. (Product birth, monopolistic market, product adjustments).
2. Expansion phase: The product has succeeded in winning a place over the market. The first benefits are realized. The product interests competitors. Products improvement have to be done taking into account the first users reactions to keep advantages in the competition.
3. Maturity phase: The transaction volume increases. The market shares decrease because production capacity and innovation are small. The profitability of the product is very good without investments.
4. Saturation phase: The market reaches a saturation point, competition increase, prices and transaction volumes decrease. Only the strongly firms can continue to compete.
5. Decline phase: Costs increase, due to the large capacity of the production and finally the product is removed from the market.

Life cycle analysis studies the phases of life, the repeated patterns that occur during life, and the causes and effects of incidences, aiming at something that can be recognized and learned from the earlier life cycles of the items under study. The representation of the life stages of all the firm products in a same diagram gives a rough estimate of the firm situation and gives mean to formulate some answers to the following questions: What products have to be replaced and how long can be the design process to do that? What could be the level of innovation of the development process? A new product (with long time development) or a redesign (to increase the life cycle of the current product)? All these questions have to be answered to facilitate the start of the design process, particularly if new scenarios are proposed like remanufacturing, servicing… Indeed, many aspects of the product are influenced by these strategic choices. However, Life cycle analysis should consider the period between birth and grave. If we consider the life cycle definition from this marketing point of view, it is only "the period of a product in the market". This interpretation emphasizes the marketing point of view and revenue planning, but excludes the impacts of product creation and disposal for life cycle profitability. So, an extension to certain defined stages, including research, development, and abandonment is needed.

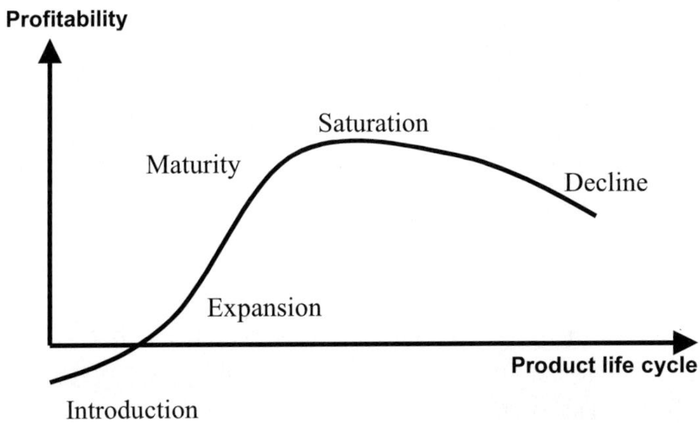


**Fig. 1.** A S-shaped curve

## 2.2 LC as a functional analysis concept: team level

To explain what is the product life cycle from a functional analysis point of view, we have to keep in mind the fact that there is not only one user of a product [4]. Designers have to consider all the users along the life cycle, to satisfy their needs (designers, manufacturers, users, recyclers…). So, when a designer speaks about product life cycle, he considers the design process, the manufacturing process, the use and now the recycling stage (figure 2). This is a non exhaustive list of all the phases that can be addressed: Design, Manufacturing and Assembly, handling, packaging, storage, Transport, distribution, sale, delivery, installation, customer use, after sales service, End of life (elimination, recycling, second life,..).

In many organizations, consideration or design of the support processes is an after-thought and many of these developmental activities are started after the design of the product is well under way (if not essentially complete). This can lead to an increase of the production costs. To integrate these considerations during the definition of the product characteristics, it is now recognized that an integrated definition of the product is necessary. Indeed, for many durable goods, there are a variety of design considerations related to the total product life cycle. For consumable products, some of these life cycle factors may be of least importance. Generally, life cycle factors that may need to be addressed during product design include: Testability / Inspectability, Reliability / Availability, Maintainability / Serviceability, Design for the Environment, Upgradeability, Installability, Safety, Product Liability, Human Factors…

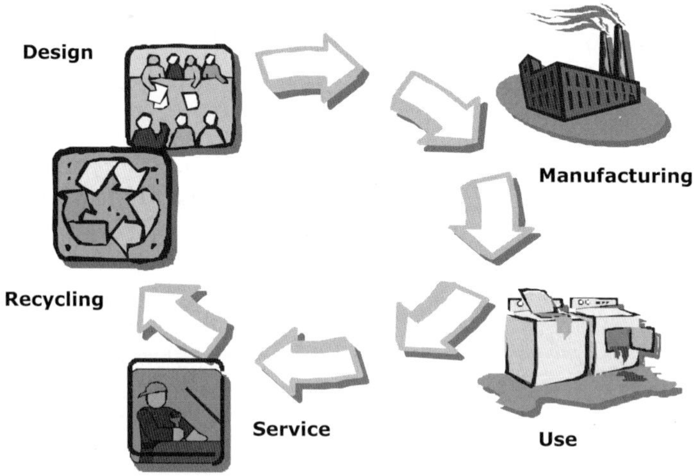


**Fig. 2.** Life cycle through the functional analysis concept

The relative importance of these factors and their orientation vary from industry to industry and product to product. Some tools and methods are proposed to guide designers when they have to define these levels of importance as QFD methods [5] or classification methods [6]. Then, general design principles for these life cycle requirements can be used to develop the parallel design of support processes with the design of the product. This parallel design requires early involvement and early consideration of life cycle factors (as appropriate) in the design process. This has lead to propose tools to help designers to manage the life cycle of the product.

### *Tools for Product Lifecycle Management*

The product life cycle management is considered as an approach to manage product data from its definition to its maintenance, while considering manufacturing aspects. The Product Lifecycle Management is an industrial subject and is born for the aeronautical and defense sectors. Now, many sectors are addressed: automobile, electronic, pharmacy, food industry, and now banks and insurance. Computer aided design and manufacturing can be considered as tools for all the product engineering. PLM is a transversal approach to create, to manage and to disseminate information to the different departments of the firm or the associated partners: engineering, marketing, purchase, logistic, manufacturing, after sales… The PLM approach is very linked to the firm Enterprise Resource Planning (ERP) and could look like a subject that want or can manage, analyze and control everything in the enterprise [7]. But PLM is really focused on the product data all along its life cycle. With a good information flow in the firm, the objectives are: to reduce the gap between engineering department and manufacturing, to take into account as soon as possible the different contingencies linked to the product evolution, to increase the frequency of new product definition, to manage resources and to control information flow (with the supply chain), to rationalize data with a unique referential

Historically, the actors that develop the large variety of tools for Product Lifecycle Management are the actors that have developed computer aided design and manufacturing tools. This can explain why all aspect of the product life cycle are not covered with PLM. Indeed, specific life cycle aspects like environmental aspects are not taken into account. The end of life of products is not really addressed in PLM tools. It is the same thing for marketing aspects, or economical aspects. So specific tools have been developed to support other expertise, less traditional than those linked to the product and to the "manufacturing" processes definitions.

## 2.3 LC to support expertise: expert level

Life cycle management is also used to help experts in their design activities. Two main expertises are presented here, linked to environment and costs.

### *Life cycle Assessments*

Life-Cycle assessment (LCA) is a systematic approach used to manage the environmental impacts of products and service systems, and it is applied at several levels:

- Conceptually as a thought process that guides the selection of options for design and improvement.
- Methodologically as a way to build a quantitative/qualitative inventory of environmental burdens or releases, to evaluate the impacts of those burdens or releases, and to identify alternatives to improve environmental performance.

In any application, LCA considers the environmental impacts along the continuum of a product's life (i.e., cradle-to-grave) from raw-materials acquisition to production, use, and disposal or recovery. The environmental impacts to consider include resource-depletion, human health, and ecological health. An important feature in this definition of life-cycle assessment is that it involves conceptual and data-intensive methodology elements. Life-cycle thinking is a way to address environmental problems from a systems or holistic perspective. In this way of thinking, a product or service is evaluated or designed with a goal of reducing environmental impacts over its entire life cycle [8]. So the product has to be taken into account with all its parts. It means: The product itself, The packaging (consumer, plant, transport, logistic…), The consumable needed for the use, The spare parts, The advertising… Indeed, these elements can have more impacts on the environment than the product itself. To have useful tools in the early design stages for estimating the environmental impacts of product alternatives, simplified life cycle assessment methodologies have been developed [9], [10]. They calculate the product environmental performance indicators by using two sets of energy-based and material-based impact drivers. This short list of drivers uses very simple input data available early when designing.

### *Product Life Cycle Cost Management*

The goal of a PLCCM system is to help employees make decisions and take actions that cause a product to be designed, manufactured, marketed, distributed, operated, maintained, serviced and disposed of in a way that creates and increases long-term competitive advantage for the firm [11]. This is accomplished by appropriately balancing the critical features of the product, including the whole life cost, the method of delivery, innovativeness and quality. The dimensions of quality are performance, features, reliability, conformance, durability, serviceability, esthetics and perceived quality. Susman [12] has presented a framework for estimating life cycle revenue and costs and discusses the correlation between the product life cycle concept and long-term corporate profits. Susman's main contribution is the integration of the two perspectives of the life cycle concept: the marketing (revenue generation) and the engineering (cost reduction) perspectives, into a single framework. The main actions from both perspectives have been illustrated in Table 1.

**Table 1.** Actions for generating revenue and reducing costs

| Revenue generation | Costs production |
|---|---|
| **Product improvement:** | **New processes** |
| • Features | **Cumulative volumes** |
| • Performance | • Experience curve |

| | |
|---|---|
| • Durability | **Average volume** |
| **Maintainability** | **Capacity utilization** |
| **Serviceability** | **Focused factories** |
| **Custom service:** | • Coordination costs |
| • Fast delivery | **Advanced manufacturing technologies:** |
| • « Hotlines » | • Less WIP, rework, inventory, floor space |
| **Product customization** | **Design for manufacturing** |
| **Expanded product line** | • Less assembly time |
| **Product warranty** | • Warranty costs |
| **New uses, new users** | • Training costs |
| **Price increases** | • Spare parts |
| **Advertising** | **Design for logistic support** |
| | **Design for maintainability** |
| | **Design for reliability** |

## 2.4 LC Integration during the design process

While examining the different life cycle concepts, approaches or tools, this stated that:

- It is important to distinguish between product life cycles and industry life cycles. Industry life cycles can last for decades, but the life cycle of particular products can be very short. These are also product classes (e.g., automotives) and product forms (e.g., modular) and product brands (e.g. Porsche). All these have different life cycles.
- It is important to clarify for whom product life cycles are managed: producers, consumers or society. Producers are interested in maximizing profit, consumers are primarily interested in product performance in a given price range, and society is interested in safety, health and environmental issues.
- The pursuit of the life cycle profits requires producers to think continuously about a product's life cycle from both the marketing and production perspectives. Strategic objectives and expense indicators change at different stages of the life cycle.
- It is important to propose tools to use product life cycle early during the design process. Indeed, it appears that it is the best way to really integrate new constraints during the product design process.
- Whatever is the expertise, every stage of the product life cycle has to be taken into account to evaluate the entire life of the products. For example, selling products at high prices to enhance revenues can lead competitors to enter the market and drive prices down.

What is needed is:

- A tool (or series of tools) that follows the design process, which embeds life-cycle information with business, financial, and technological measures, and that is simple to use. These tools will integrate what can be called the product life cycle and the environmental life cycle. It could be entitled "Product models and product development processes» like one of the objectives of the VRL-KCiP network of excellence and will focus on the development of life cycle controlling models to analyze the economic, environmental and social impacts and on defining e-services with regard to industrial benefit.
- New approaches and new expertise to allow designers to use the product life cycle scenario issued from the life cycle analysis at the beginning of their design. Indeed, the aim is to use the chosen scenario to design the product **with** the associated life cycle data. This approach is completely different of the current approach that lead to establish some priorities between the life cycle phases and then to design only **for** these stages without regarding the other impacts.

So, as a starting point of the design, the data identified during the product life cycle scenario definition have to be extracted. Then, designers have to use them to guide and evaluate their design proposal. Two levels can be concerned by these data: The project team level, and the expert level. The next sections illustrate with particular case, how these data can be used to guide designers all along the development process. Section 3 illustrates how a remanufacturing scenario can be validated by the project team, while examining product life cycle data and how this validation conduct to specify some of the technical characteristics of the product. Section 4 illustrates how experts, to think product disassembly as soon as the design project begins, can use product life cycle data.

## 3 Project definition linked to the product lC scenario: the remanufacturing case

Our first approach to design products, which are never discarded, was to examine a specific life cycle scenario based on remanufacturing. We have analyzed what are the life cycle data that are necessary to characterize a remanufacturable product and then we have proposed the concept of product profiles to guide designers during the definition of some technical product characteristics. The objectives here, are to show that strategic aspects of the firm can be linked to technical aspects of the product, to define the limits and goals of the project and to guide designers.

## 3.1 The remanufacturing strategy

Remanufacturing is an end-of-life strategy that reduces the use of raw materials and saves energy while preserving the product value added during the design and manufacturing processes [13]. But, in most of the cases, the remanufacturing processes must be adapted to existing products because products had not been designed to be remanufacturable [14]. However, the process adaptations increase costs and this can lead to reconsider the overall benefits obtained with the remanufacturing process.

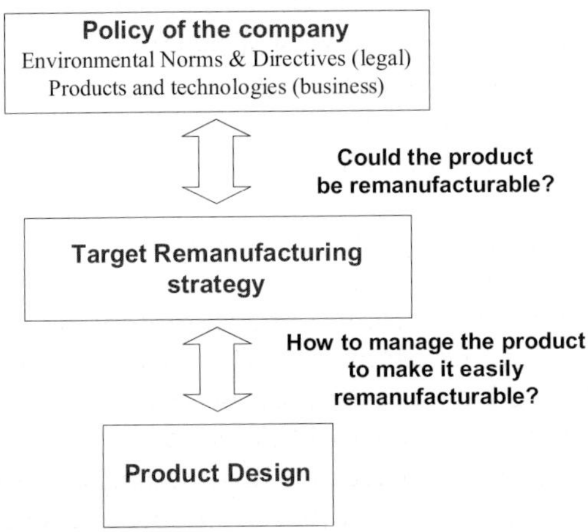


**Fig. 3.** From company context to the design of a remanufacturable product

So, the issue is now to develop new products optimizing remanufacturing and reuse strategies. The principle is to create a design situation addressing life cycle issues (and particularly end of life issues) instead of to react to a current design situation [15]. According to the policy of the company (Figure 4), designers have first to target a remanufacturing strategy before the product design. Their first question is: Could the product be remanufacturable? The second question is: How to manage the product to make it easily remanufacturable? These two levels of analysis are necessary to better integrate the remanufacturing during the design process while establishing a link between a strategically view and a technical view.

## 3.2 LC for a remanufacturable product

Many aspects have to be considered to design a remanufacturable product. It is becoming clear that designers need to go beyond product and process design considerations alone. To take into account this new kind of scenario, designers have to consider non-technical factors such as economical models, consumer needs, and market aspects. A lifecycle approach is needed to integrate understanding of technical aspects such as product design with external aspects including non-technical ones. So, to characterize a ' remanufacturable product ' a sample of existing products successfully remanufactured was studied. These products were characterized by external criteria describing the remanufacturing context of the product and by internal criteria describing the technical characteristics of the product.

### *External criteria to describe the remanufacturing context*

According to [13, 16], four categories of factors defining the remanufacturing context have been detailed: costs, technology, environment and market. The external criteria linked to each category were formulated characterizing the product remanufacturing capacity from an external point of view.

**Economic aspects:** The definition of economic aspects is useful to validate the economic viability of remanufactured products, while showing differences between a manufactured product and a remanufactured product. Thus, designers have to value the profitability of the remanufacturable product, the level of added value kept for products in end of life and savings achieved concerning the consumption of energy and raw materials.

**Technological aspects:** The definition of technological aspects is useful because technological innovation influences products remanufacturability. Thus, designers have to evaluate in what way the changes of technology will impact the life cycle of remanufacturable products and so their consequences on the product definition.

**Market aspects:** The definition of the market aspects is useful to define the sector to which the product focuses and therefore to specify needs of purchasers and users. Today many remanufactured products share the market with the original products. For example, designers have to evaluate if the consumer's interest is on the service or on the acquirement of the product.

**Environmental aspects:** The definition of environmental aspects is useful to identify environmental profits that could be done with a remanufactured product while taking into account the used resources and wastes generated.

### *Internal criteria to describe technical aspects of the product*

According to [17-19], four categories of criteria defining the product were detailed: criteria linked to the product structure, to the product quality, to the product refurbishment and to the product valorization.

**Structure of the product:** The set of criteria that characterizes the ability of the product to be collected and disassembled is described: number of parts, architecture, joining techniques, etc…
**Quality tests:** The ability of the product and its parts to be tested and inspected before being remanufactured or a new utilization is characterized.
**Refurbishment:** The ability of parts and joints to be cleaned, repaired and restored is characterized.
**Valorization:** The ability of the parts to be valorized (materials and energy) is characterized.

43 external criteria and 39 internal (or technical) criteria were pointed out to describe a remanufacturable product.

## 3.3 Remanufacturable product profile to link technical and non technical factors

To establish a link between technical and non-technical factors, clusters of products were defined. The most meaningful variables of the previously constructed characterization were extracted by a method of descriptive statistical analysis: a principal component analysis (PCA). Then, remanufacturable products profiles (RPP) were proposed as key features to structure engineers' rationale by both ensuring the economic relevance of the project and guiding the designers toward easily remanufacturable products [20]. RPP can be used when designing:

- To address the context of the end-of-life scenario alongside with the product life cycle properties
- To assist designers during the technical definition of the product
- To assess the end-of-life scenario capability throughout the design process.

The methodology has been implemented in a tool REPRO[2] (figure 5).

It can be used in the earliest stages of the design process based on the external criteria defining the context of the project on one hand, and the internal criteria defining the constraints (or objectives) on the product properties to be met for a nice remanufacturable product.RPP mainly helps designers clarifying the objectives and requirements for the end of life of the product. The methodology proposed is pro-active, giving the opportunity

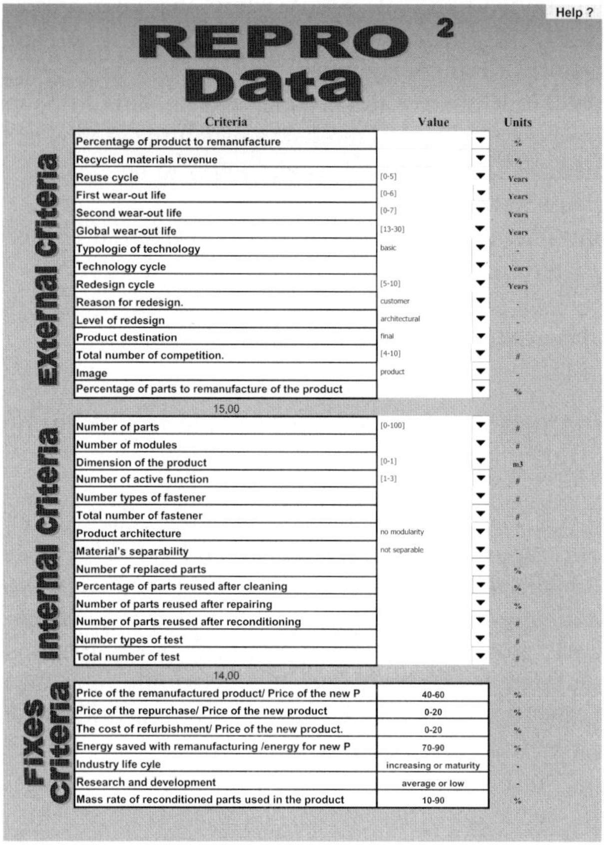

Help ?

REPRO²
Data

| Group | Criteria | Value | Units |
|---|---|---|---|
| External criteria | Percentage of product to remanufacture | | % |
| | Recycled materials revenue | | % |
| | Reuse cycle | [0-5] | Years |
| | First wear-out life | [0-6] | Years |
| | Second wear-out life | [0-7] | Years |
| | Global wear-out life | [13-30] | Years |
| | Typologie of technology | basic | - |
| | Technology cycle | | Years |
| | Redesign cycle | [5-10] | Years |
| | Reason for redesign. | customer | - |
| | Level of redesign | architectural | - |
| | Product destination | final | - |
| | Total number of competition. | [4-10] | # |
| | Image | product | - |
| | Percentage of parts to remanufacture of the product | | % |
| | 15,00 | | |
| Internal criteria | Number of parts | [0-100] | # |
| | Number of modules | | # |
| | Dimension of the product | [0-1] | m3 |
| | Number of active function | [1-3] | # |
| | Number types of fastener | | # |
| | Total number of fastener | | # |
| | Product architecture | no modularity | - |
| | Material's separability | not separable | - |
| | Number of replaced parts | | % |
| | Percentage of parts reused after cleaning | | % |
| | Number of parts reused after repairing | | % |
| | Number of parts reused after reconditioning | | # |
| | Number types of test | | # |
| | Total number of test | | # |
| | 14,00 | | |
| Fixes Criteria | Price of the remanufactured product/ Price of the new P | 40-60 | % |
| | Price of the repurchase/ Price of the new product | 0-20 | % |
| | The cost of refurbishment/ Price of the new product. | 0-20 | % |
| | Energy saved with remanufacturing /energy for new P | 70-90 | % |
| | Industry life cyle | increasing or maturity | - |
| | Research and development | average or low | - |
| | Mass rate of reconditioned parts used in the product | 10-90 | % |

**Fig. 4.** Data used to compare the product under design to the proposed RPP

to innovate in an environment limited by constraints due to the expected properties of a remanufacturable product. Some improvements have to be realized, in particular linked to the eco-design of remanufacturable products. Indeed, the different cases analyzed were essentially focused on the end of life and didn't precisely consider the whole life cycle of the product (use phase, logistic…) also if the examined products were well adapted to this kind of end-of-life. These first results have convinced us that there are certainly other product life cycle scenarios for which it exists links between internal and external criteria. In future researches, we will focus on these scenarios to try to establish such kind of database that are really helpful for designers at the beginning of their projects.

# 4 Product definition linked to its life cycle scenario: the disassembly case

Either if there is not a specific scenario addressed for the entire product, it is possible to apply data issued from the product life cycle analysis in the early phases of the project. Let us demonstrate this aspect focusing on a specific point of view: the disassembly of the product. Indeed, disassembly is certainly the main technical aspect to consider diminishing costs and technical problems for products that will have several use life. But to really improve disassembly aspects, a real effort has to be done to think disassembly during the conceptual design stage.

## 4.1 Product disassembly during its LC

Tomiyama et al. [21], showed that the way the product life cycle is envisaged impacts a lot product design activity. It is of course true for disassembly because the disassembly situations vary depending on the life cycle under consideration. Figure 6 presents the main objectives of disassembly in the life-cycle stages of a product. The large variations in disassembly situations create contradictions among the persons involved in a design process. It implies to determine and examine all the situations before choosing product solutions and optimizing them. So, the designers have to define a product disassembly strategy, addressing the whole product life cycle activity and not a unique point of view. It should lead to develop evaluation criteria common to the different disciplines involved in disassembly activities, to define a good disassembly (evaluating when designing), enabling a global optimization of the product. It is why we recommend using life cycle considerations to define the product life cycle scenario before anything else. Then we propose to use methods and tools assisting designers on assembly/disassembly issues in the early phases of the design of products and systems. It is a preventive approach that integrates disassembly constraints as soon as possible and a representation of the life-cycle disciplines that involves disassembly activities.

## 4.2 Product disassembly in the early design phases

During the Conceptual design stage a basic solution is proposed by designers through the elaboration of a functional representation of the product. This representation supports the definition of the functions and of their structures and can integrate some technical solutions.

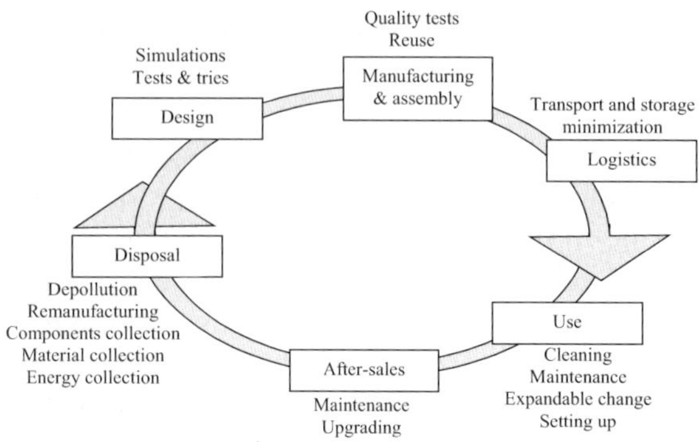


**Fig. 5.** Objectives linked to the disassembly during the product life cycle

We believe that the conceptual design stage, must also integrate some information to take into account the disassembly constraints. To do that, a first objective is to translate customer, legislative and manufacturers / recyclers requirements for the designers. Then a classical functional representation to support these requirements has to be used for their real integration during the conceptual design stage. A nice tool is the Functional Block Diagram tool, because it is very flexible and can accommodate a wide range of systems. It requires a minimum of details on the product, but gives a lot of information to start a study on the disassembly solution. So the disassembly work was based on this functional representation of the product while adding product life cycle disassembly constraints to guide designers to design product easy to be disassembled and recovered. The concept of Conceptual Disassembly Model (CDM) was developed, that is the model of the product that enables to generate a first disassembly sequence in the conceptual phase, to guide designers in their future developments [22]. With the sequence, a first evaluation of disassembly can be got. The key concept, here, is the use of a simplified BDF (figure 7). This kind of BDF can be realized during the conceptual design stage, when designers discuss on principles and try to identify the main components of their products. If these discussions are supported by a simplified BDF, data available at this stage are: the environmental requirements and the disassembly objectives, the main components of the product and their functional links, a global estimation of the weight and material of the main components, some components treatment. With these data, it is first possible to validate the rate of recyclable material and to identify the life treatment of all components. Moreover, based on a first draft of the future solution (an "intermediary object"), it is possible to generate disassembly sequences corresponding to the disassembly objective. Then, designers are assisted to improve the

assisted to improve the disassembly of components targeted by the identified sequence.

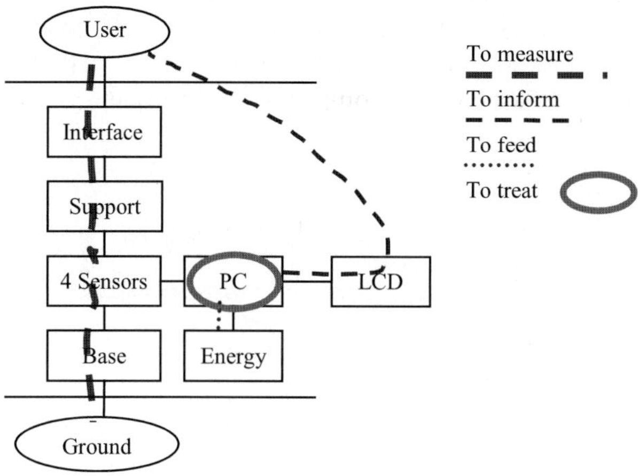


**Fig. 6.** The simplified BDF for a bathroom scale

## 4.3 The transition between functional structure and technical structure

The first "sequence" can now be used to guide designers who have to make a link between the functional structure and the technical structure of the product. On figure 8, the different stages of a design process were represented to situate the place of the CDM during the design process. In fact, the CDM has to be used during the conceptual design stage, and is a guide to allow designers to define the structure of the product, materials and fastenings. To help designers, the concept of product views has been developed. These views give means to designers to evaluate progressively the product design from the material, structure and fastening point of view. In fact, designers have to work with a goal well defined: the first proposed sequence, and the different views give indications to control the validity of their propositions. For example, if you need to maintain a component during its life cycle and that the material of this component need to be recover in the end of life stage (an electric engine for example), you need to evaluate its accessibility (structure view), its ability to be assembled and disassembled (fastening view) and to examine the compatibility of the materials that will be disassembled (material view). Our objective with this approach, is to give a mean to make rapid and reliable choices between several design options while taking into account the life cycle requirements:

functional, economical, end of life recovery requirements...This is possible during the conceptual design stage because a minimal number of data permits designers to evaluate roughly but efficiently the disassembly of a product. This approach can be particularly efficient if there are contributions of different experts like recyclers, disassembly experts and technological experts to make the best compromises during the conceptual design phase.

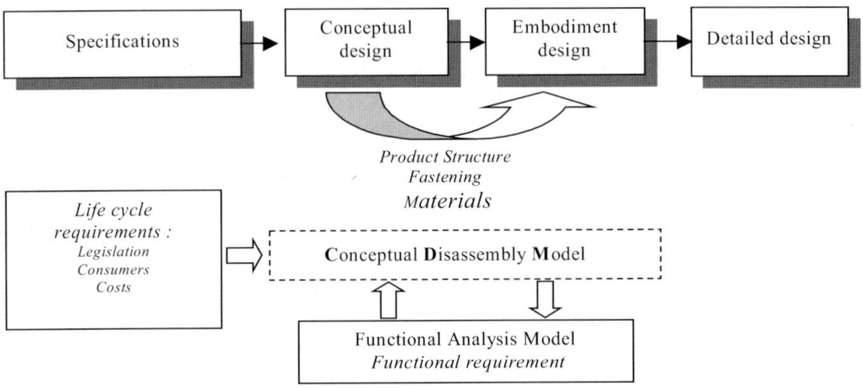


**Fig. 7.** The Conceptual Disassembly Model during the design process

## 5 Conclusion

The future of manufacturing in Europe will be very dependent on citizen's needs, on sustainable development and on new upcoming science and technology opportunities. Product life cycle planning is one of the key factors identified to guide new design approaches [23]. In the paper, we illustrated that lifecycle options need to be evaluated to meet the defined objectives by business strategy. Lifecycle options include not only choices made with regards to the post consumption stage such as 'reuse' and 'recycle' but also those that influence other life stages such as design changes to increase life and repair of a product. Life cycle approach is certainly the best one to think about "products, which are never discarded". But product life cycle covers many aspects and it was noticed that common researches only start to develop a tool able to support all these aspects during a design process (VRL-KCiP). However that may be, it is time to make first propositions to use results to be issued from a well supported life cycle approach during the design process. It was illustrated by two levels of proposals:

- The first one is guided by the choice of a specific scenario for the product. In this case, the objective is to construct specific expertises for different scenario, to give designers what are the most significant product criteria and what are their value, depending on the technical but also strategic aspects.
- The second one consists in the use of the life cycle data as soon as the conceptual design stage starts. Indeed, usually life cycle data are used during the detailed design phase to evaluate a product that is well defined (DFX, LCA or LCC approaches). Yet, it is during the conceptual design stage that it is easier to define a real compromise between all the product constraints. So it was stated that a product life cycle analysis give enough information to begin a thought in the conceptual design stage and this for many design domains as illustrated with the disassembly case and other cases [24]. What to do now is to develop tools and methods to support these new approaches.
- So, the designers have to define a product disassembly strategy, addressing the whole product life cycle activity and not a unique point of view. It should lead to develop evaluation criteria common to the different disciplines involved in disassembly activities, to define a good disassembly (evaluating when designing), enabling a global optimization of the product. It is why we recommend using life cycle considerations to define the product life cycle scenario before anything else. Then we propose to use methods and tools assisting designers on assembly/disassembly issues in the early phases of the design of products and systems. It is a preventive approach that integrates disassembly constraints as soon as possible and a representation of the life-cycle disciplines that involves disassembly activities.

Designing products, which are never discarded, could be a new concept that should appear in a near future. For example, extended products, services, … traduce a real interest of manufacturers to keep the property of their products and to manage their entire life cycle. New design approaches have to be quickly efficient to answer these new designer's needs.

## References

1. Zwolinski, P., Prudhomme, G., Brissaud, D., 2003, Environment and Design. Methods and Tools for Integration and Co-operation, In Methods and Tools for Cooperative and Integrated Design, Kluwer Academic Publishers
2. Kimura, F., 2001, Comprehensive product modeling for designing life cycle scenarios, EcoDesign, IEEE computer Society, p1-6.

3. Pesonen, L.T.T., 2001., Implementation of design to profit in a complex and dynamic business context, Academic dissertation, University of Oulu, Finland, 133p.
4. Prudhomme, G., Zwolinski, P., Brissaud, D., 2003, Integrating into the Design Process the Needs of those Involved in the Product LC, J. of Eng. Design, 114/3, 333-353.
5. Zhang, Y., Wang, H.P, Zhang, C., 1999, Green QFD-II: To Life Cycle Approach Environmentally for Conscious manufacturing by integrating LCA and LCC into QFD matrixes. Int. J. Prod. Res., Flight. 37, no 5.
6. Lagerstedt, J., 2003, Functional and environmental factors in early phases of product development - Eco functional matrix" PhD thesis, KTH, Stockholm.
7. Gao, J.X., Hayder, A., Maropoulos, P.G., Cheung, W.M., 2003, Application of product data management technologies for enterprise integration, IJCIM, 16/7-8, 491-500.
8. Eisenhard, J.L., Wallace, D.R., Sousa I., De Schepper, M.S., Rombouts, J.P., 2000, Approximate life-cycle assessment in conceptual product design, DETC'00, ASME 2000.
9. Sousa, I., Wallace, D., 2001, Product Classification to Support Approximate Life-Cycle Assessment of Design Concepts, 5th int. conf. on technology, policy and innovation..
10. Kaebernick, H., Sun, M., Kara, S., 2003, Simplified lifecycle assessment for the early design stages of industrial products, Annals of CIRP, Vol 52/1, 25-28.
11. Shields, M, Young, M; 1991, Managing product life cycle costs, An organizational model, Journal of cost management, Vol5, Autumn, p 39-52.
12. Susman, G, 1989, Product life cycle management, J. of cost management, Vol3, p8-22.
13. Kerr, W., 1999, Remanufacturing and eco-efficiency, M.Sc.Thesis Environmental Management and policy, IIIEE Lund University, Sweden.
14. Sundin, E., 2001, Enhanced product design facilitating remanufacturing of two household appliances A case study, Proceedings of ICED-01, Glasgow, 645-652.
15. Brissaud, D., Zwolinski, P., 2004, End-of-Life-Based Negotiation Throughout the Design Process. Annals of the CIRP, 53/1:155-158.
16. Rose, C., 2000, Design for Environment: A Method for Formulating Product End-of-Life Strategies, Ph.D. dissertation. Stanford University.
17. Amezquita, T., Hammond, R., Salazar, M., Bras, B., 1995, Characterizing the remanufacturing of engineering systems, Proc. of design automation conf., Boston, 271-278.
18. Bras, B., Hammond, R., 1996, Towards design for Remanufacturing-metrics for Assessing Remanufacturing, Proc. of the 1st Int. Workshop on Reuse, 5-22.
19. Sundin, E., 2001, Product Properties Essential for Remanufacturing, Proceedings of the 8th Int. CIRP Seminar on Life Cycle Engineering, Varna, 171-179.
20. Lopez Ontiveros, M.A., Zwolinski, P., Brissaud, D., 2003, Profile of products for the creation of remanufacturable products during the conceptual design phase. CIRP seminar on life cycle engineering, Copenhagen.
21. Tomiyama, T., Umeda, Y., Wallace, D.R., 1997, A Holistic Approach to Life Cycle Design, in F.-L. Krause and G. Seliger (eds.): LC Networks, Chapman & Hall, 92-103,.
22. Haoues, N., Zwolinski, P., Brissaud, D., Cornier, A., 2004, How to Integrate End of Life Disassembly Constraints in the Early design stage?, Int. Sem. CIRP LCE, Belgrade.
23. Geyer, A., Scapolo, F., Boden, M., Döry, T., Ducatel, K., 2003, The Future of Manufacturing in Europe 2015-2020 : The challenge for sustainability scenario.
24. Ishii, K., 1997, Modularity: A key concept in product life-cycle engineering, Frontiers of Engineering: Reports on Leading Edge Engineering, pp 17-22.

# Guidelines in ecodesign: a case study from railway industry

J. Lagerstedt and C. Luttropp
*Bombardier Transportation,Mainline MLN/ESE Västerås, Sweden,KTH/Machine Design, SE-100 44 Stockholm, jessica.lagerstedt@se.transport.bombardier.com*

Abstract: Designers have a key role in adapting products to the environment and a sustainable society and many attempts have been made to establish universal DfE guidelines that can be used for these design situations. However, most attempts have failed due to the level of information; too generic makes no sense for the user, and too specific means that the guidelines are too private to be useful for everyone. Furthermore, designers and the acquisition department as well as marketing and production have different background knowledge and different primary tasks. Depending on the complexity and type of product as well as on the cultural background in different companies and countries, the product development process looks different. It is therefore impossible to establish generic DfE guidelines that are valid for any product in any company. The 10 Golden Rules are a set of generic Design for Environment aspects (first presented by Luttropp (2000)). The "10 Golden Rules" are broad principles, which establish a common basis for a product development team, giving the team members a mutual perspective on the sustainability issue. The 10 Golden Rules are preferably individually customized to be of most use in direct design and product development process. This paper presents the customisation of the 10 Golden Rules at Bombardier Transportation, which is one of the worlds leading companies manufacturing rail vehicles.

Keywords: DfE Guidelines, 10 Golden Rules, Customisation of DfE Guidelines

## 1 Introduction

Designers are widely believed to have a key role in adapting products to the environment and by this promote a sustainable society. However, the time and attention designers can give to sustainability is often limited. Industrial products are complicated compromises where sustainability is just one out of many "product aspects". Figure 1 is an attempt to visualize



*D. Brissaud et al. (eds.), Innovation in Life Cycle Engineering and Sustainable Development,* 245–254.

the variety and diversity of relevant product aspect and trade-offs that have to be considered in design process. Understandably, it is unrealistic to establish a tool for every day design work on sustainability in the light of its relative importance in design work. Even if sustainability is a major concern, the product must still fulfill a functional need to defend its presence on the market. Consequently, sustainability tools for every day design work must be simple and robust.

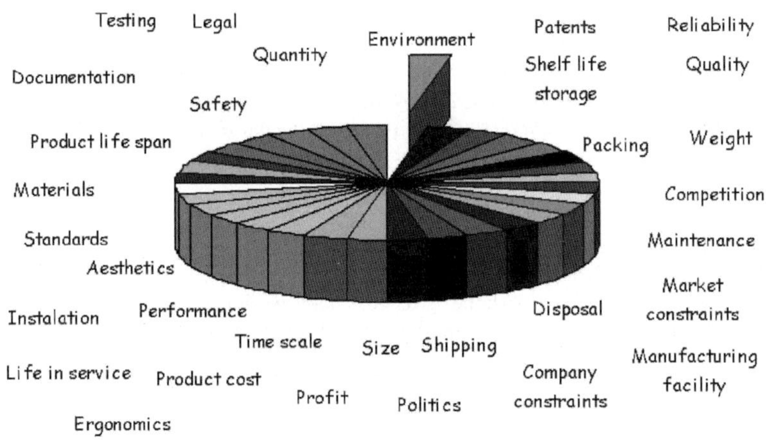


**Fig. 1.** A picture of all product aspects that must be addressed in design work

To fulfil this need a set of simple generic rules are developed at KTH. These rules or sustainability aspects are a simplification and a generic version of very many of the today present guidelines from all kinds of sources. To help designers and engineers accomplish this task during the design process, the Center of Competence for Design for Environment (CoC DfE) at Bombardier Transportation has recently released the leaflet - *Design for Environment Guideline [2]*, which is based on the10 Golden Rules first presented Luttropp [3].

## 2 The 10 Golden Rules

### 2.1 Background

First it must be stated that these 10 Golden Rules are not an invention made by the authors. It is a pedagogic summary of many of the guidelines that can be found in company guidelines and in handbooks of different origins. The Golden Rules, as presented in this chapter, are very generic and must be transformed and customized to be of real use in product development work. Guidelines in a company must be company and product specific and must be implemented by the product developers. The Golden Rules, as they are presented further down, are NOT listed in any preference order and was first presented in a 12 rule version by Luttropp (Luttropp, 2000). The basic idea is to use the 10 Golden Rules as a fundament for customizing task specific and individual guidelines.

## 2.2 The original KTH version of the 10 Golden Rules

GR 1. Don't use TOXIC substances and arrange closed loops for necessary but toxic ones.

GR 2. Minimise energy and resource consumption in production phase and transport through HOUSEKEEPING.

GR 3. Use structural features and high quality materials to minimise WEIGHT ..in products…if not interfering with necessary flexibility, impact strength or functional priorities.

GR 4. MINIMISE energy and resource consumption in the usage phase, especially for products with most significant aspects in the usage phase.

GR 5. Promote repair and upgrading, especially for SYSTEM dependent products

GR 6. Promote LONG LIFE for products especially for products with most significant environmental aspects OUT of usage phase.

GR 7. Invest in better materials, surface treatments or structural arrangements to PROTECT products for dirt, corrosion and wear, giving long life and minimised maintenance.

GR 8. PREARRANGE upgrading, repair and recycling trough access ability, labelling, modules, breaking points, manuals.

GR 9. Promote upgrading, repair and recycling by using few, SIMPLE, recycled, not blended materials and no alloys.

GR 10. Use as FEW joining elements as possible and use screws, adhesives, welding, snap fits, geometric locking etc. according to the life cycle scenario.

## 2.3 Customizing guidelines

The rules are generic and most of the times too general to be used as they stand. They must therefore be customized to be of real use in product development process. The 10 Golden Rules form a common basis for the

DfE work. However, some designers have specific use for only a few of the rules, and some of the rules can even be contradictory, which calls for a compromise in personal rules depending on individual tasks. In most cases one cannot assign a winner. Depending on the situation, the rules may have a slightly different meaning to different people. What is the right thing to do in one situation might be totally wrong under other circumstances. A rule may work well in one case, and not apply in another therefore the designer should select the rules applicable in their design work and customize them to fit their situation. For example, the first golden rule "Don't use toxic", could be customized as followed:

- Identify which are the toxic substances inside your responsibility!
- Try to find a nontoxic substitute available not jeopardizing functionality and economy!
- Consider if closed loops be arranged or are there any established collecting-recycling systems for actual substance!
- In this case, one could establish a link to rule number 8 "information":
- Label material A and try to make it as an easy accessible module!

Furthermore, rule number 9 ("mix") and 10 ("structure") are also important to consider in this context. Substituting a toxic substance might lead to a more complicated structure to achieve the modules and labels asked for, which could generate the following rule:

- Provide instructions for the disassembly of the toxic and scarce substances in the product

To establish as a comprehensive view as possible, the rules are preferably customized in corporation with environmental expert as well as managers and market expertise. A parallel can be made to living rules present in many cultures and religions, e.g. "Thou shall not steal!". This is easy to state in words, but of no real use as guidance for a thief! For a kleptomania, who would like to stop this habit, one could customize the rule like:

- Sew together your pockets
- Carry bags that can be seen thru
- Leave your overcoat at the shop entrance.
- Ask someone from the shop to accompany you during your stay in the shop.

# 3 Bombardier

Serving a diversified customer base around the world, Bombardier Transportation is the global leader in the rail equipment manufacturing and servicing industry. Its wide range of products includes passenger rail vehicles and total transit systems. Bombardier Transportation also manufactures locomotives, freight cars, propulsion & controls and provides rail control solutions. The headquarter is in Montreal, Canada, with European headquarters in Berlin, Germany. Bombardier Transportation employs some 30,000 people and has an international manufacturing presence with production facilities in 23 countries.

## 3.1 DfE at Bombardier

Bombardier Transportation, as a company devoted to continuous improvement of environmental performance, has a high ambition level regarding Design for Environment (DfE) and in the framework of the environmental management system is striving to develop and continuously adapt our products and vehicles in a lifecycle perspective. Bombardier Transportation has since 1992 worked with Design for Environment as an integrated part of delivery projects and has developed a number of tools in this area. The most important group for DfE expertise is the Center of Competence in DfE (CoC DfE), which comprises six persons. DfE contact persons have also been appointed at all divisions, which constitute the DfE Group Team network. The main objective of DfE at Bombardier is to develop vehicles that contribute to a sustainable society. This is achieved by minimizing the environmental impact during the whole life cycle already in the development phase. Furthermore, designers have a key role in the work to continuously develop and environmentally adapt our products. In order to help the engineers accomplish this task during the design process, the Center of Competence for Design for Environment (CoC DfE) at Bombardier Spring 2004 released a leaflet - *Design for Environment Guidelines*, which is based on the 10 Golden Rules (Luttropp). The DfE Guidelines provides a selection of some "quick and easy" guidelines on how to move towards more sustainable products and services.

# 4 Customising the 10 Golden DfE Rules for bombardier-the process

May 2003, CoC DfE started a project to update the existing DfE guidelines. At the same time Lagerstedt introduced the 10 Golden Rules to the group, and it was decided that a new set of DFE guidelines was needed. The project followed an iterative process, which continued until May 2004, when the Design for Environment guidelines leaflet was released to the whole company. The basic idea of this work was to establish DfE guidelines that the designers could use in their daily work at Bombardier, and thereby give the designers some general design advice to use when performing DfE. Together with earlier versions of Bombardier DfE guidelines, the 10 Golden Rules established the platform for developing a new set and layout of the DfE guidelines i.e. a customisation of the 10 GR. At first, CoC DfE performed a brainstorming session to elaborate some basic questions like: who are the potential users of this information, what is their background, understanding and pre-knowledge in the DfE area, what should the guidelines be used for, what type of information is needed for efficient communication, layout, languages, how do we present it and how and when should we distribute the guidelines? Major conclusions from the brainstorming were:

- How instead of why. Main focus should stay on how to achieve environmental improvements of Bombardier's products, rather than spending time and space on motivating why to perform DfE.
- Time and space. Information should be quick and easy to understand. As designers do not have much time to spend on DfE, the information should not exceed four pages (A4).
- Language. The idea is that the guidelines should be used by all engineers at all levels, a on a day-to-day basis, however some countries and engineering departments are many times not comfortable with English. To avoid the risk that the DfE guidelines are not used, due to lack of English understanding, it was decided to make the information available in German and French as well.
- Wording. Some of the phrases in the Golden Rules should be changed to better fit the processes and words used at Bombardier e.g. Toxic has been changed to Hazardous.
- General specific approach. Designers are experts in their own area, but normally lack environmental knowledge. Even if designers at Bombardier work on the same or similar products, they have different responsibility areas (bogie, interior, car body etc.). Consequently, it would make no sense to formulate exactly same specific DfE guidelines

for everyone; they would not be valid in all situations. For example a designer on interior would need specific DfE guidelines on fabrics, while a designer on bogies would need specific DfE guidelines on metals and corrosion preventing agents. Therefore, we choose a general specific approach.

- Remembrance. The guidelines should be easy to remember and come back to. The pictures should help people remember the leaflet. Another idea was to establish a version that could be kept as a poster on the wall, rather than keeping an electronic version, which may be lost among all other information.
- Other important information to be included in the leaflet was: coordinates to: contact persons, CoC DfE and EBOK for easy accessibility to information as needed, lists of prohibited and restricted substances, a relation to function should be included.

Thus, the DfE guidelines should be product oriented with generic guidelines with space for individual company's guidelines. The DfE guidelines are now based on three levels, see Fig. 2. The first level is basically based the 10 Golden Rules and the second level is a customisation of the 10 Golden Rules to fit Bombardier and is pretty much rail vehicle related. This level is merged with earlier DfE guidelines used at Bombardier. At the third level the designers have to write their own guidelines or action plans of what to do. Furthermore, it is important to stress that the engineers only have to select the guidelines that are suitable for their work.

## 4.1 The first version - first aid in design for environment

The first version, which was named "First aid in Design for Environment, 10 Golden DfE Rules", was tested at a DfE training session in Derby, UK. Feedback showed that the majority thought that the DfE guidelines were good or very good, informative, they stimulate ideas and are helpful. Furthermore, the DfE guidelines were considered as a good tool for daily work, but some designers expressed the point that they had problems to write their own guidelines.

## 4.2 The second version

The second version was presented to the DfE Group team. At this stage it was found that some words caused confusion e.g. Rule, which is used in the army and therefore understood as a must, and Gold, which was related to a rare metal. To cause as little confusion as possible, in introducing new names, it was decided that we should to stay with the old name, "Design for Environment guidelines". After several revisions with DfE group team on layout, wordings, and with the printing/design agency concerning layout, the leaflet was finalised in six months. The leaflet was translated into German and French.

| Design for Environment Guidelines | | My own action plan |
|---|---|---|
| **Hazardous**  Don't use hazardous substances and arrange closed loops for necessary ones | ☐ Don't use materials on BT's lists of Prohibited and Restricted substances<br>☐ Try to find solutions involving non hazardous substances, which does not jeopardise the functionality and cost limitations of the product<br>☐ If a hazardous substance cannot be substituted consider if closed loops can be arranged i.e. recycled and taken care of at end-of-life | . . . . . . . . . . . . . . . . . . . |
| **House-keeping**  Minimise energy and resource consumption in production phase and transport through housekeeping | ☐ Reuse parts and components if they can still guarantee the same quality<br>☐ Optimise and plan procurement and logistics e.g. no half empty trucks, choose less energy consuming distribution, optimise packaging<br>☐ Reduce use of consumables e.g. spill of oils<br>☐ Sort waste in recycling bins | . . . . . . . . . . . . . . . . . . . |

**Fig. 2.** Snap shot of the Design for Environment Guidelines at three levels; general, Bombardier specific and own action plan

## 4.3 Distribution

Finally, the DfE guidelines leaflet was launched March 2004 via Bombardier non-stop online, which is an email addressed to the whole company, and stored on Bombardier's intranet. Bombardier DfE guidelines are found in English, German and French. The leaflets are distributed at internal DfE training courses, but also via DfE group team. Moreover, a clickable version was established on the EBOK, (Bombardier Engineering Book of Knowledge on the intranet), see Fig. 3, where it is possible to click on each icon in the picture to get more detailed information about that specific DfE guideline. There is furthermore more information and advices behind function and design, as well as a whole set of DfE guidelines for selecting materials. Most guidelines are accompanied with practical DfE examples. To inspire designers and to help them develop their own DfE guidelines, a gallery of examples is also kept together with the guidelines in the EBOK. A printable version of the leaflet in all three languages is also included.

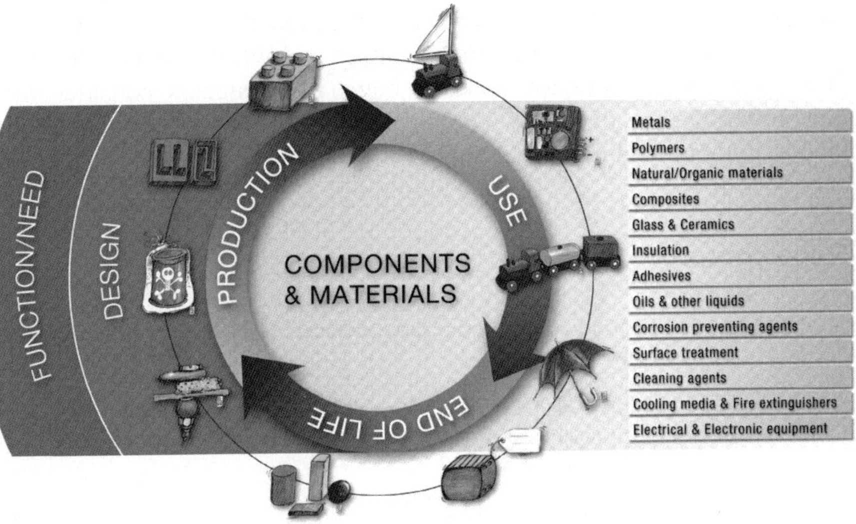


**Fig. 3.** A Snap shot of the click able version of the Design for Environment Guidelines on the Bombardier Engineering Book of Knowledge (intranet). Each picture and word is click able with more information on a new page

## 5 How are the DfE guidelines used today at Bombardier?

The basic message in the Bombardier DfE leaflet is: consider how the product interacts with the environment throughout its lifecycle. Identify activities in your daily work related to the DfE guidelines, select the guidelines that are relevant and customize them to match your work. You are the expert in your own situation. Adapt the DfE guidelines to suit your needs. Put your own DfE guidelines on the wall as a reminder! To avoid sub-optimization, one should preferably try to obtain as comprehensive a design view of the problem as possible. Moreover, resource saving must not jeopardize the function of the product.

For example lightweight cars can be motivated by decrease in fuel consumption, but lightweight should not compromise vehicle safety. The designers are therefore recommended to discuss compromises with their division contact and groups involved in the problem and should communicate the resulting view. With today's gigantic flow of information, it is difficult to make a lasting impression; a footprint.

However, the comic images used in the DfE guidelines, each representing a specific guideline, are easy to remember. Many people within Bombardier still, half a year after the release of the document, comment that they've seen the leaflet and remember it, and that they also like what they see. Follow up activities from courses and writing their own action plans is another important issue. It helps the designers consider what they've written and discuss it with an environmental expert, but also some kind of commitment must be established.

## References


1. Luttropp, C. (1999), Eco-design in early product development, Recycling R99, Geneva, Switzerland.
2. Lagerstedt, J. et al. (2004) Bombardier DfE Leaflet
3. Luttropp, C. (2000), The Dilemma of Eco Effective Products-Trade off between Contradictory Environmental Targets, CIRP 7th international seminar on Life Cycle Engineering, Tokyo, Japan.

# Identifying and assessing environmentally benign modules

M. Voß and H. Birkhofer
*Product Development and Machine Elements, Darmstadt University of Technology, Germany, voss@pmd.tu-darmstadt.de*

Abstract: Step by step, this paper presents a streamlined product structuring method which includes environmental structuring criteria. The so-called EcoDesign Structure Matrix (EcoDSM) is based on the conventional Design Structure Matrix approach which allows highlighting interdependencies between elements of the product structure. A particular concern has been put on streamlining the method, since the additional effort for the product developer analysing environmental effects of the product structure should be restricted to a minimum. Thus, this study aims at formulating (obvious) interdependencies by algebraic expressions instead of taking record of them manually. The article is completed by a case study of a household product which should rather illustrate the procedure of structuring products with the EcoDSM than proving its power to cope with complexity.

Keywords: Design for Environment (DfE), Product Structuring, Design Structure Matrix (DSM)

## 1 Product structuring

The close interrelation that exists between modular product design on the one hand and environmentally conscious product design on the other hand emerges implicitly in various scientific approaches [1]. Strategy-based EcoDesign approaches and design guidelines [2] often advise the product developer to chose modular product architecture in order to enable upgrading, component reuse, material recycling, etc.

The environmental impact of a product is more than the sum of the environmental impact of its components. The way the components are related among each other significantly determines the outcome. In general, product structuring approaches try to find a general optimum of the product characteristics, i.e. quality, cost, environmental impacts. The following



*D. Brissaud et al. (eds.), Innovation in Life Cycle Engineering and Sustainable Development,* 255–270.

three sections describe fundamental research areas in product structuring on which the EcoDSM relies.

### *Modularity Drivers*

A common basis of many product structuring approaches is the use of a set of evaluation criteria. These criteria reflect reasons for grouping components into modules that generally raise the necessity for modularity. Therefore, they are often referred to as modularity drivers, or module drivers [3]. Figure 1 contains a non exhaustive list of modularity drivers. Among other 'classic'modularity drivers aiming at the optimisation of product development or production processes other drivers concerning the use phase, such as cleaning, or end-of-life phase, such as landfill, figure in the list.

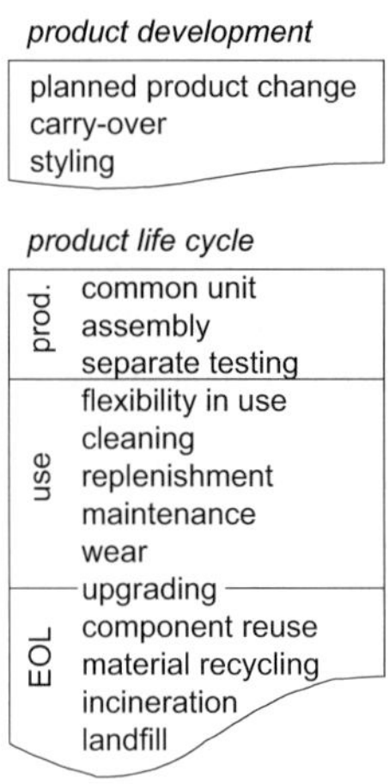


**Fig. 1.** Amended list of modularity drivers

Figure 2 presents a systematisation of environmentally relevant drivers that can trigger the module definition process. These approaches pivot upon different points in product life. Modularisation optionally helps: to prolong the initial use phase, to reuse the product or parts of it, to optimise the retirement of products.

A functional alteration is a concomitant of these life cycle approaches. In order to encourage the user to keep a product longer than the reference product, functionality has to be enriched. But the longer a product remains in use the less economic it is to keep up functionality. All things considered, life cycle approaches just work well within a small functional 'corridor' (shaded in grey).

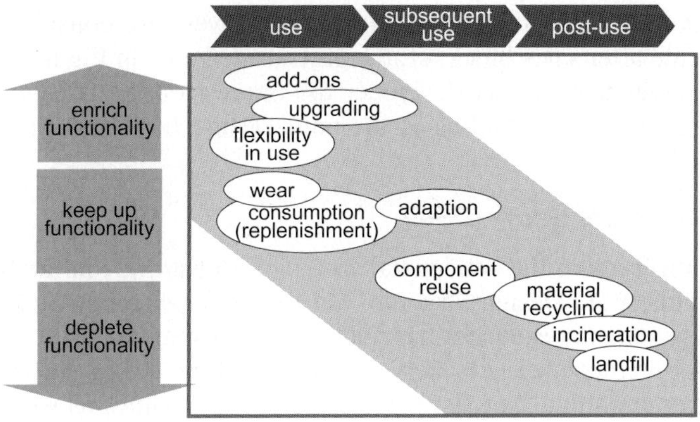


**Fig. 2.** Systematisation of environmental modularity drivers

### *Modular Function Deployment (MFD)*

Following the principles of the Quality Function Deployment (QFD) which aims at transforming the voice of the customer into design requirements Erixon's Modular Function Deployment sets up a methodology for the development of modular products. The "heart" of that Modular Function Deployment (MFD) [4] approach is the Module Indication Matrix (MIM), displayed in Figure 3. It serves to generate and to select systematically modularisation alternatives.

| Module drivers | Technical solutions | Technical solution | TS2 | TS3 | TS4 | TS5 | TS6 |
|---|---|---|---|---|---|---|---|
| Development | Carry-over | | | ○ | | | |
| | Technology push | | | | | ● | |
| | Planned product change | ● | | | | | |
| Product variants | Technical specification | | | | | | |
| | Styling | | | ● | ● | | |
| Production | Common unit | | ○ | | | | |
| | Process / organisation | | | ● | ○ | | |
| Quality | Seperate testing | | | | | | |
| | Supplier available | | | | | | |
| After sales | Service / maintainance | | ○ | | | | |
| | Upgrading | | | | ● | | |
| | Recycling | | | | | | |
| | Sum | 9 | 4 | 21 | 21 | 9 | |
| | Module candidates | | | ✓ | ✓ | | |

**Fig. 3.** Module Indication Matrix (MIM)

Environmentally relevant modularisation drivers are considered exclusively in the after sales block, cf. the grey shaded area in Figure 3. But this possibly gives the product developer the impression that Design for Service, Upgrading or Recycling is the only Design for Environment (DfE) option he has.

### *Design Structure Matrix (DSM)*

Reaching back to the conceptual works of Steward [5] in the beginning of the eighties, the Design Structure Matrix, or Dependency Structure Matrix as it is sometimes called, has evolved to a commonly used method for analysing interdependencies between structural elements. All approaches have in common that they list these elements in the lines as well as in the columns. The type of information that is analysed with this matrix based approach can be : task based, parameter based, team based and component based.

Within this article the term DSM always refers to the last category, the component based type. Literature discerns different forms of interaction between the components. Pimmler and Eppinger [6] propose a taxonomy comprising spatial, energetic, informational and material interaction. The strength of interaction often is represented by scales, mostly considering positive and negative effects.

## 2 Objectives

The objectives of the present study are twofold:

1. **Including environmental criteria into the set of modularity drivers.** Unfortunately, the state-of-the-art product structuring methods do not enable the product developer to select modules according to environmental criteria. A promising way to include an environmental perspective is to enhance the set of conventional modularity drivers.
2. **Streamlining the structuring approaches in order to minimise the additional effort.** In the development of products engineers already have to focus on a lot of optimisation aspects (time, cost, quality, etc.). Paying attention to the environment can too easily be seen as an additional burden which complicates the daily work of the product developer. Therefore, methods must be streamlined in terms of the time required for their application as compared with conventional approaches.

## 3 Approach

The developed approach can be described in four main steps as shown in Figure 4.

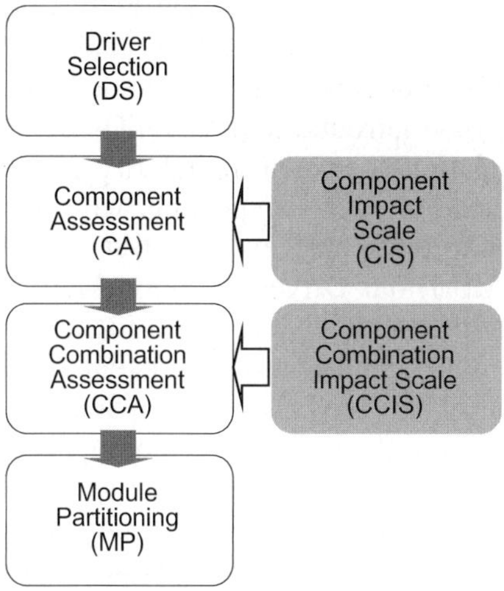


**Fig. 4.** The four main steps of the EcoDSM approach

Firstly, relevant modularity drivers are selected with help of a matrix that relates the components to the modularity drivers (Driver Selection, DS). The selected drivers are optionally listed in order of significance in a Driver Profile (DP).

Subsequently, the Component Assessment (CA) assigns the contribution of every single component to the accomplishment of the modularisation aims expressed by the selected modularity drivers. The assignment rests on a Component Impact Scale (CIS) which must be chosen from case to case. Furthermore, the assignment can be weighted with a linear, progressive or degressive scale.

Once every component has been assessed separately with respect to the relevant modularity drivers, the Component Combination Assessment (CCA) considers any pair of components in a component-based Design Structure Matrix (DSM). Based on a Component Combination Impact Scale (CCIS) the cells of that matrix state if and to which degree it is beneficial or detrimental to group the components of the according rows and columns in a module. These scales can be influenced by choosing a sensitivity factor.

Finally, the Module Partitioning (MP) aims at identifying appropriate modules with maximised internal relations (within the modules) and minimised external relations (between different modules).

### 3.1 Driver Selection (DS)

As stated before, the list of modularity drivers has been amended from an environmental point of view. But not every driver is applicable to a given product. Thus, the list must be condensed to a set of relevant drivers. For simple products this selection can still be made verbal-argumentatively whereas more complex products require methodological decision support.

On that matter, Erixon introduces in his QFD-like approach [4] a matrix that considers the relations between modularity drivers and subfunctions. In the practical application often components act as functional carriers and replace the subfunctions in that matrix. But unlike Erixon's Module Indication Matrix (MIM) our Driver Selection Matrix, which is presented in the following paragraph, solely serves to identify a profile of relevant modularity drivers. The identification of potential module candidates is postponed to the following steps. The Driver Selection Matrix (see figure 5a) relates the drivers Dy (y=1..m) to the components Cx (x=1..n) of the given product. The entries in the cells of the matrix mark if and to what degree a component is concerned by a driver. For indicating the strength of the relations e.g. a gradation of values between 0 and 9 can be used as shown in the legend of figure 5c.

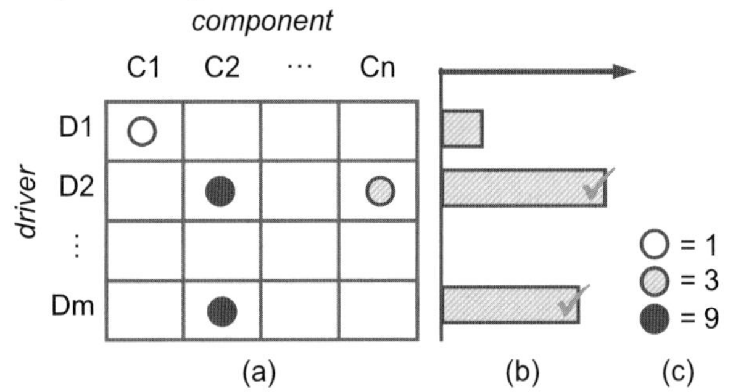


**Fig. 5.** (a) Driver Selection Matrix, (b) Driver Profile (DP), (c) Legend

The ranking of the modularity drivers is based on the following hypothesis: The more components a driver strongly affects the more important it becomes for the modularisation of the product.

Therefore, the impacts are summed up by the line. The Driver Profile (figure 5b) visualises the totalised impacts in a bar chart. Additionally, the lines of the matrix can be reordered according to the determined significance of the drivers. This gives the DP a clearer shape when considering a great number of drivers. In the further steps of the approach only the drivers with the strongest totalised impacts are considered, mostly not more than four.

### 3.2 Component Assessment (CA)

For every selected driver the contribution of the components to the modularisation targets must be assessed separately. The Component Assessment Vector (CAV) collects the single assessments and contains positive and negative values of different strengths (figure 6a).

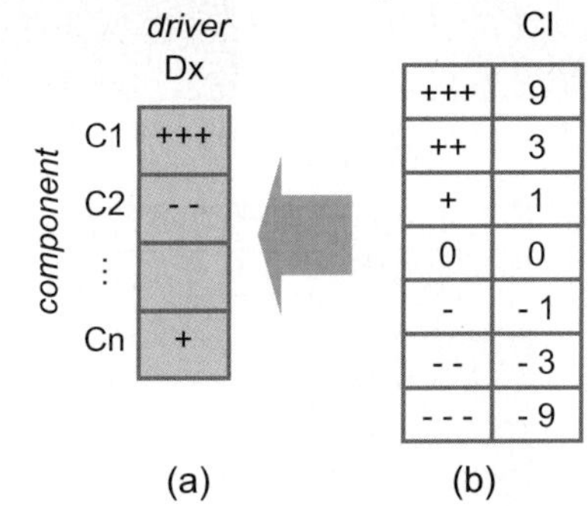


**Fig. 6.** (a) Component Assessment Vector (CAV), (b) Component Impact Scale (CIS)

In the application users can sufficiently differentiate the impacts with a trinomial scale for positive (+++, ++, +) and negative (- - -, - -, -) impacts including a neutral value (0). Filling out the CAV with symbols in a first step and translating these symbols with help of a Component Impact Scale (CIS) into values afterwards proved to be timesaving since the users did not get sidetracked by discussing numbers. The CIS (figure 6b) allocates discrete values to the impact instances that have been expressed provisionally by symbols. In dependence on the distribution of the components to the impact classifications different scales can be chosen. If the majority (~60 %) of components is slightly affected (+ or -) while only a minority (~20 %) is exceptionally affected (+++ or - - -) a progressive scale should be chosen (figure 7a). Thus, on a scale with a maximum of 9, a progressive scale takes the absolute values 1, 3 and 9. The progressive scale is widespread in metrics based methods, e.g. Quality Function Deployment (QFD) or Failure Mode and Effect Analysis (FMEA). In the opposite case where the components have a majority membership to the highest impact class the scale should have a degressive growth (figure 7c) corresponding to the absolute values 6, 8 and 9. A rather linear distribution (figure 7b) can be represented by the absolute values 3, 6 and 9.

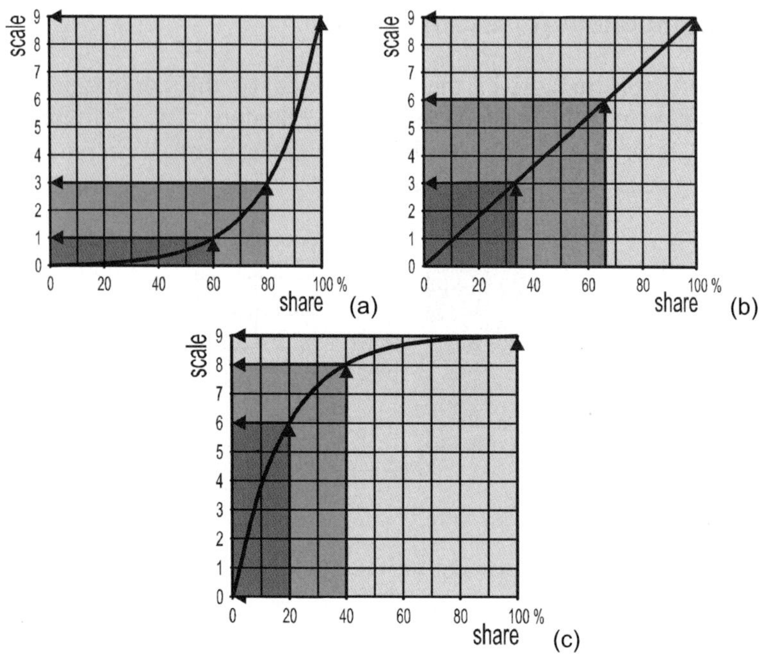


**Fig. 7.** Component Impact Scales (CIS) with (a) progressive, (b) linear, (c) degressive growth

## 3.3 Component Combination Assessment (CCA)

After having assessed the impact of a modularisation strategy separately on each component the effect of combining components to modules must be determined. The Component Combination Assessment (CCA) considers any pair of components in a component based Design Structure Matrix (DSM), cf. [6]. But unlike other approaches that let the user fill in that matrix cell for cell, the Component Combination Assessment Matrix (CCAM) is generated directly from the Component Impacts (CI) determined in the Component Assessment (CA), see figure 8a. The Component Impacts (CI) are listed in the first column and as a transposed vector in the first line of the matrix.

The Component Combination Impacts $CCI_{i,j}$ for any pair of components (Ci, Cj) can be found, eg. with the algorithmic expression:

$$CCI_{i,j} = \sqrt[n]{\frac{CI_i^{\,n} + CI_j^{\,n}}{2}} \quad . \tag{1}$$

That symmetric and linear function depends on the factor n which indicates the sensitivity of the modularity driver. The sensitivity factor must be determined from case to case, considering the special circumstances of the modularity driver and the product.

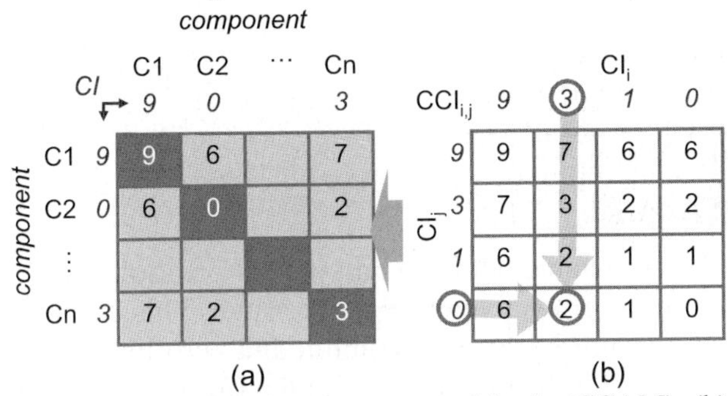


**Fig. 8.** (a) Component Combination Assessment Matrix (CCAM), (b) Component Combination Impact Matrix (CCIM)

The Component Combination Impact Matrix (CCIM) in figure 8b groups the rounded results from equation (1) with n=2. For the combination of component pair (C2, Cn) the $CCI_{2,n}$ must be determined as a function of $CI_2$=0 and $CI_n$=3. The result ($CCI_{2,n}$=2) can be read from the CCIM and is filled in the corresponding cell of the CCAM.

Figure 9 shows a graphical representation of the CCIM in a diagram, the Component Combination Impact Diagram (CCID). The diagram (n=2) can be read as follows: Starting from the impact of the first component (eg. CI=3) on the abscissa, the intersection with the curve corresponding to the impact of the second component (e.g. CI=6) shows the combination impact of the two components on the ordinate (here CCI=5).

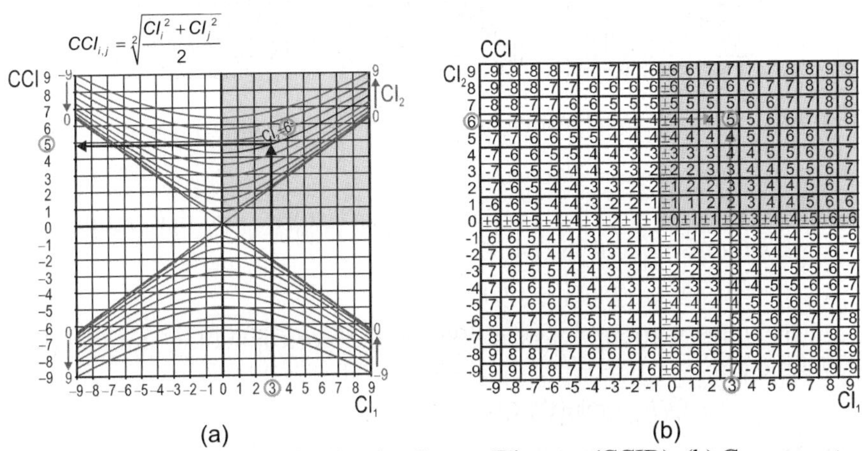


**Fig. 9.** (a) Component Combination Impact Diagram (CCID), (b) Component Combination Impact Matrix (CCIM)

Different signs ($CI_i$<0 and $CI_j$>0, or $CI_i$>0 and $CI_j$<0) are treated like in a multiplication as shown in Table 1.

**Table 1.** Combination rules

| | + | – |
|---|---|---|
| + | (+,+) → + | (+,–) → – |
| – | (–,+) → – | (–,–) → + |

For n<0, the zero value holds an ambivalent position. Dependent on the context it can act as a positive or negative value. This ambivalence is represented by the ± sign in the zero column and zero line of the CCIM and by the gap between the two zero curves in the CCID respectively.

Figure 10 assembles CCIDs for different sensitivity factors n.

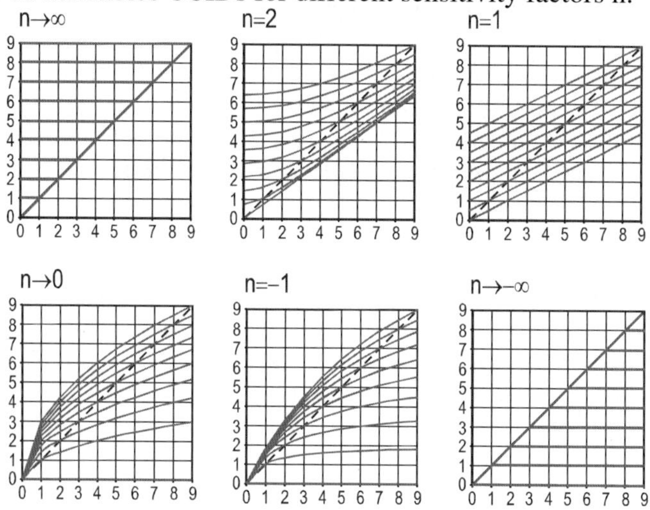


**Fig. 10.** Component Combination Impact Diagrams with different sensitivity factors n

1. The extreme case n→∞ leads to an expression that returns the maximum value of the two CIs:

$$n \to \infty:\ CCI_{i,j} = \max\left(CI_i, CI_j\right) \tag{2}$$

2. The other extreme case n→–∞ is the corresponding minimisation function:

$$n \to -\infty:\ CCI_{i,j} = \min\left(CI_i, CI_j\right). \tag{3}$$

3. n=1 furnishes the arithmetic mean of the two values:

$$n = 1:\ CCI_{i,j} = \frac{CI_i + CI_j}{2}. \tag{4}$$

## 3.4 Module Partitioning (MP)

For identifying appropriate modules the CCAM must be partitioned. The aim of the Module Partitioning is finding sub matrices that contain a maximum of internal relations and a minimum of external relations (figure 11).

maximise internal relations

| CI | Ⓐ 9 | Ⓑ 9 | Ⓒ 0 | Ⓓ 0 | Ⓔ 0 | Ⓕ -3 | Ⓖ -3 | Ⓗ -3 | Ⓘ -9 |
|---|---|---|---|---|---|---|---|---|---|
| Ⓐ 9 | 9 | 9 | 6 | 6 | 6 | -7 | -7 | -7 | -9 |
| Ⓑ 9 | 9 | 9 | 6 | 6 | 6 | -7 | -7 | -7 | -9 |
| Ⓒ 0 | 6 | 6 | 0 | 0 | 0 | -2 | -2 | -2 | -6 |
| Ⓓ 0 | 6 | 6 | 0 | 0 | 0 | -2 | -2 | -2 | -6 |
| Ⓔ 0 | 6 | 6 | 0 | 0 | 0 | -2 | -2 | -2 | -6 |
| Ⓕ -3 | -7 | -7 | -2 | -2 | -2 | 3 | 3 | 3 | 7 |
| Ⓖ -3 | -7 | -7 | -2 | -2 | -2 | 3 | 3 | 3 | 7 |
| Ⓗ -3 | -7 | -7 | -2 | -2 | -2 | 3 | 3 | 3 | 7 |
| Ⓘ -9 | -9 | -9 | -6 | -6 | -6 | 7 | 7 | 7 | 9 |

minimise external relations

**Fig. 11.** Partitioned Component Combination Assessment Matrix (CCAM)

In order to measure the cohesion of different components grouped into modules an index is needed. The Module Strength Indicator (MSI), cf. 8, consists of two terms. The first term measures the internal dependencies within the selected components in a module (SCM):

$$MSI_i = \frac{\sum_{i \in SCM} \sum_{j \in SCM} CCI_{i,j}}{n_i^2 \, CCI_{max}} \tag{5}$$

where $n_i$ = number of components included in the set SCM,
and $CCI_{max}$ = absolute maximum value (= 9 in Figure 11).

The second term measures the external dependencies, i.e. the dependencies between the components that are part of the set SCM and those which are not:

$$MSI_e = \frac{\sum_{i \in SCM} \sum_{j \notin SCM} CCI_{i,j} + \sum_{i \in SCM} \sum_{j \notin SCM} CCI_{i,j}}{2 \, n_i \, n_e \, CCI_{max}} \tag{6}$$

where $n_e$ = number of components not included in the set SCM.

Subtracting (6) from (5) provides a metric for the suitability of components to be grouped into a module:

$$MSI = MSI_i - MSI_e \,, \tag{7}$$

The MSI takes values of the interval $-2 > \text{MSI} \geq 2$, where MSI=2 represents the ideal modularisation of the product.

## 4 Case study

The presented method has been validated with product developers in several workshops at Darmstadt University of Technology and at the Technical University of Denmark. This case study describes very briefly the application of the method to a battery powered household appliance, a milk foamer (figure 12).

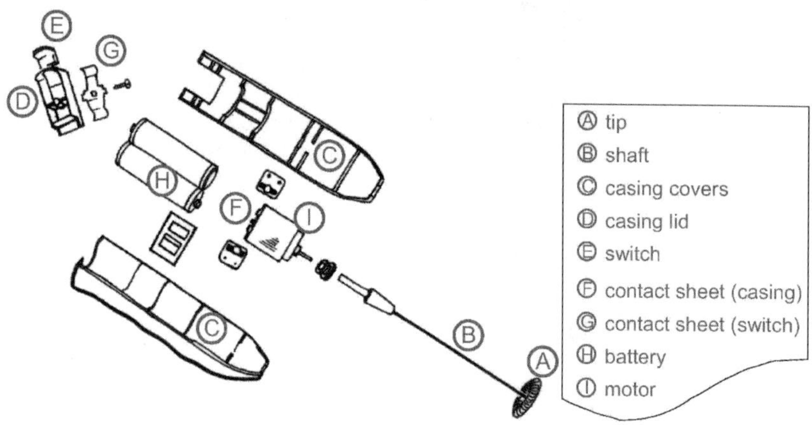


(a) (b)

**Fig. 12.** (a) Product structure and (b) parts list of the milk foamer

The main environmental impacts of the milk foamer arise from the use phase while the production and end-of-life of the product play a subordinate or even negligible role (figure 13). Having a closer look at the use phase, it is the cleaning of the milk foamer which is responsible for the most environmental impacts. These impacts result from the habit of most users to rinse the tip and shaft of the milk foamer under hot running water while letting the motor turn for a moment.

Based on the environmental assessment, the team of product developers has decided in the workshop to choose the cleaning of the product as the main driver for modularising the milk foamer. Concerning the necessity to be cleaned they assigned the highest value (+++) to the tip, followed by the shaft (++) and the casing components (+). According to the assessment the switch does not need (0) to be cleaned and the rest of the interior electrical components must not be cleaned (negative impacts). Further on, they decided to use a linear Component Impact Scale (CIS) and chose n=2 as sensitivity factor.

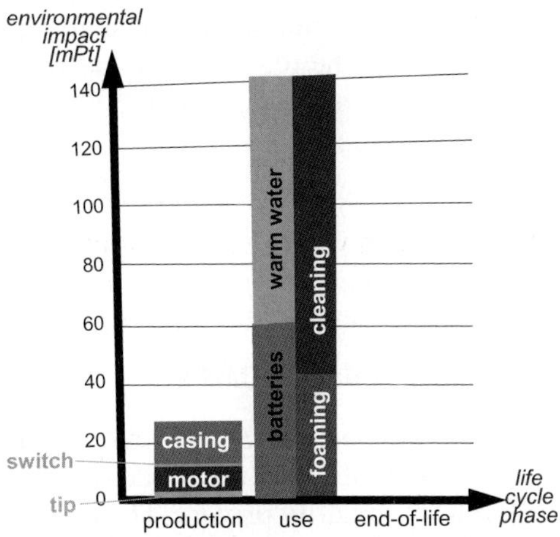


**Fig. 13.** Rough environmental assessment results with Eco-indicator 99 based on benchmarking data of a dismantled commercially available milk foamer and on a user inquiry

Supported by a spreadsheet application which calculates the Module Strength Indicators the team identified relevant modules (figure 14).

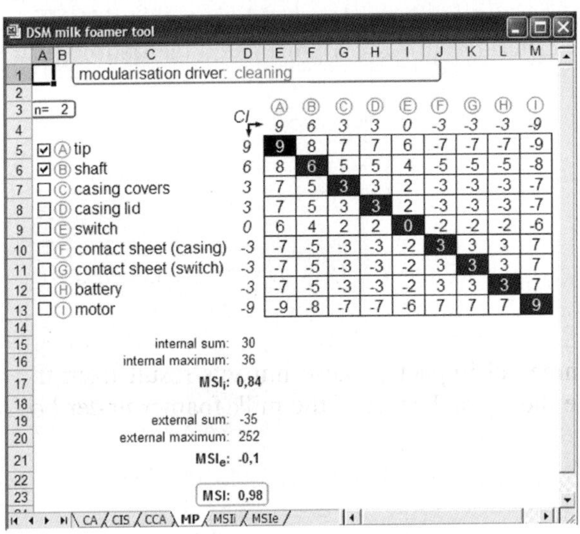

DSM milk foamer tool

modularisation driver: cleaning

n= 2

| | CI | Ⓐ 9 | Ⓑ 6 | Ⓒ 3 | Ⓓ 3 | Ⓔ 0 | Ⓕ -3 | Ⓖ -3 | Ⓗ -3 | Ⓘ -9 |
|---|---|---|---|---|---|---|---|---|---|---|
| ☑ Ⓐ tip | 9 | 9 | 8 | 7 | 7 | 6 | -7 | -7 | -7 | -9 |
| ☑ Ⓑ shaft | 6 | 8 | 6 | 5 | 5 | 4 | -5 | -5 | -5 | -8 |
| ☐ Ⓒ casing covers | 3 | 7 | 5 | 3 | 3 | 2 | -3 | -3 | -3 | -7 |
| ☐ Ⓓ casing lid | 3 | 7 | 5 | 3 | 3 | 2 | -3 | -3 | -3 | -7 |
| ☐ Ⓔ switch | 0 | 6 | 4 | 2 | 2 | 0 | -2 | -2 | -2 | -6 |
| ☐ Ⓕ contact sheet (casing) | -3 | -7 | -5 | -3 | -3 | -2 | 3 | 3 | 3 | 7 |
| ☐ Ⓖ contact sheet (switch) | -3 | -7 | -5 | -3 | -3 | -2 | 3 | 3 | 3 | 7 |
| ☐ Ⓗ battery | -3 | -7 | -5 | -3 | -3 | -2 | 3 | 3 | 3 | 7 |
| ☐ Ⓘ motor | -9 | -9 | -8 | -7 | -7 | -6 | 7 | 7 | 7 | 9 |

internal sum: 30
internal maximum: 36
$MSI_i$: 0,84

external sum: -35
external maximum: 252
$MSI_e$: -0,1

MSI: 0,98

CA / CIS / CCA / MP / MSIi / MSIe

**Fig. 14.** Spreadsheet based calculation of modularisation configurations for the milk foamer

The most interesting finding for the team was that a simple change in the product structure (in all probability) reduces the environmental impacts considerably. Grouping tip and shaft in a module obtained the highest sensible MSI value. This made the team think of a detachable module containing those two components. Thus, the tip and the shaft can be cleaned more easily by the user together with the regular dish washing. Technically, this detachable module does not set a problem with a sleeve-mounted connection to the motor shaft.

## 5 Future work

### *Multi Criteria Module Partitioning (MCMP)*

The approach presented in the paper does not yet deal with the problem how to derive an optimum overall structure based on the modularisation alternatives for the different modularity drivers. Thus, future research will be directed towards algorithms which superpose the different partitioned matrices (figure 15), cf. [9].
In this context two aspects are of special interest:

- The different modularity drivers are of varying importance, thus they must be balanced with weighting factors.
- Some modularity drivers 'evade' a mutual weighting since they imperatively must be considered. Mixing this MUST criterion with other CAN criteria requires a sophisticated methodological founding.

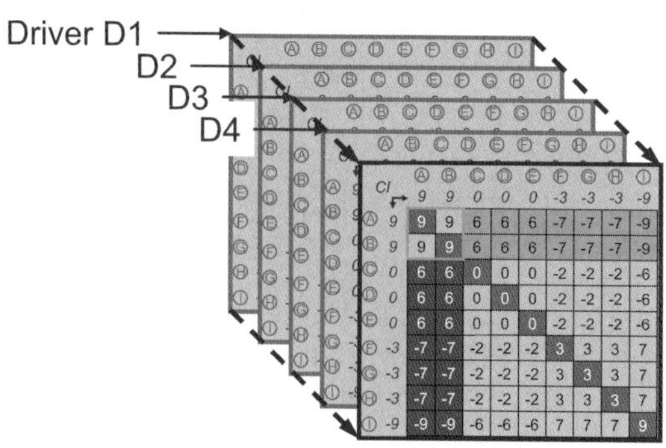


**Fig. 15.** Concept for a Multi Criteria Module Partitioning (MCMP)

### *Transferring Environmental Knowledge*

The quality of the assessment of the proposed method does largely depend on the knowledge of the product developer that executes the method. Therefore, a consistent support in form of specific knowledge units has to be included, describing e.g. the suitability for recycling of different materials for the product developer.

### *Software Implementation*

After the final establishment of the methodological procedure, of course, the approach has to be implemented in a software tool which directly accesses data of other tools. In other words, the additional effort for the product developer has to be reduced to a strict minimum, by integrating this system seamlessly into his 'digital workbench'. In the case of a redesign project the components can be derived, for example, from the existing parts structure of a CAD system.

## 6 Summary

The paper presented a streamlined DSM approach that serves to identify and assess modules according to holistic, i.e. technical, economic and environmental criteria. The methodological optimisation resides in the fact that the component based DSM has not to be filled out by hand any more. The interrelations between components in the DSM are calculated instead by a simple algebraic term on basis of the assessment of the individual components by the product developer. In addition, the set of commonly considered modularity drivers has been amended by various environmentally relevant aspects.

## Acknowledgement

The present study on modularity was a part of the research project 'Integration of Environmental Knowledge into Design Methodology'. The project was embedded in the Collaborative Research Centre (CRC) 392 'Design for Environment. Methods and Tools' at Darmstadt University of Technology. The CRC 392 was funded between 1996 and 2004 by Deutsche Forschungsgemeinschaft, the main funding body for public research in Germany. The transfer of the research findings into the industrial practice is now the focus of the recently established Transfer Unit 55 'Life Cycle Design of Active Products for the Markets of Tomorrow'.

## References

1. Dick (Voß), M., Birkhofer, H., 2003, Product Modularisation in Design for the Environment. In: Seeliger G, ed, Proceedings Colloquium e-ecological Manufacturing.
2. van Hemel, C., 1997, Ecodesign Strategies. In: Brezet, H., van Hemel, C., eds, Ecodesign. A Promising Approach to Sustainable Production and Consumption. UNEP.
3. Stake, R. B., 2000, On Conceptual Development of Modular Products. Stockholm: Royal Institute of Technology.
4. Erixon, G., 1998, Modular Function Deployment. A Method for Product Modularisation. Stockholm: Royal Institute of Technology.
5. Steward, D. V., 1981, The Design Structure System. A Method for Managing the Design of Complex Systems. IEEE Transactions on Engineering Management, Vol. EM-28, 3.
6. Pimmler, T.U., Eppinger, S.D., 1994, Integration Analysis of Product Decompositions. In: Proceedings of the ASME Int. Conf. on Design Theory and Methodology.
7. Blackenfelt, M., 2001, Managing Complexity by Product Modularisation. Balancing the Aspects of Technology and Business During the Design Process. Stockholm: Royal Institute of Technology.
8. Whitfield, R. I., Smith, J. S., Duffy, A. H. B., 2002, Identifying Component Modules. In: Proceedings of the 7th Int. Conf. on AI in Design. Cambridge, UK.
9. Cantamessa, M. Rafele, C., 2002, Modular Products and Product Modularity. Implications for the Management of Innovation and for New Product Development. In: Proceedings of the 7th International Design Conference DESIGN 2002. Dubrovnik.

# Strategies and material flow in ecodesign

C. Luttropp
*KTH/Machine Design, SE-100 44 Stockholm, Sweden, conrad@md.kth.se*

Abstract: Industrial products, green or not, are all compromises between many demands and constraints. Functionality and business performance are of course the dominants. Some products can easily be transformed in a green direction by changing a toxic substance over to something harmless others face a much more complicated situation. In order to face new challenges in product development it is essential to have a strategy for the integration of EcoDesign into "normal/classical" product development. Depending on the type of product and its impact on environmental performance, different strategies are possible.Very important is materials flow and information but also to have a procedure for getting different stakeholders inside the company together in forming requirements for new product with EcoDesign criteria integrated. Eco cycling of materials is a well known metaphor for the sustainable technical development ahead. With new concepts as Product-Service Systems and Producers responsibility a linear model can be useful. In this case the material can be looked upon as a working capital along a time axis where functionality of products and benefits to the customer is a profit and the recycling loss of materials can be seen as a cost. This paper will discuss this situation and put light on a possible new way of looking on EcoDesign.

Keywords: Business Strategy, Ecodesign, Sustainability

## 1 Introduction

The need to achieve sustainable development represents an enormous challenge to society. It means that within a few decades we must learn to deal much more efficiently with raw materials and energy. According to some estimates, within the next 50 years, the burden on the environment will have to be reduced to one half of the current levels, putting demands on the industrialised world for an increase in eco-efficiency by a factor ten, giving the developing countries a chance to grow in well fare. The industry has already started to increase the average eco-efficiency of products. However, product development efforts can only improve products in a



*D. Brissaud et al. (eds.), Innovation in Life Cycle Engineering and Sustainable Development,* 271–280.

"green" direction. Almost every product has more or less an environmental impact. The basic goal with "EcoDesign" is to achieve products better adapted to the environment; greener products. The main task for designers is to design products, which fulfils a need or provides a benefit to the customer/user. Even if welfare and growing economy in Western countries have changed the scope of design, one can say, "the role of the designer has changed from meeting needs to stimulating desires", [1]. The companies fulfil not just functional needs with their products but also image and desire. Creating the desire is sometimes an integrated part of the product design. Companies are not trying to sell green product besides their old ordinary ones. Companies try to show themselves as green companies, focusing on their brand rather than on special products. This makes the companies more vulnerable for public opinions and this is also something to adapt to in green design. In many cases this has been called a paradigm shift or The Next Industrial Revolution, [2]. This also calls for a change of life style since population is growing and this population has a tendency to allocate more and more resources also on a per capita level. Over population and over consumption are buzzwords in this area; see e.g. [3]. Eco Industrial methods can be defined as methods used to achieve products better adapted to environmental demands. These methods can be analysing tools, decision support tools or help for product development teams to visualise and support the process.

## 2 Material flows

The ownership of material in products is not consistent. Outsourcing of production and /or marketing and sales give opportunities to put non-sustainable activities out of the "brand-mark-umbrella". Traditionally companies produce products at a manufacturing plant, importing materials and exporting products to the market, see figure 1. This situation has changed dramatically, concepts as just in time has led to a situation where materials, parts and subassemblies are on the move to the final manufacturing/assembling process. Products are at the same time delivered to the market in a complex network. Regarding the company, as an organisation with a virtual materials network it is obvious that materials flow concern the company and the connected brand labels in a very direct way; see figure 1. Public opinion is a strong force and it is today not accepted to argue that toxic materials etc. are somebody else's concern. The company "brand-label-umbrella" has a wide boundary covering every material aspect of the product, inside or outside the core company. Cramer has stated,

"The amount of manoeuvring room depends largely on the influence of the company in the supply chain. Far-reaching environmental improvements can usually be made only with the co-operation of other partners in the chain. These partners must be willing to implement the desired changes in their products. The more power a company has, the more pressure it can put on others in the chain to adapt their products. Co-operation in the supply chain often plays a role when recycling systems are established.

**Fig. 1.** Traditional stationary companies with materials flow in to the company and products delivered out to the market are transformed into a more virtual appearance, with suppliers and just in time as a major concept. In this later case materials can be seen as an alternative capital with inflow and outflow of materials. However, even if the company does not own the material, due to just in time or other concepts the responsibility is virtually connected to the product brand label

No matter what form of co-operation is chosen, the agreement will be successful only if the company is willing to turn environmental threats into opportunities. This requires a proactive instead of a defensive attitude toward environmental issues. A company's flexibility also plays an important role in determining how much room it has to manoeuvre. For instance companies that have invested in large facilities may find it more difficult to change their processes or products than companies with a more flexible process. Much will depend on the willingness of management to take risks." [4]. When looking upon a company's manoeuvring room and flexibility in the context of material flow, the company, add material and loose material but where the main body of material is a part of the company. Hence it will be of great importance to evaluate which phase of the product that has the largest environmental burden. Is it pre use phase, use phase or after use phase? For a product, which has the use phase identified as the dominant environmental burden, can an indicator that presents production and disposal issues still be valid but will not likely be critical. Here it is the use phase, which needs to be sustainable [5]. Global attributes such as temperatures, compositions and concentrations are not predictive indicators. Links are also required which can validate company/product/service-based indicators and tie sustainability thresholds to global conditions. Until indicators can be validated, we collectively run the risk of moving towards non-sustainability [5]. "One could question if, and if so how one can measure the environment, the sustainability of species as well as individual and collective welfare, all the while recognising that the very act of deciding what we measure is subjective" [5].

## 3 Ecodesign

Many design activities in modern product development are carried out in a parallel way. Experts on different fields are combining their competence in teams in order to shorten product-developing time. However, there must always be a balance between all demands on the forthcoming product. All these demands, "Design Core", as they are presented by e.g. [6], have to exist together; many of them at the same time. The circle diagram in figure 2 illustrates all demand connected to a forthcoming product. Environmental matters are visualised with the new piece, environment, trying to merge the classical compromise. Even if environmental demands are important and crucial, there are a lot of other demands to be considered as well, [7].

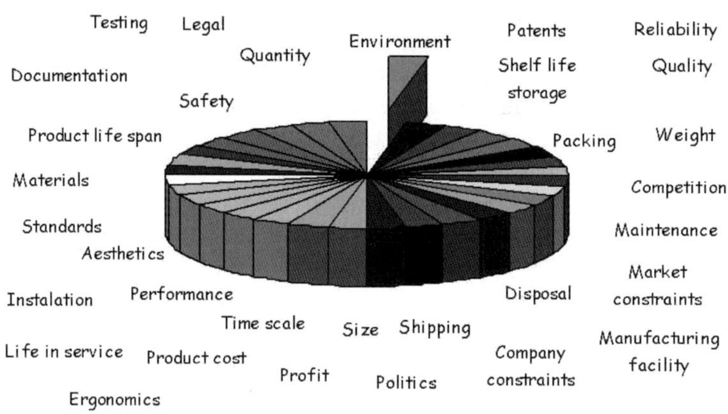


**Fig. 2.** A picture of all product aspects that must be addressed in design work

Functionality of the product and profit to the company are examples of two aspects, which always have very high priority, higher than environmental demands. Without customers buying a function and companies making profit there will be no market, no matter the environmental issue. The environmental issue is fairly new as subject with a short tradition which is a major reason why these questions have a problem to penetrate design work. With focus on environmental issues time will allow these matters to merge into normal design work. Environmental knowledge is present today as facts and formulas on specialist levels. It takes some time before this knowledge has transformed into the core competence of designers. Recycling and other environmentally imposed actions have to be related to all the elements in the design core without taking over, since environmental demands can never have top priority.

### 3.1 Life style elements

Life style has a large impact on sustainability. In Western societies brand labels are growing elements. For young people the label brand has taken over from functionality. A product carries functionality but the functions are today more of a constraint that is taken for granted. The products are today getting more attention than function, which is the legitimacy of the product. This must be taken care of in sustainability related design. A basic issue is the balance in products. How much environmental impact can be related to different functions and how much is related to the function carrier; the product? This balance is visualised with figure 3 [8]. Elaborating with this balance one can observe a handful situation.

**Fig. 3.** Balance between product and the related environmental impact is essential

## 4 The impact/function balance

### 4.1 Win-win

In this case there is a mutual interest between environmental demands, economy and other beneficial factors. Reducing fuel consumption of automotives is good for the environment and the wallet of the owner. However the SUV concept overrides this consumer desires "makes it worth" raised fuel costs in order to have a very trendy car. Sometimes a good compromise is possible on economic, environmental and technical matters. In some cases, housekeeping leads to both better products and benefits to the environment. Car washing facilities with a closed water system is one example. The closed system provides a possibility to use more chemicals and more water with a cleaner car as result, being still very environmental. Every one wins. When implementing ISO 14001 the first survey of the company often shows that housekeeping measures can save money already during the first year. Unfortunately this low hanging fruits are positive harvests, which, cannot easily be repeated; see figure 4.

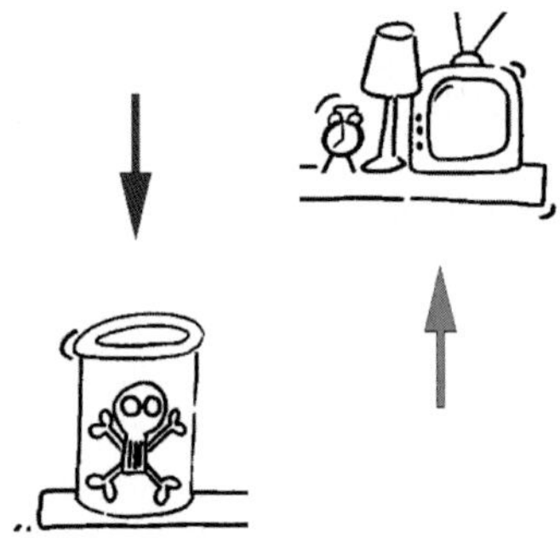

**Fig. 4.** Win win strategy: Functionality goes up; environmental impact goes down

## 4.2 Unnecessary

Another rewarding situation is eliminating unnecessary functions. A good example is the food served in short flights they are on their way out with a lower ticket price and less resource as result. However this can lead to rebound effects in the sense that more people can afford flying and the total amount of flying might raise; see figure 5.

**Fig. 5.** Sometimes unnecessary functions can be eliminated

## 4.3 Green fix

Life Cycle Analyses have found materials with high environmental impacts in products where the impact substance very easily can be replaced. The main function of the product is not very close connected to the specific substance or can be changed into a less harmful one; see figure 6. In

Sweden reductions have been made concerning toxic pigments or solvents in paint. In a student project at a course in EcoDesign, students found 12 grams of lead (Pb) as ballast in an older shaving handle. In new ones, present in shops today, the ballast is changed into 12 grams of stainless steel with much less environmental impact as result. Packaging is also diminishing and packaging materials are changed into more environmentally friendly materials. These green fixes are also a matter of low hanging fruits and will soon be completed, [9]. After some time one can believe that all this accidental mistakes as far as materials concern will be corrected. The environmental lesson is spread to designers and products that can easily be changed, are changed to a large extent.

**Fig. 6.** Green fix: harmful material removed without affecting basic functionality

### 4.4 Linear down

Too often common strategy is trying to just avoid a harmful substance or exchange it for some other less harmful substance. However this strategy has its dangers and obstacles. It is easy to fall into two different pit holes. The first is when the cause of the environmental impact is carrying the functionality. Typical example is some modern water-soluble paint, which is so environmentally friendly that the protection of houses does not work; being toxic is so to speak the business idea. New promising research point out other ways of protection and then we are in the win/win segment! (figure 7).

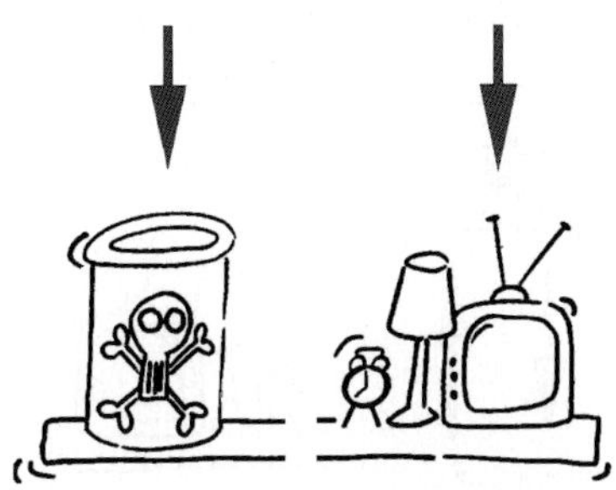

**Fig. 7.** Functions eliminated with connected environmental impact

### 4.5 Green washing

Another pit hole is the risk of doing "green washing". The new exchange substance can be bad from other point of view or other perceived improvements can be at a closer look worse than the first one. Changing from PVC into some other plastic with a shorter product life as result might not be a total gain; see figure 8. Some products are not greener than other products without the claimed green features. Putting labels on plastic parts don't mean that recycling really takes place. Using environmental labels without agreed constraints and measures are of course not beneficial to the environment. Implementing ISO 14001 don't mean by default that the actual company is very green. ISO 14000 is a toolbox for companies wanting to be greener, which is something else!

**Fig. 8.** Green washing: environmental impact is raised and functionality is reduced

### 4.6 Complex

Many products have a complex environmental situation. If we improve one parameter another might get worse. Changing one substance might lead to other problems not foreseen. A toxic surface treatment might lead to longer life and the answer of this equation is not trivial; see figure 9.

**Fig. 9.** A complex situation that often occurs with trade off situations as result

## 4.7 Linear up

In many situations it is the usage phase that is the predominant in environmental impact. In this case the product owner has a key role. So called active products as cars, refrigerators, houses etc. are examples of this. The more we use the car the more impact and especially how we drive. Inside temperature winter and summer has a large impact on the consumed energy etc.; see figure 10.

**Fig. 10.** Strong link between product user and resulting environmental impact

# 5 Company strategy

In order to handle this strategy dilemma it is essential that product development understands this. The first step must be to identify where on this map the company or it's products are and from this point try to find a rewarding strategy or start a program to handle and communicate a possible complex situation. Some companies have begun to translate the environment from a minor management concern into a core issue that relates directly to efficiency and competitiveness. Probably this will boost even more for companies to go beyond environmental performance and begin to discuss sustainability, which will further more challenge companies to consider the environmental impact of the materials they select, their social implications of their products and operations, and in some case the need for their product. One large challenge will be to integrate environmentally and sustainability concerns more systematically into the companies decisions and working routines. The company have to decide whether and to what degree they want to be "green". The minimum level of strategy is compliance with environmental regulations. The challenge will include development of a common vision and strategy, the selection of management systems and tools, and collection of data. Setting appropriate goals will be a critical matter in the integration of environmental and sustainability issues and developing performance measures to identify if the implementation is moving in the right direction [10]. Environmental management systems can be of help to provide the change. Industries need to manage their environmental impacts and sustainability issues, so it realise not only benefits for the environment but for business as well. Companies try to show themselves as green companies and this way focus on their brand rather than one product. This makes the companies more vulnerable for public opinions and this is also something to adapt to in green design.

## 6 Summary

It is important for companies to position on the sustainability map. The vulnerability of the brand label makes it necessary to take control of the total material life cycle, whatever the company boundaries are. The situation can be concluded:

- Search for win-win situations and don't forget the option of unnecessary functions, see chapter 4.1-2
- Watch out for possible green fixes chapter 4.3 and especially watch out for pit holes like in chapter 4.4-5
- Communicate complex situations as presented in chapter 4.6-7

A few statements can be made:

- Functionality must be regarded as environmental income.
- Functions must be balanced against total economy and LC technology.
- To control and monitor the "company materials body" allocating principles and standardised system border descriptions must be developed.

## References

1. Cooper, T., 1999, Creating an economic infrastructure for sustainable product design, Journal Sustainable Product Design, January 1999, UK
2. McDonough, W. Braungart, M.,1998, The NEXT Industrial Revolution, The Atlantic Monthly, October,
3. Schmidt-Bleek, F., 1994, Wieviel Umwelt braucht der Mensch?, Deutsche Tachenbuch Verlag, Germany (In German)
4. Cramer, J., 1999, Toward Sustainable Business: The Eco-efficiency challenge, Journal Environmental Quality Management vol. 9, no. 2, pp. 53-63. ISSN: 1088-1913
5. Hunkeler, D., 1999, Ecometrics for Life Cycle Management. A Conflict Between Sustainable Development and Family Values?, Int. J. of LCA, 4(5), 291-298
6. Pugh, S., 1991, Total Design, Addison-Wesley Publishing Company, UK
7. Luttropp, C., 1999, Eco-design in early product development, Recycling R99, Geneva, Switzerland
8. Lagerstedt, J., 2003, Functional and environmental factors in early phases of product development-Eco Functional Matrix, PhD thesis KTH Machine Design.
9. Luttropp, C., 2000, The relevance of Eco Industrial methods in the Context of Producers responsibility, Electronics goes green 2000+, Berlin
10. Brady, K., Henson P., Fava J.A., 1999, Sustainability, Eco-efficiency, Life-Cycle Management, and Business Strategy. J. Environmental Quality management, 8/3, 33-42

# Screening life cycle modelling for sustainable product design

M. Fargnoli and F. Kimura
*University of Rome "La Sapienza", Italy, The University of Tokyo, Japan,*
*mario.fargnoli@uniroma1.it*

Abstract: The present research work concerned the study of eco design methodologies able to lead designers in planning the entire life cycle of products in the early stages of design and development activities, also including marketing and management aspects. In order to make the use of traditional eco-tools more efficient we tried to develop a streamlined life cycle approach, based on the Life Cycle Assessment criteria, aimed at investigating the end of life of products, throughout the development of several different scenarios which take into account the main phases of the product life cycle. With this aim in mind, an attempt was made to integrate the Life Cycle Modelling approach in the first phases of the Methodical Design process by means of the development of a design technique, called "Screening Life Cycle Modelling" (SLCM).

Keywords: EcoDesign, Life Cycle Modelling, Life Cycle Assessment

## 1 Introduction

The attention paid to Sustainable Development problems has certainly helped in making the industrial world more aware, due to the new specific standards and regulations oriented to the environmental impacts reduction, and by the market request for more and more environmentally sound products. The necessity to evaluate the environmental performances of a product before deciding its production has become significant both because of expenditure of the resources which its introduction on the market requires, and for the influence that it can have on the "image" of the company. The goal of this research is to examine the influence that decisions concerning the product life cycle can have on the product's sustainability. This paper shows the benefit of both applying material/energy flow models and developing life cycle scenarios during the early stages of the product



*D. Brissaud et al. (eds.), Innovation in Life Cycle Engineering and Sustainable Development,* 281–292.

development process, with the aim to define an optimal "environmental product profile". In particular, in order to make the use of traditional eco-tools more efficient, we tried to develop a streamlined life cycle approach, based on the Life Cycle Assessment criteria, aimed at investigating the end of life of products, throughout the development of several different scenarios which take into account the main phases of the product life cycle. With this aim in mind, an attempt was made to integrate the Life Cycle Modelling approach in the first phases of the Methodical Design process by means of the development of a design technique, called "Screening Life Cycle Modelling" (SLCM). The first results showed that the use of SLCM can help designers in achieving a better design in a more effective and efficient way, defining environmental design requisites of the product, identifying critical design options, optimizing at the same time the marketing strategies and adapting the traditional design concepts to the producers' bottom line.

## 2 Background and motivations

The latest research work carried out by scientific institutes and companies has provided some important results that show key design schemes, which can be followed for a wider action aimed at solving problems related to the environmental impacts of industrial products [1-3]. A significant effort in this direction has been made also by international bodies and public authorities which have been issuing many regulations and standards aimed at indicating the new "green approaches" for product design and development. In this ambit, particular attention is paid to the assessment of the whole product life cycle, enlightening the role played by the last phases of the product's life: indeed, the publication of the recent EU norms concerning recycling (e.g. RoHS and WEEE Directives), as well as the issue of similar regulations in non European countries (e.g. the Japanese "Law for Recycling of Specified Kinds of Home Appliances"), clearly show how aware public opinion is regarding this problem [4]. In accordance with this trend, many authors in the field have shown the importance of planning the entire life cycle of products since the initial stages of design activities, as well as the marketing and management aspects.It is clear, then, that the valuation of the whole life cycle of the product is necessary, from raw material acquisition, through production, use and disposal, since each life cycle stage can have a significant influence on the products' environmental performances. The theoretical basis of the research work carried out consisted in a detailed study of design tools which take into account environmental aspects throughout the product life cycle. Moreover, a survey carried out among several industrial firms has pointed out that generally their design activities neglect the first stages of the design process, both to reduce the time-to-market since it is seen as time consuming to "re-think" the product, and because it is considered more difficult to "start from scratch" rather than directly introducing minor improvements to already existing products, reducing the designers' freedom.

Even if the greatest hindrance of operating in initial stages of the product design and development is the lack of detailed information concerning the product's characteristics and in particular information regarding its life cycle, it is indeed at this point that the rethinking of the product should be considered fundamental in order to optimize the product's characteristics in the most effective and efficient way, as well as deploying a feasibility study aimed at defining the product's life phases (i.e. considering market and sales strategies, reusing and recycling options, etc.). In this field, several tools and design approaches have been developed with the aim of analysing and improving all phases and processes which characterize the life cycle of a product. Among all of these, Life Cycle Design (LCD) and Life Cycle Assessment (LCA) represent the best known and the most used tools. In particular, LCA identifies and quantifies both direct and indirect environmental impacts, as well as consumption of energy and resources of a product [5–6]. LCD applies this concept in the design activities: it seeks to minimize the environmental burdens associated with the product life cycle and leads the designer to correctly define the product's environmental requirements in the design process [7].

Beside these, recent studies have shown that the development of a product life cycle simulation (Life Cycle Modelling, LCM) allows the integration of technical issues and product flow management, which enable the evaluation of the product's environmental performances, helping designers in both understanding the potential benefits of environmental improvements and managing the overall product development [8–9].

## 3 Methodology

With the aim of following this approach and at the same time trying to overcome the above mentioned difficulties concerning the use of the Life Cycle Design tools, as well as optimizing their use with the traditional Eco-design tools, a Life Cycle Modelling technique, called Screening Life Cycle Modelling (SLCM), has been developed. It consists of a streamlined life cycle simulation, the results of which can be considered as a possible support for the further application of design and management methods.

## 3.1 The screening life cycle modelling

### *Main characteristics*

The Screening Life Cycle Modelling can be considered as an assessment technique aimed at evaluating the main environmental impacts related with a product life cycle in the first phase of the Design Process [11, 12] (usually called "Clarification of the task" or "Task Analysis"), providing at the same time options for improvement throughout the development of a series of life cycle alternatives. The assessment criteria are based on the ones which characterize the traditional LCA; on the other hand, the level of information needed in order to apply the SLCM is very low compared with the traditional LCA. The final aim of the method is to provide an easy to use tool for the evaluation of the most significant directions that designers can follow in carrying out the project. The application procedure of the SLCM consists of four main steps:

1. Base Scenario (BS) definition: in this phase, the base life cycle model is defined, stating main parameters and/ or indicators.
2. Alternative Scenarios (ASs) definition: on the basis of the preliminary study concerning the problem which is being studied (i.e. a particular need of the customer, a specific law requirement, etc.) or of precise company needs, several models are developed.
3. Simulation: in this step, the life cycle analysis of both the BS and the ASs is carried out.
4. Analysis of results: the results of the simulation are critically analyzed in order to deploy a feasibility study and to single out possible modifications and alternatives.

The development of life cycle models can be carried out using different software languages/tools for modelling, such as UML [10]; similar considerations can be made regarding the simulation, which can be carried out using different calculation software.

## 3.2 The integration into the design process

As mentioned above, the method has been developed in order to use it during the first stages of the design activity. In particular, in accordance with the Methodical Design theory, if we consider the design process divided in four main phases (Analysis of the task, Conceptual design, Embodiment design, Tests and validation), it is possible to correctly integrate the SLCM method in the first phase, as shown in Figure 1.

However, such a method can also be used in different moments of the design process, depending on the need of designers: for example during the second phase, in order to choose the best concept, as well as with the aim to operate a bottom up verification.

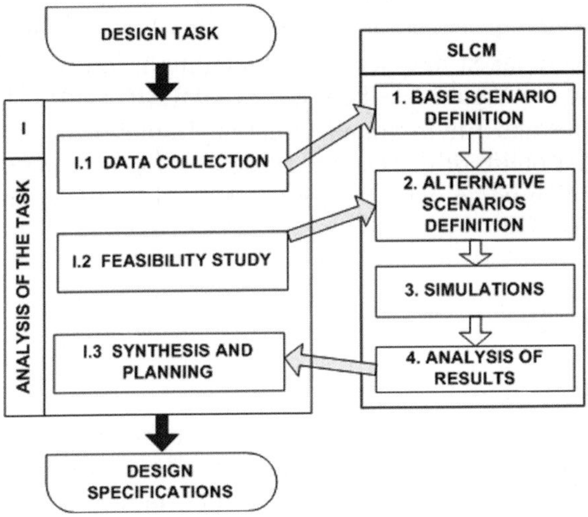


**Fig. 1.** Moment of use of the Screening Life Cycle Modelling: procedure of integration in the first phase of the Design Process

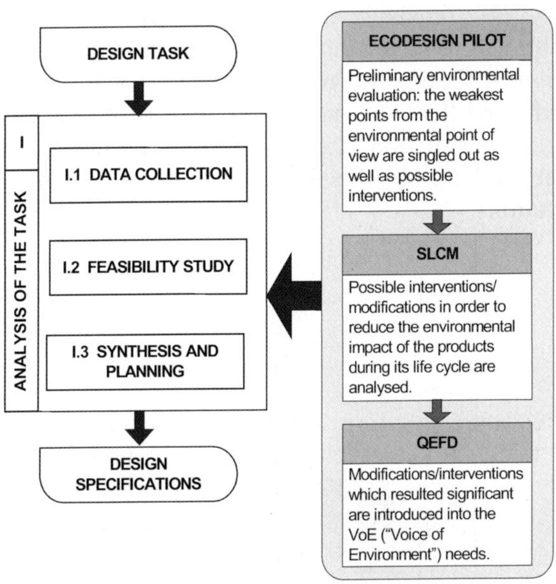


**Fig. 2.** The development of a "Design Module" concerning the coordinated use of the SLCM method together with the Ecodesign PILOT and the QEFD

### *Co-ordinated use of SLCM*

In order to obtain more effective results from the application of SLCM, its coordinated use together with other design tools has been studied. More in detail, considering its use in the first phase of the Design Process, several options were taken into account developing a series of "Design Modules", as shown in Figure 2, where the ECODESIGN PILOT method [13] and the Quality and Environment Function Deployment [14] are considered in the first phase of the Design Process. Of course, depending on the specific design situation, the use of other different methods and techniques can be possible.

## 4 Case study

In this section an example of the application of the SCLM is shown. In particular the design approach developed has been applied to the redesign of a refrigerator for domestic use. In order to identify the weakness aspects of the product's environmental profile, the Eco-Indicator 95 technique [15] was used: in Figure 3 the most important results are shown. Since from this analysis it emerged that the use phase is the most influential on the product's environmental performances, the ECODESIGN PILOT method (Product Type "D": intensive use) was applied to the already existing product: the aim was to understand which aspects require more effort to be improved: the most significant results are summarized in Table 1. Such an application indicated that the main interventions in order to improve the product regard: the reduction of energy during the use phase, the increase of the product's reliability, the improvement of dismounting and disassembling operations.

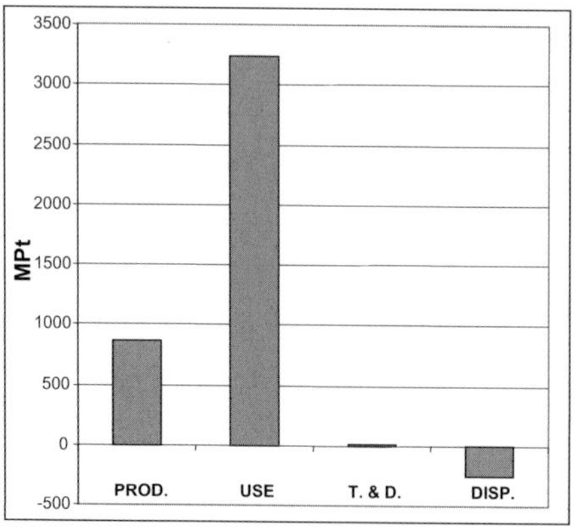


**Fig. 3.** Main results from ECOINDICATOR 95 application to the existing product

**Table 1.** Main results of ECODESIGN PILOT

| ASPECTS | Priority |
|---|---|
| Prefer the use of recycled materials | 40 |
| Preferably use single material components and/or reduce number of different materials | 40 |
| Ensure high reliability of product | 40 |
| Design product for possible upgrading | 40 |
| Minimize energy consumption at use stage by increasing efficiency of product | 40 |
| Make possible use of process materials from renewable raw materials | 30 |
| Ensure reversibility of assembly procedure | 30 |
| Design product structure for easy disassembly | 30 |
| Minimize time and paths for disassembly | 30 |
| Use easily detachable connections | 30 |
| Ensure easily visible access to connections for disassembly | 30 |
| Make possible separation of materials for recycling | 30 |
| Ensure simple extraction of harmful and valuable substances | 30 |
| Design product for adjustment and adaptation at use stage | 30 |
| Minimize energy demand at use stage by choosing an adequate principle of function | 30 |

## 4.1 Base Scenario definition

**Table 2.** Main parameters of the Base Model

| BASE MODEL (BM) | |
|---|---|
| Compressor | 1 X 150 W |
| Life span | 12 years |
| Working hours per day | 18 |
| Efficiency improvement ratio | 0,5 per year |
| Internal Recycling (plastics) | 0% |

The first step of the application of SLCM is to define the base scenario model, whose estimated lifetime is 12 years, on the basis of data provided by the company, Merloni s.p.a. (Table 2). In particular we focused the analysis on the use phase and considered as an indicator the one provided by the following formula (1):

$$Use\ \mathrm{Im}\ pact = P \times h \times d \times y \times e \qquad (1)$$

where: **P** represents the compressor power, **h** is the number of working hours per day, in this case 18, **d** is the number of working days per years, **y** is the number of years of use, in this case 12, **e** represents the energy consumption indicator, whose value of 0,385 *mPt/kWh* has been attributed, in accordance with experts from the company.

## 4.2 Alternative Scenarios definition

**Table 3.** Main parameters of the Alternative Scenarios

| SCENARIO 1 (M1) | |
|---|---|
| Compressor | 1 X 150 W |
| Life span | 6 years |
| Working hours per day | 18 |
| Efficiency improvement ratio | 0,5 per year |
| Internal Recycling (plastics) | 5% |
| **SCENARIO 2 (M2)** | |
| Compressor | 2 X 120 W |
| Refrigerator Life span | 4 years |
| Freezer Life span | 6 years |
| Refrigerator working hours per day | 10 |
| Freezer working hours per day | 6 |
| Efficiency improvement ratio | 0,5 per year |
| Internal Recycling (plastics) | 5 % |

On the basis of results obtained from the application of ECODESIGN Pilot, the most significant interventions regard: the modification of the energy consumption during the use phase; the possibility to recycle plastic materials in order to use them (or a part of them) to produce new refrigerators The increase of the product's reliability. Another aspect that emerged from the analysis regards the need to improve the dismounting and disassembling operations. It has to be underlined that these aspects also correspond to the new EU directives requirements.

Of course, in accordance with these considerations and in order to reach homogeneous results in the simulation, we assumed that identical refrigerators are produced during the whole simulation period. On the other hand, in order to take into account natural technological improvements, an eco-efficiency improvement following the indications provided in the "Eco-efficiency Indicator Handbook for Products", by the Japan Environmental Management Association for Industry (JEMAI) [16];

mental Management Association for Industry (JEMAI) [16]; the improvement value adopted corresponds to half of the one promoted by JEMAI. The possibility to produce the refrigerating part and the freezing one separated both to improve the energy consumption and the disassembling operations, and due to maintainability and reliability considerations (in particular, the internal components of the refrigerator part need to be repaired or replaced very frequently). Main characteristics of the Alternative Scenarios' models are summarized in Table 3.

## 4.3 Simulation

The simulation carried out was based on the data collected about the base model and on the following assumptions: constant production of machines per year (10.000 units); the price of the refrigerator doesn't change during the whole simulation period (24 years) and includes the programmed maintenance (first three years). The eco-efficiency improvement ratio is applied for new machines after the first life cycle (i.e. for the M1 model after the sixth year and for the M2 model respectively after the fourth and the sixth year). The evaluation of the environmental performances was based on the ECOINDICATOR 95 criteria, and in particular for the use phase the formula (1) was applied. It has to be underlined that in all three scenarios the main structure of the machine was not changed. In order to consider the higher request of materials and processes in the case of the scenario M2, the value of the environmental impact in the production phase was increased on the basis of data provided by the company (such an increase mainly regards the presence of an additional compressor). It was not possible to complete the study of the end of life phases (collecting, sorting, reusing recycling etc.) because when most of data were collected a take back policy was not yet applied.

## 4.4 Analysis of results

Even though it might seem, from an external point of view, that it would be more efficient to extend the life of this kind of electrical appliances, first results of the simulation (Figure 4) show that considering a shorter life cycle of the machine, can be very convenient both from the general environmental impact of the product and from the particular point of view of the energy consumption. Such an aspect also positively characterizes the behaviour of the scenario M2, in spite of the higher costs and impacts during the production phase. In fact, as shown in Figure 5, the environmental impact (measured in MPt) of the whole life cycle of the scenarios developed (i.e. material acquisition, manufacturing, distribution, use and disposal phases) shows that the shorter life alternatives are very competitive from the environmental point of view. Moreover, it has to be underlined that the simulation concerned a 24 year period: actually, according to both company data and market analysis, a refrigerator is usually replaced after 15-16 years. Hence, we can see that a shorter life cycle appears more sustainable from the environmental point of view.

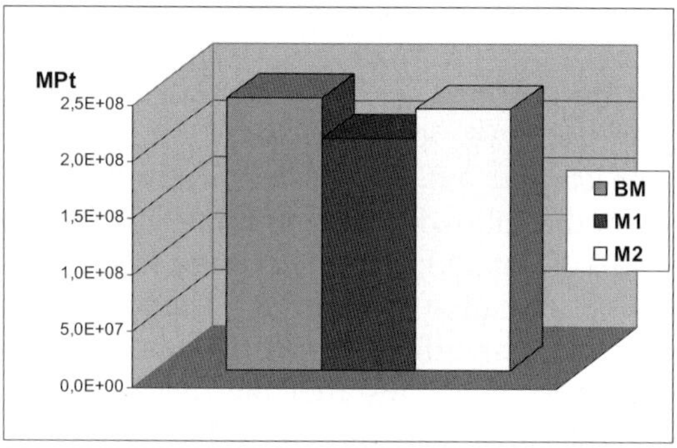


**Fig. 4.** Main results of the simulation concerning the whole life cycle

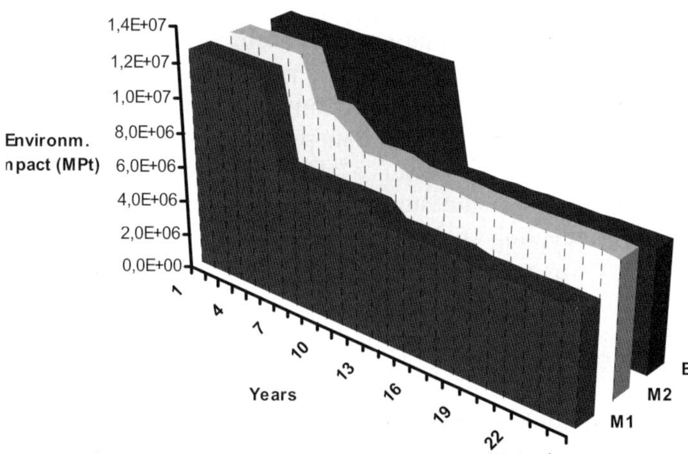


**Fig. 5.** The behaviour of the environmental impact during the whole life cycle of the three models considered during the whole simulation period

The results obtained allowed us to define a series of design specifications and priorities very useful in order to further develop the product design, which so far has been carried out until the conceptual phase (in Figure 6 the two main concepts of the product are shown).

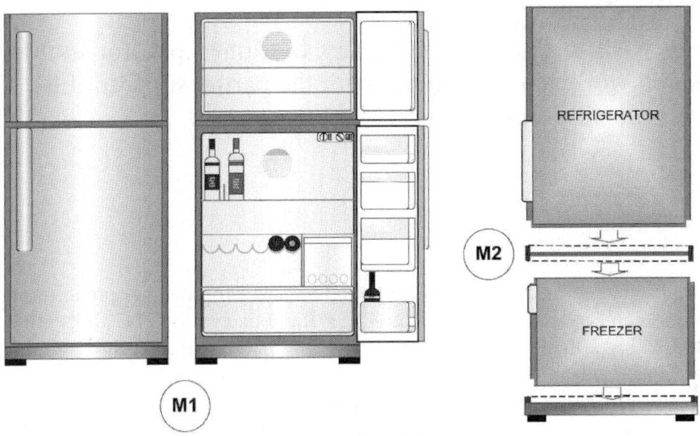


**Fig. 6.** General concepts of the 2 models developed

## 5 Conclusions and summary

Preliminary results confirmed the importance of considering the entire life cycle of products during the first stages of the design activity, and underlined the importance of the improvement of the environmental performances of industrial products. The coordinated use of the Ecodesign tools allowed us to optimize the environmental analysis of the system providing at the same time feasible indications for its modification and improvement. A further study concerning the reliability of the machine has been foreseen with the aim to better choose among the design options, as well as to define the opportunities of material recycling.

In this paper an approach to life cycle design and management is discussed by means of the introduction of an easy-to-use design tool, which can be applied to the design of any kind of industrial products. The life cycle simulation carried out has shown the possibility to improve product environmental sustainability since the first stages of the design activity by a rather unusual application of design tools in a coordinated way.

## Acknowledgements

Authors wish to thank the company Merloni s.p.a. for its helpfulness. The research work was carried out with the partial support of the JSPS Research Fellowship Program.

## References

1. Brezet, H., Hemel, C., 1997, "ECODESIGN – A promising approach to sustainable production and consumption" a joint UNEP/Rathenau Institut/TU Delft publication.
2. Alting, L., Legarth, J.B., 1995, "Life Cycle Engineering and Design", Annals of CIRP, vol.44, no.2: 569-580.
3. Gruner C., Birkhofer H., (1999), "Decision Support for Selecting Design Strategies in DFE"; Proceedings of ICED 99, Munich, Vol. 2, 1089-1096.
4. EU, 2003, Directive 2002/96/EC on Waste Electric and Electronic Equipment (WEEE) and Directive 2002/95/EC on "Restriction of the use of certain Hazardous Substances in electric and electronic equipment" (RoHS), Official Journal of the European Union, http://europa.eu.int/environment/.
5. Wenzel, H., Hauschild, M., Alting, L., 1997, "Environmental Assessment of Products - Vol. I: Methodology, tools and case studies in product development" Chapman & Hall.
6. Tomiyama, T. et al., 1997, A Holistic Approach to LCD, Life Cycle Networks, 4th International Seminar on Life Cycle Engineering, CIRP, Berlin.
7. Kimura, F., Suzuki, H., 1995, Product Life Cycle Modelling for Inverse Manufacturing, Life Cycle Modeling for Innovative Products, Chapman & Hall.
8. Fukushima, Y. Hirao, M., 2002, A structured Framework and Language for Scenario-Based Life Cycle Assessment, Int. Journal of LCA, n. 7.
9. Kimura, F., 1999, Life Cycle Design for Inverse Manufacturing, Proceedings of the EcoDesign '99, Tokyo.
10. Fowler, M., 2003, UML Distilled, Addison-Wesley, Pearson Education Inc.
11. Pighini, U., Fargnoli, M., 2001, Engineering Design in the Development of Sustainable Products, Proceedings of ICED01, Glasgow.
12. Hubka, V., Eder, W.E., 1988, Theory of Technical Systems, Springer, New York.
13. Wimmer, W., Züst, R., 2003, ECODESIGN PILOT, Product-Investigation, Learning- and Optimization-Tool for Sustainable Product Development, Kluwer A.P.
14. Masui, K., Sakao, T., Kobayashi, M., Inaba, A., 2003, Applying Quality Function Deployment to Environmentally Conscious Design, International Journal of Quality and Reliability Management, Vol. 20, No.1, pp. 90-106.
15. Eco-indicator 95 Manual – Prè Consultants, 1995, www.pre.nl.
16. JEMAI, 2004, "Eco-efficiency Indicator Handbook" for Products, http://www.jemai.or.jp

# Using design for environment for redesigning a household appliance

S. Kara, H. Kaebernick and S. Ibbotson
*Life Cycle Engineering and Management Research Group, School of Mechanical and Manufacturing Engineering, The University of New South Wales, Sydney NSW 2052 Australia, S.Kara@unsw.edu.au*

Abstract: Over the last decade, growing consumer awareness of the environment, and the stringent government regulations have forced manufacturers to consider Design for Environment (DFE) during product development. DFE is a systematic approach of incorporating environmental attributes into the design of a product by considering the potential environmental impact of a product throughout its life cycle. Although DFE assessment is generally carried out during the early design phase, it is also a useful exercise to assess the environmental impact of an existing product, and to provide the design suggestions. In this paper, a DFE assessment of a household appliance is presented with the aim to assess the environmental and economical impact of an existing design. Later, the result of the initial assessment is used to provide design suggestions for improving the existing design. The outcome of the study suggests that DFE assessment allows manufacturers to compare different design configurations in order to reduce the environmental and economical impact of a product.

Keywords: Design for Environment, Environmental Impact, Cost Impact, Redesign

## 1 Introduction

The emerging legislations in Europe, the United States, Japan and Australia are forcing companies around the world to implement "polluter pays" policies [1–3]. Two important reasons behind these legislations are to force producers to take responsibility for waste disposal, and the increased awareness of the public, which is demanding access to information about the environmental performance of companies. In addition, shareholders, consumers and employees are increasingly seeking improved environmental and social performance from business.This is consistent with global moves by business, governments and communities



*D. Brissaud et al. (eds.), Innovation in Life Cycle Engineering and Sustainable Development,* 293–302.

towards achieving sustainability [1–4]. The World Business Council for Sustainable Development coined the term 'eco-efficiency' to describe the efforts by businesses to take the environment into consideration in their operations.

For a typical product, 70% of the cost of development, manufacture and use is determined in its design phase. By integrating environmental considerations into the upfront product design, a company can increase efficiency, reduce waste of materials and energy, and take cost into consideration in their operations [4–5]. There are different approaches suggested in the literature to include the environmental impact of a product during the design phase [6]. One of these approaches could be utilised by first setting strategic considerations, then focusing within the product, and finally implementing an appropriate Design for Environment (DFE) tool [7]. DFE is a systematic approach of incorporating environmental attributes into the design of a product by considering the potential environmental impact of a product throughout its life cycle. DFE aims to avoid or minimise significant environmental impacts and increase resource efficiency at all stages of a product's life cycleraw material extraction and processing, manufacturing, packaging and distribution, product use, and end-of-life. One of these tools is the DFE software tool developed by Boothroyd and Dewhurst Inc and TNO Industry Centre in Delft, the Netherlands [8–10]. This software simulates the end-of-life disassembly of the product for the designers and quantifies the economic and environmental affects. In this paper, as part of an industry based project, a DFE assessment of a household appliance, in this case a washing machine is presented by the aforementioned DFE software with the aim to assess the environmental and economical impact of an existing design. Later, the result of the initial assessment is used to provide design suggestions for improving the existing design.

## 2 Utilisation of DFE software

Figure 1 shows the structure of the DFE program. The initial disassembly list of the product can be created in the Disassembly Worksheet window. In addition, it can also be obtained from a previous Design for Assembly (DFA) if available. The materials and manufacturing processes used to produce each item are then selected from the software database.

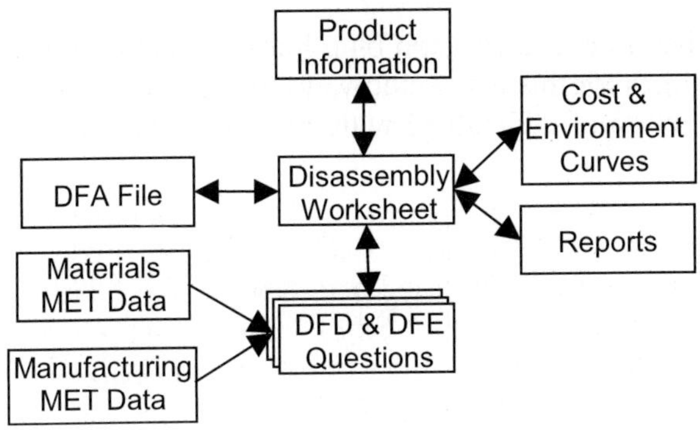


**Fig. 1.** Structure of the DFE software

The end-of-life destination (re-use, recycle, landfill) for each item can be entered as a next step. At this point, the precedence relations for each item need to be entered in order for the software to decide which items must be removed prior to an item. Having entered all these information, the program can determine the best disassembly sequence and associated financial and environmental analysis results.

## 3 DFE analysis of a washing machine

As part of an industry based project, a DFE analysis of a top loading washing machine was carried out in order to find out the environmental efficiency of the existing design. As shown in Figure 2, the washing machine consists of five subassemblies, namely the control panel subassembly, the top cover subassembly, the inner bowl/outer bowl /agitator subassembly, the motor and transmission subassembly, and the wrap cover sub assembly. The washing machine materials are mostly carbon steel and thermoplastic. Therefore recycling is the best option for the majority of the components. However, an assessment methodology developed in this project, suggests that there are five components with reuse potential. These are the solenoid brake motor, gearbox, transmission assembly, electric motor and bearings, which are found in the motor and transmission subassembly [11]. A manual disassembly operation was carried out. During the disassembly, a time-motion study was conducted in order to find out the quantitative disassembly data required by the software. The material and the manufacturing processes data required for the analysis were acquired from the appliance manufacturer. As shown in Figure 2, the initial disassembly sequence, which is simply the reversal of the assembly process, was used for the selected components of the washing machine. It should be noted that only the essential components are included in the Figure 2 for the purpose of clarity. The financial (the top blue line) and environmental (the bottom green line) lines for this disassembly are given in Figure 3. As can be seen, a financial return of AUD80.82 with end-of-life options assumed for each item [12].

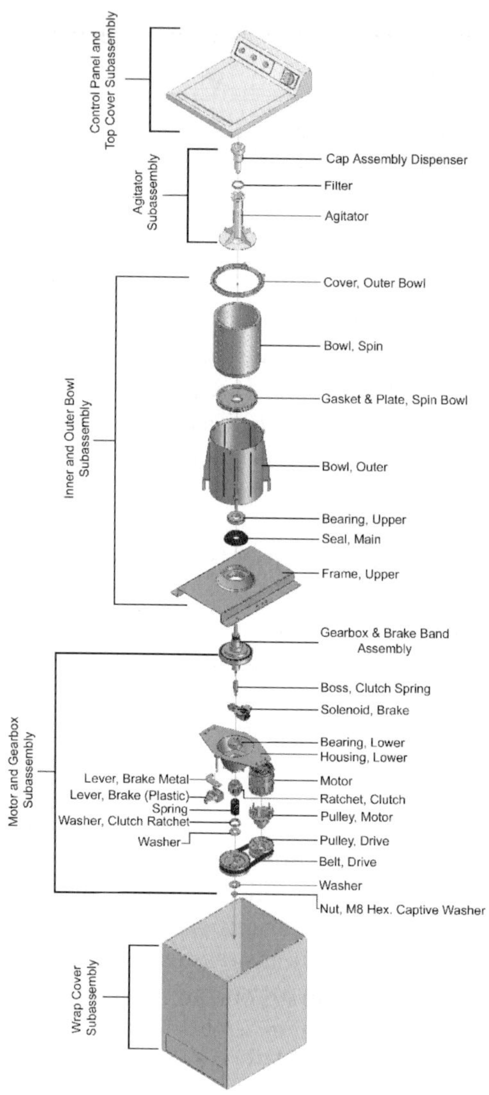


**Fig. 2.** Initial disassembly of the washing machine

The disassembly time was estimated to be 580 seconds, which was obtained by completing the whole disassembly process. The corresponding environmental line shows that the least environmental impact can be achieved by disassembling the whole product, which is consistent with the literature [9].

In order to optimise the initial disassembly sequence, the financial line needs to be investigated carefully. There are three items (solenoid break motor, motor, and the gearbox) with high manufacturing value as indicated by the large steps as shown in Figure 3, which are disassembled at position 32, 42 and 48. These items as well as those with the reuse potential need to be disassembled as early as possible as long as the precedence relations permit.

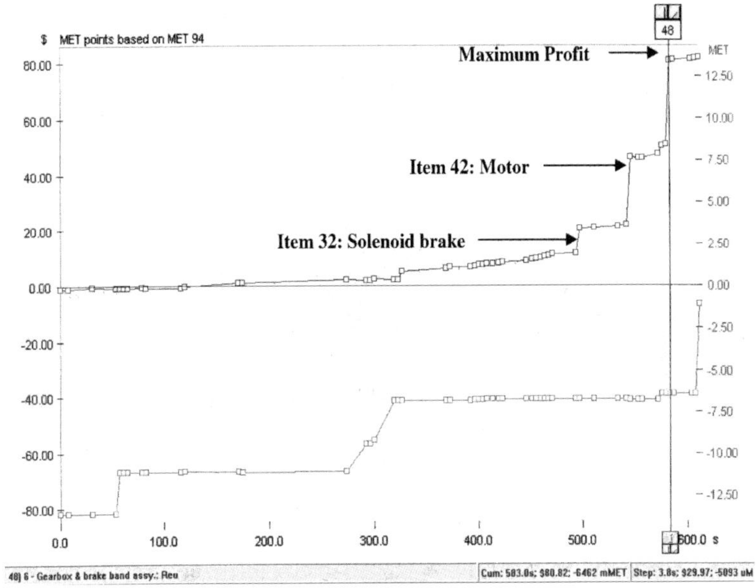


**Fig. 3.** Results for the initial disassembly of the washing machine

The corresponding new disassembly sequence is given in Figure 4. In this case, the disassembly is carried out from the bottom up in order to remove the electric motor, solenoid brake motor and other components. However, in order to remove the gearbox assembly, a top-down disassembly is required since the gearbox shaft is connected to the agitator assembly. The financial and environmental lines for the new disassembly sequence are shown in Figure 5.

**Fig. 4.** Results for the optimized disassembly of the washing machine

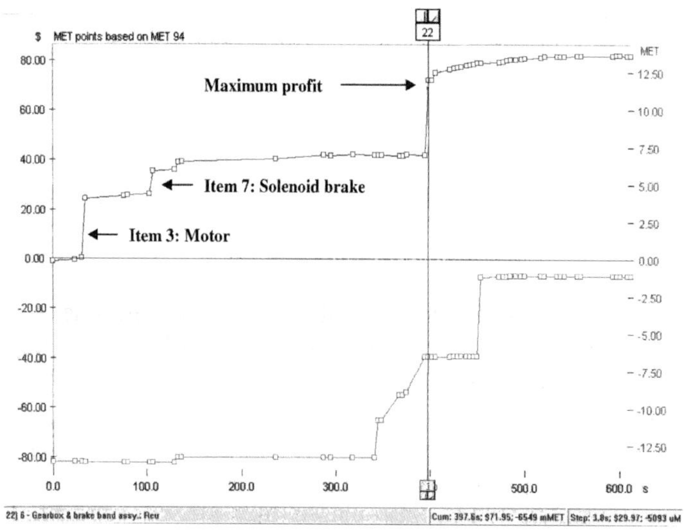


**Fig. 5.** Results for the optimized disassembly of the washing machine

According to the new sequence, the aforementioned components are removed at positions 3, 7 and 22 respectively. The new cumulative disassembly time is estimated 397.6 seconds with the same financial return. As indicated by the green line, the MET point is the same as for the previous sequence since the end-of-life options stay the same.

## 4 Design suggestions

The DFE results suggest that the precedence relations allow the disassembly of the electric and solenoid break motors at positions 3 and 7. However, further investigations revealed that these disassembly sequences are not viable since the existing design does not allow a bottom up disassembly, and there is no clear access to the selected components in an upright position. Furthermore, the disassembly needs to be carried out in an upright position due to safety and ergonomic reasons. That means, the precedence relations alone do not provide sufficient information in order to determine a realistic disassembly procedure. Therefore, the DFE results need to be critically analyzed with the aim to identify valid sequences and to consider possible design improvements.

The key problem for the realization of the optimized disassembly sequence is the restricted access to the selected components. Therefore the first of the simple design improvements is to change the size of the existing back plate, which allows access to these components. By increasing the size of this plate as shown in Figure 6, the optimised disassembly sequence as suggested by the DFE software can be realised. This will allow disassembly of the components with high manufacturing value such as the electric motor, and the solenoid-break motor as early as possible. The other important component with high financial value as indicated by a big step of the financial line in Figure 3 is the gearbox, which represents roughly 25% of the overall manufacturing value of the total product. In the original design, as shown in Figure 7, this component is assembled with the upper frame, sealing, bearing and agitator. Therefore, the control panel and top cover subassemblies, the agitator, the inner/outer bowl subassembly, bearing, sealing, and the upper frame need to be removed in order to access the gearbox at the optimised position 22 as shown in Figure 7. Although, the precedence relations allow the gearbox to be disassembled at position 22 as indicated by the optimised disassembly sequence during the DFE assessment, further analysis reveal that a clamp-bracket system is used to assemble the gearbox to the upper frame as shown in the original design in Figure 7. By considering both the DFE and DFA principles at same time, a new design solution was found as shown in the proposed design section of the Figure 7. The clamping mechanism in the original design was located above the upper frame, not accessible from the bottom of the machine. In the new design, the clamping mechanism is located below the upper frame that means outside the bowl assembly. This new design allows disassembly of gearbox at position 15 without removing the aforementioned components, consequently saving significant disassembly time.

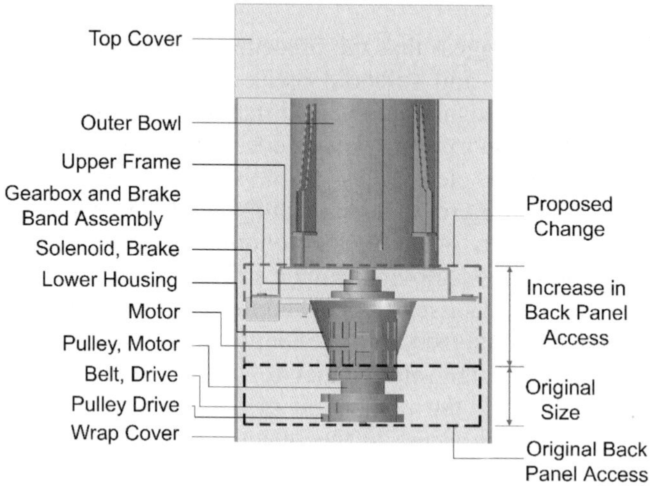


**Fig. 6.** Rear cutaway view of the washing machine

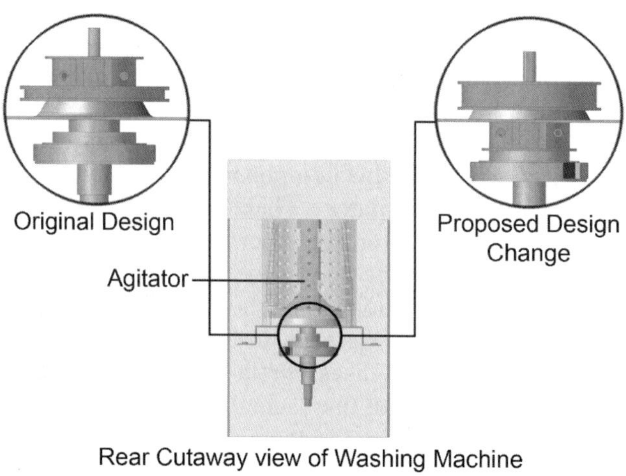


**Fig. 7.** Original and proposed design for the gearbox assembly

## 5 Discussions and conclusions

In this paper, a DFE assessment of a household appliance was presented with the aim to assess the environmental and economical impact of an existing design. Although the DFE assessment is generally carried out during the early design phase, it is also a useful exercise to assess the environmental impact of an existing product, and to provide suggestions for design improvements. The outcome of this case study suggests that the precedence relations alone do not provide sufficient information in order to determine a realistic disassembly procedure. Furthermore, a design solution can be found by considering both the DFE and DFA principles at the same time without compromising any assembly and disassembly processes. Since the washing machine is an active product, the majority of the environmental impact is generated during the usage phase. Therefore, further reduction of the environmental impact requires assessing the functional performance of the washing machine. The washing machine used for this case study is an old generation machine with a belt drive system. The new design concept, which is about to be released by the manufacturer, has a direct drive system. In this design concept, the gearbox will be driven by the electric motor without a pulley system. This will allow two of the most valuable components, namely the electric motor and the gearbox, to be integrated, hence making the disassembly easier.

## Acknowledgement

Authors of this paper kindly acknowledge the Coopeative Research Centre for Intelligent Manufacturing Systems and Technologies for its finacial contribution and Mrs. J. Hoya's contribution to the content of this paper.

## References

1. Lagerstedt, J., 2000, Company International Communication in DFE – A Survey in Germany and Swedish Industry, Proceedings of the 7th CIRP International Seminar on Life Cycle Engineering, 27-29 November, Tokyo, Japan, 203-210.
2. Ishii, K., 1998, Design for Environment and Recycling: Overview of Research in the United Sates, Proceedings of the 5th CIRP International Seminar on Life Cycle Engineering, 16-18 September, Stockholm, Sweden, 191-200.
3. Murtagh, N., Bamba, T., 2000, "A DFE Support Tool Using CAD Model Files for Data Input, Proceedings of the 7th CIRP International Seminar on Life Cycle Engineering, 27-29 November, Tokyo, Japan, 188-195.

4. Kara, S., Sun, M., Kaebernick, 2002, A Trade-off Model for Sustainable Product Development, Proceedings of the 9th CIRP International Seminar on Life Cycle Engineering, 9-10 April, Erlangen, Germany, 103-110.
5. Yarwood, J.M. and Eagan, P.D., 1998, Design for the Environment Toolkit: A Competitive Edge for the Future, Minnesota, USA.
6. Brissaud, D., Zwolinski, P., 2004, End-of-Life Based negotiation throughout the Design Process, Annals of CIRP, Vol. 53/1, 155-158.
7. Hauschield, M. Z., Jeswiet, J., Alting, L., 2004, Design for Environment – Do We Get The Focus Right?, Annals of CIRP, Vol. 53/1, 1-4.
8. Boothroyd Dewhurst Inc., 1996, DFE User Guide.
9. Knight, W., Curtis, M., 1999, Design for Environment Software Development, The Journal of Sustainable Product Design, April, 36-44.
10. Sweatman, A., Simon, M., 1996, "Design for Environment Tools and Product Innovation", Proceedings of the 3rd CIRP International Seminar on Life Cycle Engineering, 18-20 March, Zurich, Switzerland, 119-126.
11. Kara, S., (2003), Reuse and Recycling of Appliances, CRC for Intelligent Manufacturing Systems and Technologies, Project SP3.2 Final Report, Sydney, Australia.
12. Hoya, J. (2002), Design for Environment Analysis of a Household Appliance, BE Thesis, The University of New South Wales, Sydney, Australia.

# Modular design of technical product-service systems

J.C. Aurich, C. Fuchs and C. Wagenknecht
*Institute for Manufacturing Engineering and Production Management, Kaiserslautern University of Technology, Germany, aurich@cck.uni-kl.de*

Abstract : Technical services (maintenance, upgrading, user training etc.) aim at enhancing the performance of investment goods. To systematically exploit their potentials the scope of traditional product engineering methodologies needs to be enlarged. A two step method will be introduced for this reason. It systematically builds up on the principle of modularization for realizing technical Product-Service Systems, i.e. customer solutions comprising both physical (product) and non physical (technical service) constituents. The method covers a process library for designing and manufacturing technical Product-Service Systems as well as a procedure for selecting, combining and adapting appropriate process modules.

Keywords: Life Cycle Management, Design Process, Product-Service System

## 1 Introduction

Life Cycle Management (LCM) aims at optimizing the product life cycle in order to realize sustainable products, i.e. products that result in minimum environmental loads, provide customer satisfaction and allow the manufacturers to make profits (Triple Bottom Line [1]). For this reason it is necessary to consider the economical and ecological effects of each life cycle phase, i.e. design, manufacturing, usage and remanufacturing. For long life investment goods as focused on in this paper the greatest part of their environmental loads results from their usage [2]. The amount of these loads is partially product inherent, i.e. predetermined during product design by the product manufacturer. However, within these product specific boundaries the individual circumstances of product usage resulting from the individual character of the product customer greatly influence the actual loads.



*D. Brissaud et al. (eds.), Innovation in Life Cycle Engineering and Sustainable Development,* 303–320.

In this context, technical -s such as maintenance, contracting, teleservice or user training represent a good means for positively influencing the product usage phase. Especially maintenance has recently been discussed as a powerful method for LCM [3]. Technical services are characterized by direct interaction between staff of the manufacturer and customer, e.g. in terms of service technicians being deployed to the customer. Thus, they provide a link of both spheres and permit information from product usage to be fed back into the manufacturing enterprise. Furthermore, technical services enable the combination of product specific manufacturer knowledge with usage specific customer knowledge in order to increase the economical and ecological product performance. As a consequence, new and advanced product benefits become possible that with respect to investment goods can be described as follows. Investment goods play a crucial role in the value creation processes of the customers e.g. as manufacturing or construction equipment. Corresponding technical services therefore mainly aim at improving these processes e.g. through providing higher equipment up-times or increased productive life times. This very often goes along with reduced energy consumption and waste [4], increased productivity of the corresponding natural resources [5] and a higher level of product benefit in general. Furthermore, technical services provide the manufacturers with a means for differentiating their products from technically similar rival products as well as for flexibly individualizing and enhancing them according to the demands resulting from the geographical or cultural backgrounds of their customers. Thus, new market potentials and higher profit margins become possible. Finally, the realization of technical services at the actual location of product operation, i.e. where required, results in the need for close-meshed service networks and highly qualified service staff. This contributes to the aims of the knowledge society in general as well as to geographically balanced work distribution in specific. Summarizing, technical services represent significant sustainability enablers in terms of life time extension and increased product performance.

## 2 Conceptual framework

Today, the turnover of technical services accounts for about 20 percent of the total turnover of the German discrete part manufacturing industry. Thereby, about half of these services are subject to indirect pricing [6], which means that they are inseparable from the corresponding products. It can thus be followed that products and technical services are together perceived as customer solutions. Such individualized product-service combinations are at the same time discussed as a means for more sustainable production and consumption [7]. Therefore, following the definition of Product-Service Systems provided by Mont [7] as well as the extended product understanding by Jagdev et al. [8] this type of solution will subsequently be referred to as a technical Product-Service System (PSS). A technical PSS consists of a physical product core enhanced and individualized by a mainly non-physical technical service shell that is realized throughout the entire life cycle of the physical product core. Thus the demands of the customers, i.e. product users in terms of a required benefit level can be met. Figure 1 illustrates that the physical PSS core and the non physical PSS shell each consist of several constituents (products, product components and technical services).

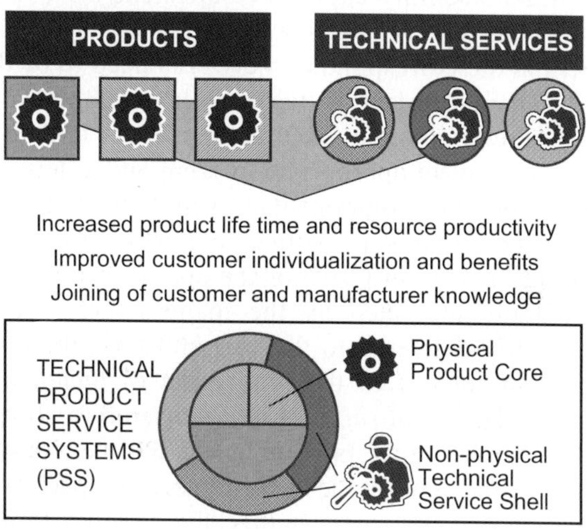


**Fig. 1.** Technical Product-Service Systems

## 2.1 Product and technical service engineering

Providing technical PSS that fulfill the described potentials requires a thorough understanding of the current practices in the investment goods industry concerning product and technical service engineering. Design and manufacturing respectively realization issues will therefore be critically assessed in the following. The success of newly developed or fundamentally adapted products in the discrete parts manufacturing industry strongly relies on the application of systematic engineering methodologies (e.g. Yoshikawa [9], Wheelwright and Clark [10]) as well as corresponding processes (e.g. VDI Guideline 2221 [11]) and methods (e.g. TRIZ, QFD or

FMEA [12]). Even though the same is true for the design and realization of technical services [12], intuitive service engineering approaches are still predominantly applied in practice. A recent study among German enterprises shows that only one third of the respondent manufacturers frequently make use of systematic approaches (e.g. processes and methods) for technical service design [13]. As a direct consequence the individual and changing customer demands cannot be flexibly met. Additionally, a lack of integration can be perceived between the technical services offered by an enterprise which results in unbalanced and incoherent service portfolios [13]. Unsystematic service design also affects product design in the sense that design activities cannot be systematically integrated and thus the mutual interrelations between products and services, e.g. in terms of serviceability and maintainability cannot be sufficiently accounted for. With respect to the realization of products and technical services the roles of the involved partners in the corresponding networks need to be discussed. The high complexity of investment goods usually results in network based manufacturing. Thereby, it is the manufacturers' task to define the product characteristics and to select appropriate internal and external suppliers for the corresponding product components. Even though the manufacturing processes of the external network partners are not directly prescribed, they are significantly influenced in terms of the specified product quality. The components are then integrated by the manufacturer and the resulting products subsequently delivered to the customers via the distribution network partners. With respect to the realization of technical services the situation is similar. Technical services being offered to complement products, the product manufacturer is responsible for their design. Based on these specifications the worldwide service network partners (contractors and independent service branches) realize the services at the location of customer product usage. As different from product manufacturing the quality of technical services cannot be measured physically but mainly depends on the quality of the underlying servicing (service manufacturing) processes [12]. Process definition and standardization is therefore of high importance.

## 2.2 Scenario for technical PSS engineering

The described circumstances of product and service design respectively realization can be used for developing an integrated framework for the design and realization of technical Product-Service Systems, in the investment gods industry (compare figure 2).

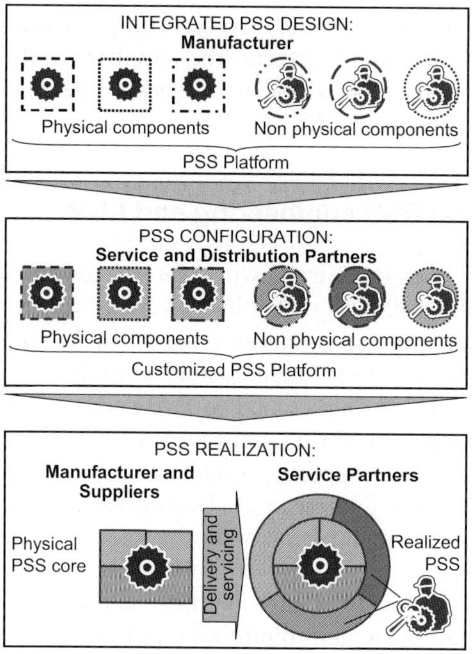


**Fig. 2**. Scenario for technical PSS design in the investment goods industry

Thereby, the manufacturer is responsible for the design of integrated PSS platforms that can be adapted according to the individual demands of the customers. As a result, different alternatives of the physical PSS components (e.g. motor alternatives) and corresponding types of services (e.g. motor maintenance) are specified. In addition to each service type standards for service realization (e.g. use of motor oil of a specific quality) are defined. For actually performing the design of the described integrated PSS platforms an integrated design process based on systematic processes for product and service design is required. Configuration of the resulting PSS platforms is then based on the customers expressing their individual demands to manufacturer representatives, e.g. the distribution or service partners. With respect to the physical PSS components on the one hand the desired alternatives (e.g. motor alternatives) are specified and product manufacturing initiated in the production network. The configuration of the non physical PSS components on the other hand is not restricted to choosing from a predefined set of alternatives. Based on the selected types of services and their corresponding process standards individual services can be designed to complement the physical product core in terms of its non physical service shell. Again, this step requires a systematic service design process to be applied within the service branches. The designed technical services can then be delivered as demanded by the customer and the PSS thus be realized consecutively. The production and service network, comprising the manufacturer, the suppliers and the service respectively distributions branches will subsequently be referred to as the extended value creation network.

## 2.3 Technical PSS engineering and LCM

Based on the presented framework for the design and realization of technical PSS the following conclusions with respect to LCM can be drawn. According to Westkämper et al. [14], LCM aims at optimizing the interaction of product design, manufacturing and life cycle activities in order to protect natural resources and to maximize the effectiveness during product usage. Therefore, LCM subsumes design for life cycle (life cycle engineering), technical support, product data management and life cycle evaluation (assessment and costing). Design for life cycle covers 'the design and manufacture of products with the goal of protecting the environment and conserving resources, while encouraging economic progress, keeping in mind the need for sustainability, and at the same time optimizing the product life cycle and minimizing pollution and waste' [15]. In the context of technical PSS, this implies addressing both the design and manufacture respectively realization of products and technical services (technical support) in an integrated way. The product life cycle is characterized by the collaboration of many different partners. In order for each of them to provide the products and services needed to fulfill the growing demands of the customers, common data is required. Therefore, in terms of product data management, standards for the integrated design and realization of technical PSS in the extended value creation network including the documentation of life cycle information are necessary. Life cycle information is crucial for evaluating the economical and ecological impacts of each life cycle phase. Only life cycle optimized technical PSS be designed and realized. Therefore, information from product usage, i.e. the product customer needs to be available for the manufacturer. Since technical services provide a good means for gathering such information and for systematically building up a corresponding knowledge base information feedback processes need to be considered during technical service design [16].

## 2.4 Resulting fields of action

On the basis of the above remarks the following conclusions for technical PSS engineering can be drawn.

- Unsystematic approaches for technical service design need to be replaced by well-structured design processes with a strong focus on information exchange between manufacturers and their customers.
- Product and technical service design processes need to be aligned in order to provide integrated customer solutions, i.e. technical PSS.
- Flexible process standards need to be established in the extended value creation network considering the allocation of engineering tasks to the partners.

In the following these fields of action will be addressed by means of an integrated concept. Firstly, the systemization of technical service design and a corresponding process for technical service design will be discussed. Secondly, an approach for the integration of product and service design processes will be introduced. Thirdly, the allocating of design tasks to the partners in the extended value creation network will be addressed.

## 3 Technical service design

Existing product design methodologies such as indicated above represent promising starting points for technical service design [17]. Firstly, they are based on a systems approach which makes them suited for technical (products) and socio-technical systems (technical services) alike. Secondly, extending the scope of run-in processes and methods to new problems supports widespread application in industrial practice. Applying product design methodologies for technical service design requires a modification thereof as follows. Starting, an existing product design process needs to be structured with respect to identifying well defined design phases with a homogeneous set of activities and outputs. Parallel, existing technical services need to be analyzed in order to specify the object of technical service design in terms of its properties. Furthermore, technical services being an important source for customer information, a systematic analysis of the internal customer interfaces (e.g. call center, service technicians) is required. Thus, available customer information (e.g. from the machine users) can be matched with demands from product engineering, marketing or other departments. Based on these analysis results the desired outcome of the service design process can be specified. Generic service product models thereby represent a valuable means based on which product design activities can be systematically modified. Thus, it can be provided that the design object (technical service) defines the corresponding design process and not vice versa. As the final step organizational standards such as project management need to be defined and established.

## 3.1 Properties of technical services

Based on the following properties technical services can be completely described. Such descriptions are not only a necessary prerequisite for the systemization of corresponding design processes but even more importantly for the actual design of technical services. Only thus can the output of a design process and its reliability be measured.

- *Objectives:* A general overview of a technical service can be provided by means of defining its objectives in line with the existing service and product strategy. Therefore, its relations with the products and technical services already existing in the portfolio of an enterprise need to be specified.
- *References:* By means of technical services socio-technical systems comprising physical products and/or components thereof as well as product users are affected. A distinction between different services can be drawn in this context. In the case of manipulative or logistical services e.g. products are specifically enhanced, undergo transport or storage whereas in the case of qualifying services persons, such as the product users are being trained. Furthermore, organizational services can address both products and persons. For specifying their effects a clarification of the corresponding service systems, i.e. all relevant partners involved in their realization as well as their interactions is helpful.
- *Functions:* Three basic service functions can be distinguished that need to be detailed for a technical service. They either mainly address the user (external) or the manufacturer (internal). The support function refers to ensuring the guaranteed use of the corresponding product. It is closely related to liability. Requirements fulfillment refers to enhancing this use by means of complementary offers. Finally, information gathering as an internal function aims at providing the manufacturer with information from product usage e.g. user experiences, expectations or suggestions. In this case the user is only indirectly addressed e.g. through resulting product improvements. Based on this distinction the corresponding measures for the realization of the functions (e.g. maintenance activities or cycles) can be specified.
- *Resources:* Both physical and non-physical resources are required for realizing a service. Physical resources can be equipment (e.g. machines, tools, vehicles), supplies (e.g. lubricants, filters) or spare parts. Non physical ones comprise detailed descriptions of the servicing and information exchange processes as well as corresponding instructions, guidelines, forms, databases etc. Furthermore, specifications of the qualification profiles necessary for realizing a service (e.g. expert knowledge, social skills) are comprised therein. Thus, suitable staff can be assigned

and trainings can be initiated. Finally, the organizational circumstances in which services are realized are to be clarified. This especially refers to the duties, decision competencies and tasks of the service staff.

## 3.2 Technical service design process

Systematic processes both for product and technical service design are a crucial prerequisite for the design of technical PSS. For this reason, based on the above approach the technical service design process described in figure 3 has been recently developed in cooperation with several discrete part manufacturing enterprises [18]. The depicted generalized version is thereby based on a reference process for product design suitably modified after Wheelwright and Clark [10]. The first phase starts with the identification of customer demands based on which service targets and requirements can be specified. Next, economical and technical feasibility analyses are to be conducted. The concept development phase starts with the specification of the key parameters of the actual technical service design project, i.e. times and budget as well as responsible persons. Then, potential solutions are to be identified suited for meeting the demands of the customers. Since the identified solutions at the same time represent restrictions for information exchange (e.g. teleservice suggests electronic information exchange) a first information exchange concept needs to be developed. Next, the best suited solution needs to be selected and together with the description of the servicing framework compiled into the service draft. The following service specification phase corresponds closely with the design of physical products. It refers to preparing all documents that describe the technical service, i.e. the service product model, service system model and servicing as well as information exchange process models. The service system model represents the ideal interactions between relevant staff of the manufacturer and customer (e.g. call-center, service manager respectively technician, warehousing, product user). The technical service product model refers to specifying the above described properties. Finally, the servicing and information exchange process models provide a detailed overview of all activities necessary for realizing a technical service. Next, realization planning needs to be conducted especially in terms of resource planning, specification of responsibilities and staff qualification. The latter is especially crucial due to technical services being realized and consumed simultaneously. Finally, the prototypical service has to be tested together with a pilot user. Methods from product design such as TRIZ, Quality Function Deployment (QFD), Failure Mode and Effects Analysis (FMEA) or Service Blueprinting are well suited to support this design process [12].

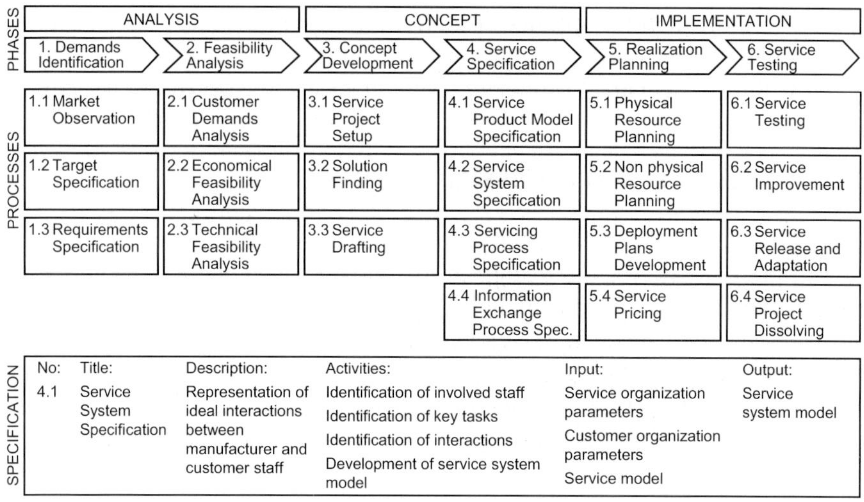


**Fig. 3.** Technical service design process

## 4 Integration of product and service design

In the underlying framework described in figure 2, PSS engineering is based on several partners, i.e. discrete part manufacturing enterprises and distribution respectively service branches collaboratively providing an individualized solution to a customer. Thereby, the physical product core is manufactured by the production network partners (manufacturer and suppliers), the non physical service shell is realized by the service network partners (manufacturer and service branches) and the manufacturer is responsible for the integration of both parts of value creation. In order for the PSS components to fit to each other a common starting point for all involved partners is required in terms of a reference process model for PSS design. This reference model has to provide a common understanding of the design and manufacturing respectively realization processes and the corresponding activities thereof appertaining. At the same time the differences e.g. resulting from the particular geographical or cultural backgrounds of the partners, their different sizes or the demands specified by the customers need to be appropriately considered. This requires the reference process model to be suitably adaptable. Therefore, practically applicable guidelines are required.

## 4.1 Process modularization and linkage

Process modularization is proposed for solving the conflicts between the need for standardized and individual technical PSS engineering processes. Thus, up a "library" of elementary process building blocks for product and service design respectively realization can be built up. On the basis of specific rules these building blocks can be collaboratively used by the network partners for compiling their own individual engineering processes as required. Thus, the manufacturer can e.g. assemble a process for integrated PSS design while the service network partners make use of the same building blocks for customer individual configuration of the resulting PSS platform. Process modularization mainly aims at coping with the complexity of a process. This is especially relevant with respect to the design of a technical PSS consisting of physical and non physical building blocks with multiple interrelations in between. Thus, the transition of traditionally function oriented industrial enterprises with product design, product manufacturing and technical service working parallel and largely independent of each other towards a more integrated form of work where all entities relevant for providing a customer solution truly collaborate is supported. Process modularization results from the decomposition of a process into corresponding modular subprocesses, i.e. process modules (figure 4). A process module is thereby defined as a logically differentiable building block of a process that possesses reference character and is therefore widely applicable. This is possible due to a process module being a "black-box", characterized by inputs and outputs respectively corresponding transformations [19]. In this definition, the inputs and outputs of process modules represent standardized interfaces that allow different process modules to be linked and reused. Linkage is possible if the output of one process module corresponds with the input of another process module. Due to the reference character of the process modules their linkage makes it possible to derive numerous process models as required.The transformations in terms of specified changes of the states of the ingoing system elements (information and resources) result from a cohesive set of activities that can be adapted as required. Due to the black-box principle these activities can be specified on different levels of detail such as appropriate for an application case.

Even though process modules can be identified on all levels of a decomposed process, it can be useful to identify elementary process modules on a rather low hierarchical level because they provide the broadest possible field of application. Based on a process module library individual processes can be designed by means of selecting, adapting and linking appropriate process modules. Thus, with respect to the intended integrated design of technical Product-Service Systems both product and service design process modules are required. Appropriate modules can then be linked by means of matching corresponding inputs and outputs, i.e. ingoing and outgoing information respectively resources. In figure 4 corresponding inputs and outputs are represented by identical geometrical shapes. Based on the described principles the above technical service design process can be modularized. As the first step the principle of process decomposition needs to be applied in order to derive a set of cohesive design activities and corresponding input and output information.

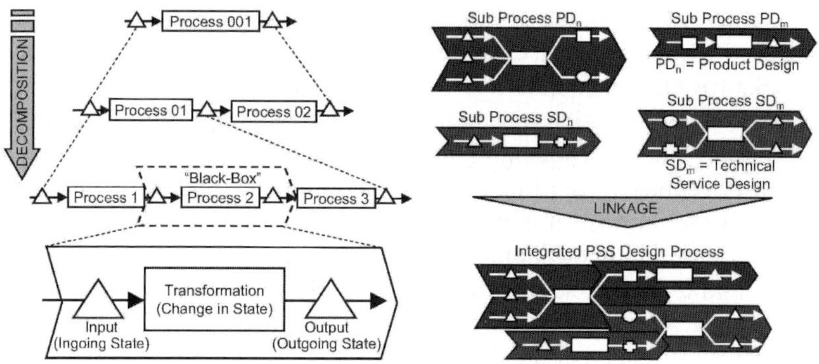


**Fig. 4.** Process decomposition and process module linkage

Since the principle of decomposition primarily refers to the segmentation of transformations, the phases in the above model for technical service design can be taken as the starting point. Input and output information results from analyzing interactions in between different design activities and especially their effects on the corresponding physical and non physical products. An exemplary process module for the specification of the service system is provided in figure 3.

## 4.2 Integrated Process Model

Next, the resulting service design process modules need to be integrated into a process library that supports the modeling of the targeted integrated PSS design process. With the Integrated Production Process Model – IPPM a well suited reference process model concept has been proposed [19]. It is based on the principles of object oriented modeling thus allowing the definition of processes based on relevant process objects, structures, relations as well as required functions and corresponding properties. In the modified version of the IPPM (figure 5) the concept of modularization has been incorporated so that predefined process modules for life cycle oriented process design can be provided. Starting with the focus on internal manufacturing activities the IPPM has already been extended to e.g. consider product design and logistics under the scope of inter organizational collaborations in product design and manufacturing. In light of successful industrial applications additional process modules for service design and service realization (servicing) are to be added in order to come up with a fully life cycle oriented process module library (figure 6).

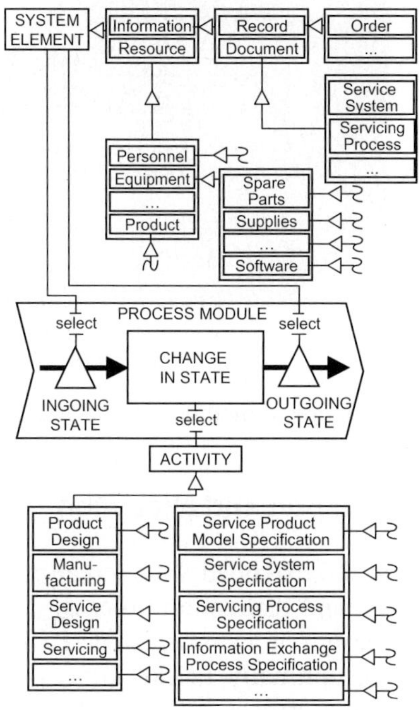


**Fig. 5.** Modified IPPM

The modified IPPM relies on three basic principles for providing the required elementary process modules [20]. The "principle of extension" is based on adding additional activities, resources and information to the predefined set of available design elements. Based on these design elements new process modules can be created thus offering an indirect way for increasing the scope of the module library. The "principle of multiple referencing" leads further as it allows to integrate already existing process models respectively modules. This is useful for flexibly integrating already existing process libraries into the IPPM. The most advanced extension of the modularized IPPM is based on the definition of additional process modules, e.g. for new application domains. The extension by service design process modules represents this approach. Summarizing, process models can be extended on different levels of details and scope. In order to obtain new process modules and to extend the available process module library the concept of modularization typically applies the concepts of extension and multi referencing. That is, both the definition of new process modules as well as the integration of existing process module libraries becomes possible.

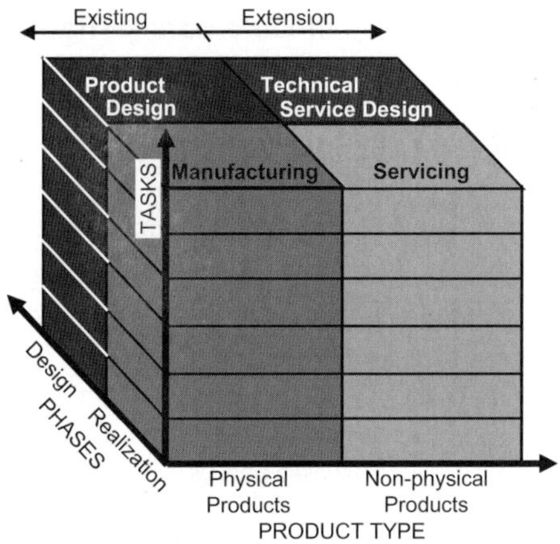


**Fig. 6.** Scope of modular PSS engineering

## 4.3 Process compilation and allocation

The different roles in the extended value creation network result in the need for different design and realization processes. The manufacturer being responsible for the design of a customizable PSS an integrated PSS design process is required, optimally linking process modules for product, service and manufacturing process design. This process can then be adapted by the (international) network partners to account for their individual demands. In detail, the production network partners require processes for designing physical product components that typically consist of process modules for product and manufacturing process design. Furthermore, the service partners require processes for configuring customer individual products and technical services comprising product and service design process modules. Both processes can be derived from the above integrated technical PSS design process. Thereby, it is likely that individual process modules have to be adapted. With respect to the manufacturing of the physical product core corresponding manufacturing processes need to be specified for application by the manufacturer and his suppliers. Thus, a coherent set of physical product components can be realized that fulfill common quality standards. Concerning the realization of the non physical technical service shell corresponding servicing processes (e.g. maintenance, user training) are required. Again, for realizing common quality standards the manufacturer can specify a basic servicing process to be adapted by the local, i.e. international service branches as required. For compiling the required design and realization processes and allocating them to the network partners the following procedure is proposed.

As the first step the design object needs to be described, i.e. the function, structure etc. of the physical product core and the service shell need to be specified. Based thereon the network partners need to agree on the targets of their collaboration as well as their individual roles in the design and realization process. Finally, the design tasks need to be described generically. The second step refers to a top down decomposition of the required design steps. Thereby the modified IPPM is to be used with the aim of modeling design tasks by means of corresponding design process modules. As the third step the process modules need to be assigned to the network partners. Several criteria can be thought of according to which the assignment can take place such as the individual role of the network partners, their market power or their engineering competencies.

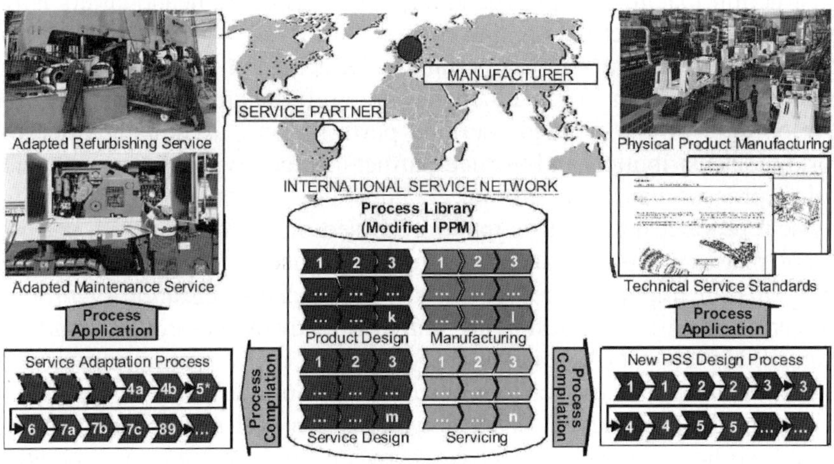


**Fig. 7.** Concept application in the international service network of an investment goods manufacturer

As the fourth step an additional adaptation of the process modules might be needed in case the process modules provided by the modified IPPM do not provide the degree of specification necessary for a practical case of application. Several possibilities exist for adapting a process module. Firstly, one process module (*N*) can be disassembled into two process modules (*Na* and *Nb*). Inversely two process modules (*N* and *M*) can be assembled into a single process module (*NM*). Thirdly, the activities corresponding to the transformations provided by a process module (*N*) can be adapted to form process module $N^*$. In all three cases the input and output information respectively resources need to be left unaltered ("black-box principle"). Finally, as the fifth step a simplification of the resulting process might be useful. However, this step is not obligatory because the linkage of the process modules by means of their corresponding inputs and outputs inherently provides an applicable process. In combination, the modified IPPM and the described steps for task allocation represent a promising method for networked design and realization of technical PSS.

## 5 Case study

Valuable insight into the requirements of investment goods manufacturers concerning the integration of product and technical service engineering could be gained from the implementation of the technical service design process described in figure 3 in a manufacturing enterprise for heavy road construction machines [18]. The enterprise's corporate headquarters is responsible for product design and manufacturing e.g. of road milling machines. Further responsibilities pertain to the design of corresponding technical services such as maintenance or refurbishing as part of the development of the strategic service portfolio and to looking after the worldwide distribution and service partners. Recently, a technical service design process has been introduced equivalent to the process used for physical product design. It is therefore based on a checklist with specific design phases, activities and measurable output. The application thereof represents an important step towards the intended integration of product and service design. The following immediate advantages could already be realized. Firstly, time to market could be reduced by performing product and service design activities jointly where appropriate. This e.g. pertains to customer demands identification as well as technical and economical feasibility studies. Secondly, synergies between products and corresponding technical services can now be systematically realized by optimized information exchange between product and technical service design. For example the design of maintenance processes and the corresponding documentation thereof necessary for supporting the service partners is now performed upon definition of the product layout (e.g. filters, tubings). Furthermore, the design of electronics such as telemetry systems is now partially based on the specified customer information to be gathered during servicing. Positive consequences could also be perceived with respect to the international service partners. One of their main tasks refers to designing locally adapted service processes on the basis of standards provided by headquarters. For example maintenance processes frequently have to be adapted according to the local climate that e.g. affects product corrosion. Refurbishing is another example for necessary adaptations. While in industrialized countries it usually implies the upgrading of certain product components a downgrading is frequently required in developing countries (e.g. exchange of electronic by electro-mechanical components). These modifications are now supported by the service design process, especially by those activities pertaining to service specification and realization planning. Equivalent demands can be observed in other branches of the investment goods industry that rely on similar structures such as agricultural machinery manufacturing.

## 6 Summary and conclusions

Starting with the potentials of technical services concerning life time extension and increased product performance this paper introduced technical product-service systems as an inseparable combination of physical products respectively product components and technical services. On this basis, a scenario for the design and realization of technical product service systems in the investment goods industry has been provided, focusing on the contributions of the manufacturer, the suppliers as well as the distribution and service branches. Furthermore, conclusions with respect to LCM have been drawn and three relevant fields of action identified. Concerning the first field, i.e. the systematic design of technical services a corresponding phase model has been introduced. Upon its modularization and the integration of the resulting process modules into a common process library the linkage with existing product design processes has been described and a promising starting point for integrated technical PSS design, i.e. the second field of action provided. Concerning the third field a compilation process for the allocation of tasks to the network partners has been addressed. Together the process library and the compilation process represent a promising method for integrated PSS design. First applications such as briefly described in a case study show the potentials of the method in the investment goods industry. The presented approach being based on the principles of process modularization and process linkage, special focus needs to be laid on specifying the input and output information of the corresponding process modules for product and technical service design respectively realization. Consequentially, there exists a need for further research pertaining to the systematic investigation of the interrelations between products and technical services as well as their engineering activities.

## Acknowledgements

The research underlying this paper has been supported by the German federal ministry of education and research, BMBF (number 01 HG 0106).

## References

1. Brundtland, G.H., 1987, Our Common Future. World Commission on Environment and Development, Oxford University Press, New York.
2. Lamvik, T., 2001, Improving environmental performance of industrial products through product service systems, Trondheim.
3. Takata, S., Kimura, F., van Houten, F.J.A.M., Westkämper, E., Shpitalni, M., Ceglarek, D., Jay Lee, 2004, Maintenance: Changing Role in LC Management, Annals of the CIRP, 53/2: 1-13.
4. Smith, L., 1999, Reducing the Environmental Impact of Driving – Effectiveness of Driver Training, Proceedings of Ecodrive, Graz: 48-55.
5. http://product-life.org/case_studies.htm
6. Stille F., 2003 DIW-Wochenbericht 21/03 Produktbegleitende Dienstleistungen gewinnen weiter an Bedeutung, Berlin.
7. Mont, O.K., 2002, Clarifying the Concept of Product-Service System, Journal of Cleaner Production, 10/3: 237-245.
8. Jagdev, H.S., Brennan, A., Browne, J., 2004, Strategic Decision Making in Modern Manufacturing, Kluwer Academic Publishing, Boston.
9. Yoshikawa, H., 1989, Design Philosophy The State of the Art, Annals of the CIRP, 38/2: 579-586.
10. Wheelwright, S.C., Clark, K.B., 1992, Revolutionizing Product Development - Quantum Leaps in Speed, Efficiency, and Quality, Free Press, New York.
11. VDI Richtlinie 2221, 1993, Systematic approach to the development and design of technical systems and products, VDI-Handbuch Konstruktion, Beuth Verlag, Berlin.
12. Eversheim, W., Kuster, J., Liestmann, V., 2003, Anwendungspotenziale ingenieurwissenschaftlicher Methoden für das Service Engineering, Bullinger, H.-J., Scheer, A.-W., Service Engineering, Springer Verlag, Berlin Heidelberg New York: 417-441.
13. Spath, D., Zahn, E., 2003, Kundenorientierte Dienstleistungsentwicklung in deutschen Unternehmen, Springer Verlag, Berlin Heidelberg New York.
14. Westkämper, E., Alting, L., Arndt, G., 2000, LC Management and Assessment: Approaches and Visions Towards Sustainable Manufacturing, Annals of the CIRP, 49/2: 501-522.
15. Jeswiet, J.A., 2003, Definition of Life Cycle Engineering, 36th CIRP international Seminar on Manufacturing Systems, Saarbrücken: 17-20.
16. Warnecke, G., Schülke, P., 2000, Design of Preventive Customer Service Processes, Production Engineering, 7/2: 75-78.
17. Bullinger, H.J., Fähnrich, K.-P., Meiren, T., 2003, Service engineering – methodological development of new service products, International Journal of Production Economics, 85: 275-287.
18. Aurich, J.C., Fuchs, C., DeVries, M.F., 2004, An Approach to Life Cycle Oriented Technical Service Design, Annals of the CIRP, 53/1: 151-154.
19. Wagenknecht, C., Aurich, J.C., 2004, Design of modular planning processes using model-based predictive control, CIRP int. Sem. on Manufacturing Systems: 447-453.
20. Warnecke, G., Stammwitz, G., Hallfell, F., Förster, H., 1998, Evolutionskonzept für Referenzmodelle. Industrie Management, 14(2): 60-64.

# Estimating the environmental profile of early design concepts

W. Dewulf, B. Willems and J.R. Duflou
*Mechanical Engineering Department, Katholieke Universiteit Leuven, Belgium,*
*Wim.Dewulf@mech.kuleuven.ac.be*

Abstract: The Eco-PaS methodology supports anticipative weak point analysis of a product's life cycle environmental impact. Using quantitative information on functional parameters available in early design stages, the order of magnitude of an environmental indicator is estimated. This result is obviously influenced by a number of uncertain factors. However, a critical analysis of these factors allows to point the designer at promising design directions at a time of the design process when many degrees of freedom remain available. This paper recapitulates the principles of the Eco-PaS methodology, and applies the method to a design concept of an electric fruit juicer. The validity of the Eco-PaS approach is assessed by comparing the results with a streamlined Life Cycle Assessment study performed on a detailed product design.

Keywords: Eco-PaS, environmental impact screening, conceptual design

## 1 Introduction

A range of product life cycle environmental assessment tools has been developed over the years. However, the major limit of most tools is the need for too much input data in terms of at least a detailed materials inventory. The latter is often not available. For example, during early design phases, only functional requirements and product concepts are available. Nevertheless, decisions taken in the conceptual design phases can influence the outcome of a design exercise far more significantly than any optimisation step later on in the design process [1]. In an eco-design approach an early recognition of favourable system component solutions is therefore of great importance. Moreover, even for fully developed products, a detailed materials inventory is difficult to obtain when many components are bought off-the-shelf. This especially holds true for small and



*D. Brissaud et al. (eds.), Innovation in Life Cycle Engineering and Sustainable Development,* 321–334.

medium-sized enterprises, which can provide little incentives to enforce suppliers to provide the requested data.

# 2 The Eco-PaS methodology

## 2.1 Basic principle

As a solution to these problems, we have introduced the Eco-PaS methodology (Eco-Efficiency Parametric Screening) [2, 3]. Eco-PaS makes use of Eco-cost estimating relationships (E-CERs), defined as mathematical expressions relating an eco-cost as dependent variable to one or more independent eco-cost driving variables. In this framework an eco-cost can be expressed in monetary units, such as external costs or willingness to pay, or by any other commonly used environmental performance indicator, such as, for ex-ample, the Eco-Indicator99 [4] that will be used in this paper. The eco-cost driving variables are functional requirements (FR) or design parameters (DP) that product developers have at hand when designing or selecting components from catalogues. Consequently, parametric expressions are used which express the environmental impact $\xi$ as a function $\xi$ =f(FR, DP). For example, the cradle-to-gate Eco-Indicator'99 score $\xi$ for electric motors can be estimated based on their nominal power $P_{nom}$: $\xi = f(P_{nom})$.

## 2.2 Deriving E-CERs

Three techniques can be used to derive E-CERs: theoretical model development, regression analysis on empirical data and growth laws [5].

### *Theoretical model development*

This first technique can be used for developing E-CER's for relatively simple components for which well-known physical laws or technical formulae lie at the basis of E-CER development. For example, for structural components, the relationship between functional parameters such as maximum loads and deformations on the one hand, and component dimensions on the other hand is known for given load conditions, while the relationship between dimensions and cradle-to-gate environmental impact is straightforward based on geometric formulas.

Consider a material selection problem for a uniformly loaded rectangular plate-like construction with a maximum allowed deformation as main concern. The E-CER will in that case be [3]:

$$\xi_{plate,i} = \sqrt[3]{\frac{\alpha.q.b^4}{y_{max}}}.a.b \;.\; \frac{\rho_i.\xi_i}{\sqrt[3]{E_i}} \tag{1}$$

for α a parameter depending on the plate proportions and the boundary conditions, $y_{max}$ the maximum allowed deformation, q a uniform load, a and b the plate length and width, $E_i$, ρ and $\xi_i$ respectively the Young modulus, the density and the environmental impact score of the plate material i. Note that the plate thickness, which is typically determined late in the design process, is not present in the formula. Moreover, since only the second factor in this formula can be influenced by the material selection, the screening of different material alternatives can be limited to a maximisation of the following ratio:

$$\frac{\sqrt[3]{E_i}}{\rho_i.\xi_i} \tag{2}$$

which only contains structure independent parameters as can be obtained from a material database. Note the similarity between this ratio and the ratios determined by Ashby [6] for minimal weight design.

### *Regression analysis*

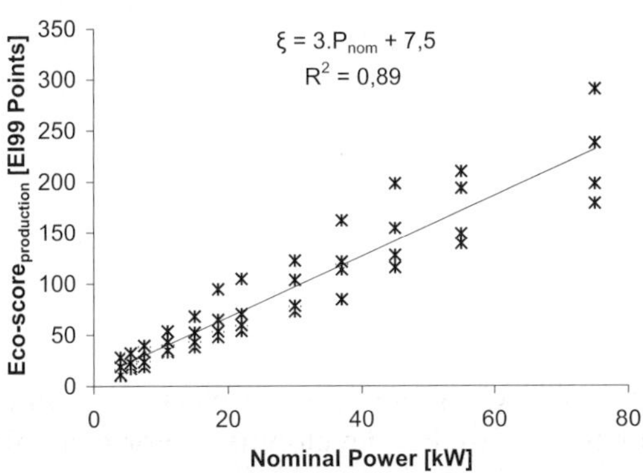


**Fig. 1.** Cradle-to-gate Eco-Indicator99 score for 3-phase induction motors (only taking into account material production impact)

Regression analysis allows deducing E-CERs from empirical data. Dominant eco-impact drivers can be determined for many system components, starting from more detailed LCA output. Purpose of this approach is to identify one or more functional parameters that are sufficiently strongly correlated with the environmental impact created by a category of components in their production and/or utilisation phase. For example, Figure 1 illustrates that for 3-phase electromotors the nominal power can be used as an independent variable to estimate the eco-impact caused during the production of the motor.

#### *Environmental Impact Growth Laws*

Finally, in analogy to so-called Cost Growth Laws [7], Voß [8] introduces the Environmental Impact Growth Law (EIGL) approach for predicting the ecological properties of size ranges. The EIGL approach is suitable when an environmental impact assessment has been carried out for a basic design $D_0$. The environmental impact of a subsequent design $D_i$ is then estimated by multiplying the environmental impact of $D_0$ with a growth factor, which expresses the size of $D_i$ compared to the size of $D_0$.

### 2.3 Uncertainties

While it is well-known that the calculation of a product life cycle's environmental performance is subject to a number of uncertain factors, it is obvious that the estimation of environmental impact using E-CERs is subject to an even higher degree of uncertainty. The E-CERs should therefore be supplemented with information about the importance of uncertain factors. We have previously shown how the different types and sources of uncertainty can be dealt with in a theoretically consistent and structured way using the appropriate statistical formulae [13]. Moreover, fuzzy sets and interval arithmetic are currently being tested as a more practical approach to deal with these uncertainties [14]. The following sections give a concise overview of the basic principles.

#### *LCA related uncertainties*

Life cycle assessment is subject to various sources of uncertainty and variability. Throughout the cause-effect chain from product and product life cycle over emissions to impact and damage, a large number of elements are variable, uncertain, or even just unknown. Huijbrechts ([10], [11]) and Björklund [12] provide in depth surveys of such factors. For example, Huijbregts [10] distinguishes three types of uncertainty (i.e. parameter uncertainty, model uncertainty, and uncertainty due to choices) and three types of variability (i.e. spatial variability, temporal variability, and variability between objects and sources) in the inventory and the impact assessment phases of LCA. In the framework of Eco-PaS, another classification of uncertainties is used: on the one hand are uncertainties that a designer or company can influence (e.g. through material selection, dimensioning, or supplier selection), on the other hand the uncertainties which they cannot influence (e.g. uncertainties due to methodological choices or uncertainties on basic ecological and toxicological models). It is our opinion that eco-design support should make abstraction of the latter type of uncertainties. With respect to the former type, three major sources of uncertainties can be distinguished.

#### *Uncertainties on E-CERs for single component types*

First, uncertainties are introduced when empirically modelling the E-CER using e.g. regression analysis: the original data points will only seldom lie exactly on the eventual regression line (Figure 1). Standard statistical formulae, known from regression analysis, can be used to model this uncertainty range (See e.g. [13, 15]).

#### *Uncertainties on combined E-CERs for different component types*

E-CERs can be defined at different levels of abstraction. On a *low level*, an E-CER can be defined covering a specific range of components, e.g. the motors of one specific type and one specific supplier. E-CERs of this type are typically used in a detailed design phase. On an *intermediate level*, an E-CER can be defined covering the motors of one specific type but including all suppliers, thus supporting embodiment design decisions. On a *high level*, a generic E-CER can be defined covering the full range of electric motors. These high-level E-CERs will be needed in the conceptual design phase. They can be derived as a weighted average of a set of low-level E-CERs. Obviously, the uncertainty on the high-level E-CERs will be significant: the broader the scope of the E-CER, the less certain we are about the eventual result of the analysis. Thus, this type of uncertainty is a result of the mere fact that many decisions have not yet been taken in early design phases. We have previously exhibited that the uncertainty on the high-level E-CERs can be calculated using statistical formula known from joining two or more populations. For a more elaborate explanation, we refer the reader to [13].

#### *Uncertainty on the independent variables*

Uncertainties on the independent variables are the third source of uncertainty covered by the Eco-PaS methodology. They are typical to early design phases, when many decisions have not yet been taken. For example, in the electric motor case, the actually required nominal power will be uncertain until embodiment design of neighbouring systems has been completed. Similarly, the exact use and disposal scenarios are often unknown to the designer. Consequently, the functional parameters or design parameters used as independent variables of the E-CER will not be crisp numbers, but ranges.

***Using fuzzy sets for representing uncertainties***

Traditional statistics can offer an opportunity to develop a theoretically and mathematically consistent framework for uncertainty handling. However, statistics are basically dealing with probabilistic problems, while many issues addressed in the previous sections – especially with respect to independent variables – are intrinsically possibilistic. Consequently, an approach using fuzzy sets and interval arithmetic is being developed as an alternative approach. A detailed description of the fuzzy set approach is outside the scope of this paper. Nevertheless, the case study presented in the next section was performed using this fuzzy set approach.

## 3 Case study: an electric fruit juicer

In order to validate the Eco-PaS methodology, a concept for an electric fruit juicer has been modelled and the results of the Eco-PaS analysis are compared to the results of a screening LCA of a commonly available model. The case study covers both the E-CER development and the use of E-CERs for design concept analysis. In real-life situations, the designer would only need to do the latter task, while method developers, consultants and academia execute the former task, as is the case for regular LCA database development.

### 3.1 Product model

The concept of the electric fruit juicer has been modelled based on five functional blocks: press tip, jug, electric motor, housing and cable (Figure 2). Moreover, the electricity consumption in the use phase is modelled. Cleaning as well as the growing and transportation of oranges are excluded from the system boundaries. The EcoIndicator'99 has been retained as the environmental indicator $\xi$[Pt].

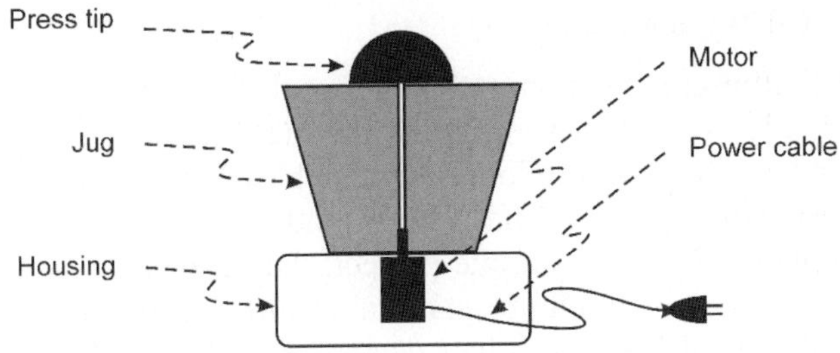


**Fig. 2.** Conceptual representation of an electric fruit juicer

### *The cable*

Due to the existence of the IEC227 standard on electrical power cables, it is relatively easy to determine the material content. An E-CER was developed for the IEC227-HO3VVH2-F type of cable, which is commonly used for houshold appliances such as radios, hand mixers and fruit juicers:

$$\xi = 0.005 + 0.025 \cdot L \tag{3}$$

where the length of the cable L[m] is the only input parameter required. Since the E-CER is valid for only one, well-standardised type of cables, the coefficients of the equation are crisp.

For our example project, the designer makes use of this E-CER to model the power cable of his fruit juicer design. He estimates that the total cable length needs to be ca. 1 meter, or between 0.8m and 1.2m. This is modelled as a triangular fuzzy number.

### *The electric motor*

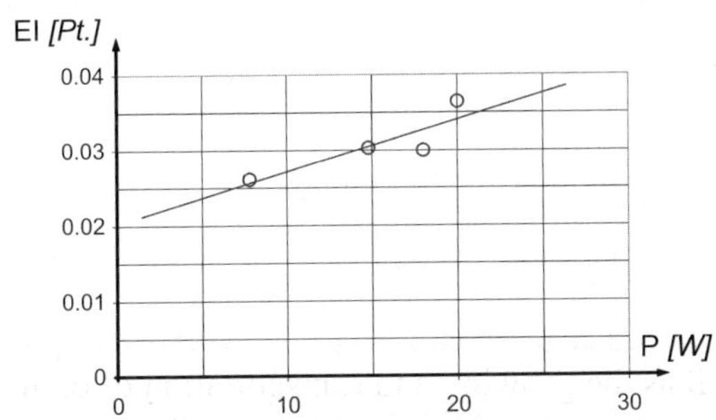


**Fig. 3.** E-CER for small electric motors expressing the environmental impact as a function of the input power [8]

Voß et al. [8] used the Environmental Impact Growth Law approach to develop an E-CER for producing small electric motors (Figure 3):

$$\xi = 0.020 + 0.00065 \cdot P \tag{4}$$

where P[W] represents the nominal power of the electric motor.

However, Figure 3 clearly indicates the presence of an uncertainty interval around this equation. Consequently, based on a variance analysis, the coefficients of Equation 4 have been modelled as fuzzy sets instead of as crisp numbers. For our example project, the designer makes use of this E-CER to estimate the cradle-to-gate environmental impact of producing the electric motor of his fruit juicer design. Based on a short overview of other kitchen appliances and of competing products, he estimates the need for ca. 20W of electrical power, being at least 15W and at most 30W. This is modelled as a triangular fuzzy set.

### *The jug*

In many design projects, recipients need to be developed that can contain a specified amount V[m³] of gas, liquid or bulk material. Other important design parameters influencing the eventual environmental impact are the material, the shape and the thickness of the recipient. We have proven that – for each type of recipient – the environmental impact can be modelled as [14]:

$$\xi = \gamma \cdot V^{0.66} \cdot t \cdot \kappa \tag{5}$$

where:

- $\gamma$[-] is a dimensionless parameter, representing the influence of the recipient's shape on the environmental impact score: obviously, a spherical recipient has a lower $\gamma$ than a cubical alternative with similar thickness and volume contents. Table 1 shows the value of $\gamma$ for selected recipient shapes;
- $\kappa$ [Pt/m³] is a material parameter, being the product of the specific environmental impact score found in many databases (e.g. the Eco-Indicator lists [4]) and the material density;
- t[m] represents the wall thickness of the recipient. Obviously, this parameter is typically not available in early design phases. Nevertheless, for a first impact screening, it doesn't seem unreasonable to ask the designer for an estimation of the wall thickness.

In our example case, the designer considers only plastics, which is modelled as a $\kappa$ value ranging from 300 to 600 Pt/m³ (Table 2). A wall thickness between 0.5mm and 2mm is supposed. Since the shape has not been determined yet – he only knows that the top side of the jug should be open – the $\gamma$ parameter ranges from 3.8 to 6.4 (Table 1).

When the designer would have selected a cylindrical shape with diameter equalling the height (which is a good approximation for many fruit juicers commonly found on the market), the $\gamma$ value would be restricted to the crisp value 4.6. Consequently, postponing the exact shape determination of the jug introduces an uncertainty of only +/- 30% on the result for the jug.

**Table 1.** Value of the shape parameter $\gamma$ for recipients

| *Shape* | *Shape parameter $\gamma$* | *Remarks* |
|---|---|---|
| Cube | 5.0 | One side open |
| Half sphere | 3.8 | One side open |
| Cylinder (One side | 4.4 | Diameter = height*2 |
| open) | 4.6 | Diameter = height |
| | 5.2 | Diameter = height /2 |
| Cone | 6.4 | Diameter = height *2 |
| (One side open) | 4.6 | Diameter = height |
| | 4.0 | Diameter = height /2 |

**Table 2.** Value of the material parameter $\kappa$ for recipients

| *Material* | *Material parameter $\kappa$* | *Remarks* |
|---|---|---|
| ABS | 428 $Pt/m^3$ | |
| PC | 600 $Pt/m^3$ | |
| PP | 300 $Pt/m^3$ | |
| PS | 420 $Pt/m^3$ | |

### *The housing*

On first sight, the housing and the jug are similar elements. However, while the functional parameter for the jug is clearly the enclosed volume, the functional parameters of the housing are the length, width and height of the volume to be enclosed. Consequently, the E-CER has the following form:

$$\xi = \gamma' \cdot (w \cdot l + w \cdot h + l \cdot h) \cdot t \cdot \kappa \tag{6}$$

where:

- $\gamma'$[-] is a dimensionless parameter, representing the influence of the housing's shape on the environmental impact score. Note that $\gamma'$ is different from $\gamma$ used in Equation 5, although it fulfils a similar function;
- $\kappa$ [$Pt/m^3$] is the material parameter;
- t[m] is the wall thickness of the housing;
- w[m], l[m] and h[m] are respectively the width, length and height of the envisaged housing.

In the fruit juicer example case, the designer again considers only plastics, modelled as a κ value ranging from 300 to 600 $Pt/m^3$. A wall thickness between 0.5mm and 2mm is supposed. Since the shape has not been determined yet, the γ' parameter can range from 1.2 to 2. Finally, the designer estimates length and width to be between 15cm and 30cm and the housing height to be between 3cm and 6cm.

#### The press tip

Since the press tip needs to fit the shape of common citrus fruits, its dimensions are the functional parameters. Consequently, the E-CER of Equation 6 can be reapplied. The press tip is modelled as an open half-spherical volume ($\gamma = 0.8$) with diameter $b = l = h/2$ of $10 \pm 1$cm. The designer knows the tip will be made of plastic ($\kappa = 300$ to $600$), with a wall thickness between 0.5mm and 2mm.

#### The use phase

Finally, the use phase is modelled. Since the environmental impact of producing one kWh of electricity equals 26mPts (average for Europe), the environmental impact of the life cycle equals:

$$\xi = P \cdot L \cdot 0.000026 \qquad (7)$$

where:

- P[W] is the electrical power consumption of the fruit juicer. As stated above, the designer estimates a need for ca. 20W of electrical power, more specifically: at least 15W and at most 30W;
- L[h] the total life time usage of the product. The designer estimates a pessimistic scenario of 75 hours and an optimistic scenario of maximum 200 hours;

## 3.2 Results and validation

Based on the above E-CERs and input parameters, a fuzzy estimation of the environmental impact of a fruit juicer is obtained. The interpretation of this fuzzy set leads to a defuzzified 'average' estimated impact of 0.35 Pts (based on a point of gravity defuzzification approach) with an uncertainty interval of [0.13 0.48] at membership level $\alpha$=0.5 of the fuzzy set. Next to the total estimated impact of the fruit juicer, Eco-PaS also provides an analysis of the result and of the related uncertainties (Figure 4).

From these figures, the designer learns that the housing is a crucial element in the eco-design of an electric fruit juicer, since it combines a high contribution to the overall impact (Figure 4a) with a very significant contribution to the overall uncertainty of the result (Figure 4b). The latter indicates that this component has a high improvement potential. In the second place, the designer will need to focus on the jug design as well as on elements influencing the total energy consumption. However, in contrast to most active products, the electricity consumption of the juicer is not the first concern in this eco-design process. Finally, the motor and the cable have important contributions to the total environmental impact (Figure 4a), but represent very little improvement potential (Figure 4b). Of course, this conclusion is only valid as long as no other, radically different solutions are selected that were not yet represented in the utilised E-CERs.

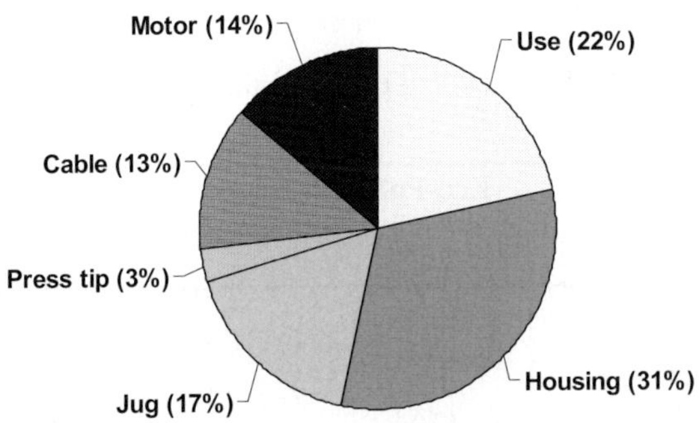


(a)

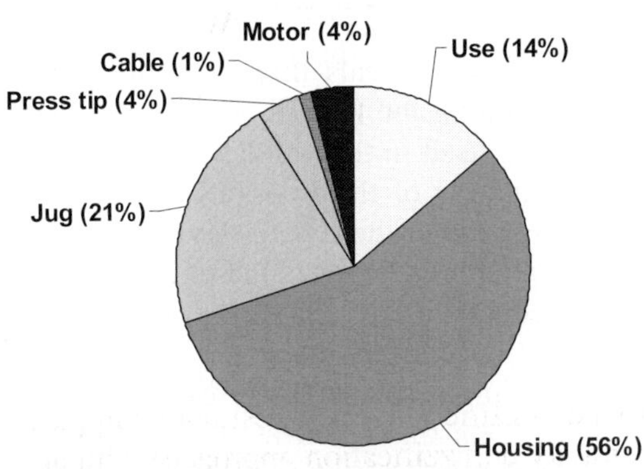


(b)

**Fig. 4.** Result of an Eco-PaS analysis of a fruit juicer design concept: (a) contribution of the building blocks to the estimated impact and (b) contribution of the building blocks to the overall uncertainty on the estimated impact

A screening LCA of a typical electric fruit juicer was made in order to validate the results of the Eco-PaS analysis (Table 3). The screening LCA score for the product's life cycle amounts to 0.315 points, which fits very well in the possibility range indicated by the Eco-PaS results. Nevertheless, the application of Eco-PaS using readily available E-CERs requires only a few minutes time and very little input data, while the streamlined LCA including full disassembly and weighting of material fractions easily requires a few hours' work as well as the availability of a fully developed product. When analysing the overall result more thoroughly, it proves that for most components, the share in the overall environmental impact indicated by the LCA and Eco-PaS is similar. Two main exceptions appear to be the housing and the electric motor.

**Table 3.** Comparison of the Eco-PaS and the screening LCA results for an electric fruit juicer

| Part | Eco-PaS<br>Estimated environmental impact | Screening LCA<br>Calculated environmental impact |
|---|---|---|
| Motor | 14% | 40% |
| Cable | 13% | 11% |
| Press tip | 3% | 4% |
| Jug | 17% | 19% |
| Housing | 32% | 12% |
| Product use | 22% | 14% |
| ***Total*** | ***0.35Pts*** | ***0.315 Pts*** |

Scrutinising the electric motor reveals that the type of electric motor used by Voß [8] when developing the E-CER used in this case study was very different from the model used in the validation product. This conclusion points at the basic weakness of the Eco-PaS approach: E-CERs are only valid for the component range they were developed for, and should not be extrapolated. Consequently, a larger range of differing motors needs to be assessed in order to deduct a more reliable E-CER. Obviously, this E-CER will have a much larger uncertainty range.

The deviation on the housing result is easily explained by taking into account the large uncertainty on this result as depicted in Figure 4b. As pointed out before, the designer has in this case the opportunity to significantly influence the environmental profile of the product by making appropriate eco-design choices.

## 4 Summary

This paper applied the Eco-PaS methodology on a conceptual representation of an electric fruit juicer. The validity of the Eco-PaS approach is assessed by comparing the results with a streamlined Life Cycle Assessment study performed on a detailed product design. This comparison shows that Eco-PaS allows to provide proper estimates and weak-point analysis based on a minor amount of data and little time investment. Moreover, analysing the remaining uncertainty on the Eco-PaS result allows identifying not only the major contributors to the overall impact estimate, but also the primary improvement potential. However, the results also point at the basic challenge of the Eco-PaS approach: E-CERs are only valid for the component range they were developed for, and should not be extrapolated. Consequently, future research should concentrate on developing a large set of E-CERs for many different component ranges.

## Acknowledgement

The authors would like to recognise the support of the Flemish government through IWT (Institute for the Promotion of Innovation by Science and Technology in Flanders) in the framework of the postdoctoral research program.

## References

1. Andreasen, M.M., Hein, L., 1987, Integrated Product Development, IFS Publications, Bedford.
2. Duflou, J., Dewulf, W., Al-Bender, F., Sas, P., Vermeiren, C., 2002, Parametric Eco-Efficiency Analysis: A DfE Support Tool, Proc. of 9th CIRP International Seminar on Life Cycle Engineering, Erlangen, April 9-10, 121-127.
3. Dewulf, W., 2003, A Pro-Active Approach to Ecodesign: Framework and Tools, Ph.D. Dissertation 2003D06, Katholieke Universiteit Leuven, Mechanical Engineering Department, Leuven.
4. Goedkoop, M., Effting, S., Collignon, M., 2000, The Eco-Indicator99 - Manual for Designers, Pré Consultants, Amersfoort.
5. Dick, M., Dewulf, W., Birkhofer, H., Duflou, J., 2004, Estimating the environmental impacts of similar products, Proc. of International Design Conference - Design 2004, Dubrovnik, May 18-21, 1515-1520.
6. Ashby, M., 1992, Materials Selection in Mechanical Design, Butterworth Heinemann, Oxford.
7. Pahl, G., Rieg, F., 1984, Kostenwachstumsgesetze für Baureihen (Cost Growth Laws for Size Ranges), Hanser, Munich.

8. Voß, M., Birkhofer, H., 2004, Using Similarity Relations for Determining Environmental Impacts, Proc. of Global Conference on Sustainable Product Development and Life Cycle Engineering, Berlin, September 29 – October 1, 399-402.
10. Huijbrechts, M., Application of Uncertainty and Variability in LCA, Part I: A General Framework for the Analysis of Uncertainty and Variability in Life Cycle Assessment, International Journal of Life Cycle Assessment 3 (5), 273-280.
11. Huijbrechts, M., Application of Uncertainty and Variability in LCA, Part II: Dealing with Parameter Uncertainty and Uncertainty due to Choices in Life Cycle Assessment, International Journal of Life Cycle Assessment 3 (6), 343-351.
12. Björklund, A.E., Survey of Approaches to Improve Reliability in LCA, International Journal of Life Cycle Assessment 7 (2), 64-72.
13. Dewulf, W., Duflou, J.R., 2004, Estimating environmental impact scores and related uncertainties using the Eco-PaS methodology, Proc. of Electronics Goes Green 2004+, Berlin, September 6-8, 447-452.
14. Dewulf, W., 2004, Verfijnen en toepassen van de pro-actieve eco-design methode Eco-PaS met bijzondere aandacht voor het werken met onzekerheden (Refining and applying the pro-active eco-design method Eco-PaS while paying special attention to dealing with uncertainties), K.U.Leuven internal working document, (in Dutch).
15. Neter, J., Kutner, M., Nachtsheim, C., Wasserman, W., 1996, Applied Linear Statistical Models, 4th Edition, IRWIN, Chicago.

# Product Life Cycle Management

# Design for environment by target life cycle costing

D. Janz, M. Hornberger and E. Westkämper
*Dept. of Product and Quality Management – Life Cycle Management Team, Fraunhofer Institute of Manufacturing Engineering and Automation (IPA), Stuttgart, Germany*

Abstract: Although many methodologies and tools have been developed to design environmentally friendly products, life cycle cost aspects have barely been included and likewise there has been no significant discussion on how to effectively combine different tools. There are specific tools supporting environmentally friendly design such as LCA (Life Cycle Assessment), and tools supporting cost optimisation such as Target Costing and Life Cycle Costing. A combination of these tools is necessary to minimise costs along the whole product life cycle while simultaneously fulfilling environmental requirements. The Target Life Cycle Costing (TLCC) method was developed by the Fraunhofer IPA, concentrating on Design for Environment and on optimising product functionalities by decreasing environmental costs. This paper analyses how designers can optimise costs and conduct product redesign by Target Life Cycle Costing, providing case studies on two products: a medical appliance and a washing machine. The analysis shows that implementing TLCC facilitates cost reduction by product redesign considering functional roles of product components from both an environmental and an economic point of view. This type of combination supports designers effectively, whereas a single tool could not achieve the same results.

Keywords: Design, Environment, Lifecycle

## 1 Initial situation

For a start, businesses are expected to create sustainable value both for their owners and for other stakeholders including state authorities and the general public. After deduction of expenses, their activities are to turn a profit. At the same time, the pressure on businesses is growing to keep the burden they place on the environment under control and even reduce it. Tougher statutory regulations (e.g. Directive on End-of-Life Vehicles and Directive on Waste Electrical and Electronic Equipment) and changing



*D. Brissaud et al. (eds.), Innovation in Life Cycle Engineering and Sustainable Development,* 337–348.

market requirements force companies to deal thoroughly with the recycling and disposal of their products. The directives and rules mentioned above aim at expanding the traditional role of the manufacturers and holding them responsible for the disposal of post-consumer products. This approach concerns the anticipated profit and cost of an organisation as well as its assessment of risks and opportunities. Producers are to develop concepts for the design and manufacture of products which ensure a maximum life span and – at the end of a product's life cycle – allow employing the most efficient recycling or disposal method. Extending the producer responsibility raises questions of how to optimally allocate costs and revenues from development, manufacture, use and recycling.

## 2 Solution concept

Target Life Cycle Costing (TLCC) takes a holistic viewpoint (life cycle perspective). It assesses the total cost of acquisition, ownership and recycling/disposal of a product and helps evaluate and optimise a product's life cycle cost, comparing it to the market-oriented target costs while satisfying specific environmental targets as well as performance, safety, reliability, maintainability and other requirements. Information relevant for the control of operations is captured and edited throughout all product life cycle stages. Recycling or disposal processes form an integral part of the perspective. A major advance of the holistic perspective is that all project-related activities can be clearly allocated, forecasted and analysed. Thus long-term cost reductions are possible. TLCC is a market-oriented cost management tool based on the strategic cost management concept. It delivers an answer to one of the most important business questions: What resources to use in order to be competitive and at the same time fulfil environmental requirements? [1]. All this requires businesses to position themselves against their competitors and use an appropriate 'product-market-resource strategy' to better focus on customers and markets. An important issue affecting market-orientation is the function of resource planning and allocation. The analysis of resources as part of product and marketing considerations has to include environmental aspects and alternatives both for the use of secondary raw material and the reuse of product components. This allows to influence the resource structure at the earliest possible point, creating a relationship to costs. The costs for delivering products and services are determined by cost drivers (e.g. factor quantity, factor quality, factor price, production programme, manufacturing process, etc.). Cost management strives for the optimisation or minimisation of cost drivers. There are three possible leverage points: cost level, cost trend and cost structure [2]. The TLCC as a strategic cost management tool focuses on cost aspects related to basic strategic questions. In the context of the producers' extended responsibility for the environmental aspects of their products, it is important to have leverage over costs and cost drivers to ensure long-term success potential throughout the product's life cycle. The major features of TLCC are its long-term perspective on resource consumption and environmental impacts. Accordingly, all costs are regarded as variable. TLCC as a strategic cost management tool is in line with the statutory and market requirements and includes a full costing approach. The aim of TLCC is to equally fulfil market and environmental objectives. Therefore, the business perspective comprises the complete value chain including the own company as well as suppliers, customers, recyclers and disposal companies. Strategic cost management requires to act within a relatively open framework, taking a great number of action parameters into account. The focus is therefore not on accuracy, but on the robust and easy use of this method.

# 3 The TLCC approach

## 3.1 Basics

The TLCC method combines the functions of resource procurement and resource consumption with resource recovery. Similar to the Target Costing approach, TLCC includes measures to impact costs at the earliest possible point in time [3]. Unlike traditional Target Costing, it includes both the companies involved in product making and those involved in the downstream phases of use and recovery and makes them part of its cost analysis. The demands these actors place on the product are accordingly incorporated. Moreover, this approach takes the requirements of environmental laws into account. The TLCC concept includes the following procedural steps:

- Definition of the product's functional structure,
- Weighting of product functions,
- Development of a conceptual product design to determine components,
- Estimation of the number of components,
- Weighting of components with regard to their functional contribution,

- Evaluation of main environmental impacts at component level and calculation of the respective costs,
- Calculation of historical product life cycle costs,
- Formulation of the target life cycle costs reduction,
- Optimisation of components (redesign) based on the integration of target life cycle costs.

The TLCC analysis provides an overview of the economic aspects of a product while considering its environmental impact. The primary objective is to provide input for decision-making especially in the early phases of a product's life cycle [4]. Unlike Life Cycle Assessment, it does not offer a detailed and overall picture of the environmental impacts of a product. TLCC helps to choose the most economic design solution while including the environmental cost in the costs of acquisition, ownership and disposal. For example, TLCC analyses how the replacement of raw materials by secondary raw materials impacts the cost of acquisition, ownership and disposal. Total costs for environmental impact activities and cost for energy consumption are calculated as separate cost categories before including them in the life cycle costs. After calculating the complete product life cycle costs, the market-based target costs are determined. An analysis of the life cycle cost allows identifying the cost drivers before deriving starting points for optimisation from the difference between costs incurred and target costs. The following practical implementation aims to minimise costs mainly by redesigning products. An important feature of this optimisation concept is that it focuses on functions [5].

# 4 Implementation of the TLCC-approach

## 4.1 Using TLCC on a medical appliance

The considered medical appliance is a portable oxygen concentrator. Its functional structure contains frequent and critical functions as well as recycling-oriented functions. Figure 1 shows the most relevant items in the functional structure as the results of a market analysis conducted by the product manufacturer. The weighting scale includes 3unit-steps, starting with 0 and ending with 9.

| Functions | | Weighting |
|---|---|---|
| Operating, frequent functions | placing the device to correct location (mouth) | 9 |
| | gripping / holding the device | 9 |
| | removing the device | 9 |
| | storing the device | 9 |
| | detecting measurement complete alarm signal | 6 |
| | reading display | 6 |
| | easy cleaning and desinfection of device | 9 |
| | easy filter changing | 9 |
| | mobility | 9 |
| Critical operating functions | electronic oxygen convertion | 6 |
| | ventilation pressure regulation | 9 |
| | relative humidity regulation, non condensing | 9 |
| | temperature regulation | 9 |
| | background sound pressure level | 6 |
| | respiratory frequency regulation | 9 |
| | convertion of electric power to mechanical power | 6 |
| | functional checks | 9 |
| | indication of defects or deviation | 9 |
| Recycling/ Reuse | easy dismantling of components | 9 |
| | easy reassembly of components | 9 |
| | use of recyclable materials, non-usage of hazardous substances | 9 |

**Fig. 1.** Most important items in the functional structure of the oxygen concentrator

Based on this structure the assemblies and components which support the listed functions have been determined. This way, the product was broken down into 293 components. According to the functional contribution of components, nine component clusters were created as shown in figure 2.

| Functional importance | Cluster | Nr. of components | Direct costs | Total costs |
|---|---|---|---|---|
| 9 | i | 4 | 519,01 € | 691,12 € |
| 9 | h | 1 | 95,00 € | 95,00 € |
| 9 | g | 2 | 54,44 € | 54,44 € |
| 9 | f | 9 | 89,84 € | 110,72 € |
| 6 | e | 8 | 42,79 € | 49,69 € |
| 6 | d | 30 | 60,05 € | 91,31 € |
| 6 | c | 27 | 35,40 € | 38,20 € |
| 3 | b | 7 | 5,13 € | 5,13 € |
| 3 | a | 38 | 6,29 € | 10,46 € |
| **Total** | **Total** | **126** | **907,95 €** | **1.146,07 €** |

**Fig. 2.** Component clusters

The next step includes the evaluation of the main environmental impacts at component level. As environmental impacts we considered the energy demand of components, the required dismantling time and the incurred recycling/recovery costs. By concentrating on mechanical components and assemblies, it was possible to determine the cumulative energy demand.

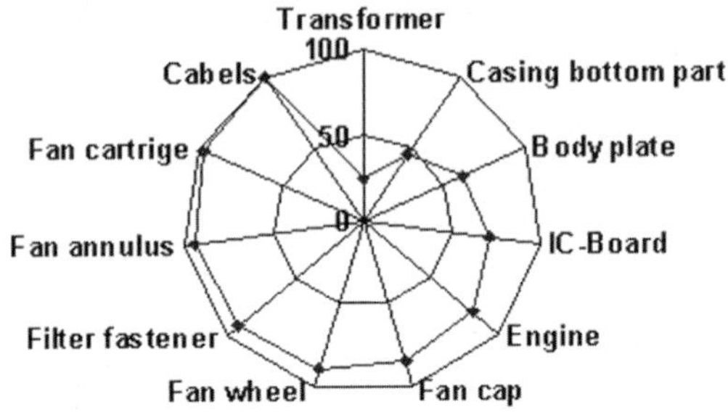


**Fig. 3.** Energy balance at component level

There is a close relation between the energy demand of components and recycling measures since the processes for making new products are substituted by treatment processes enabling their reuse [6]. In order to comply with environmental requirements the goal was to raise the recycling rate through product redesign without increasing the engineering and economic efforts and the energy demand. The determination and analysis of the energy demand helped evaluate the reuse, recycling und recovery options and select the most appropriate alternative. It also provided a basis for product redesign requirements. At component level, the dismantling processes and the most appropriate recovery and disposal alternatives were analysed and the respective time and costs presented in a matrix. Thus, both the relevant material fractions contained in components and the allocated costs could be displayed (Figure 4). The next analysis phase focused on cost aspects. Based on the above mentioned nine component clusters cost data including purchase, production and replacement costs (direct costs) as well as dismantling and recycling costs were considered. The cost evaluation identified cost drivers (corresponding to the component clusters f, g, h, i).

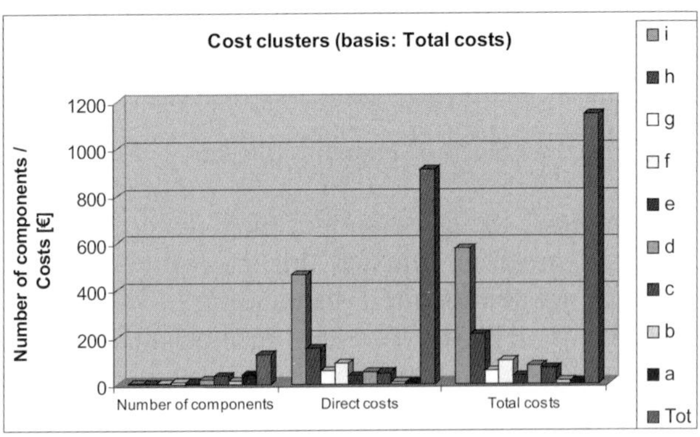


**Fig. 5.** Component cost clusters

| Material fraction | Way of recovery/ disposal | Recovery benefits [€/kg] | Disposal costs [€/kg] |
|---|---|---|---|
| Fe-Stainless Steel | Stainless steel mill | 1,05 | |
| Fe-Steel (ordinary steel) | Steel mill | 0,11 | |
| Fe-Steel/Plastic | Small shredder | 0,07 | |
| Nf-Metall-Compound | Remanufacturing | 1,00 | |
| Aluminium | Aluminium plant | 1,50 | |
| Brass | Copper mill | 1,45 | |
| Cabel recovery | Cabel shredder | 1,25 | |
| Thermoplasts_ABS | Re-granulation_ABS | 0,35 | |
| Thermoplasts_PP | Re-granulation_PP | 0,27 | |
| Thermoplasts_EPP | Re-granulation_EPP | 0,25 | |
| TPE | Cement works, Combined heat and power plant (CHP) | | 0,08 |
| Duroplasts | Cement works, Combined heat a. power plant (CHP) | | 0,08 |
| Thermoplasts:PA, PC, PE, POM, PS, PET | Cement works, Combined heat and power plant (CHP) | | 0,08 |
| PVC | Waste incineration plant | | 0,25 |
| Elastomers | Waste incineration plant | | 0,25 |
| Plasticcompound | Waste incineration plant | | 0,25 |
| Plastic-Metal-Compound | Waste incineration plant | | 0,25 |
| Other materials, combustible | Waste incineration plant | | 0,25 |
| Other materials, non-combustible | Domestic waste landfill | | 0,20 |
| PCB with capacitors | PCB-recycling + Undergroung landfill | | 0,20 |
| Capacitors | Underground landfill | | 1,25 |

**Fig. 4.** Recycling matrix medical appliance

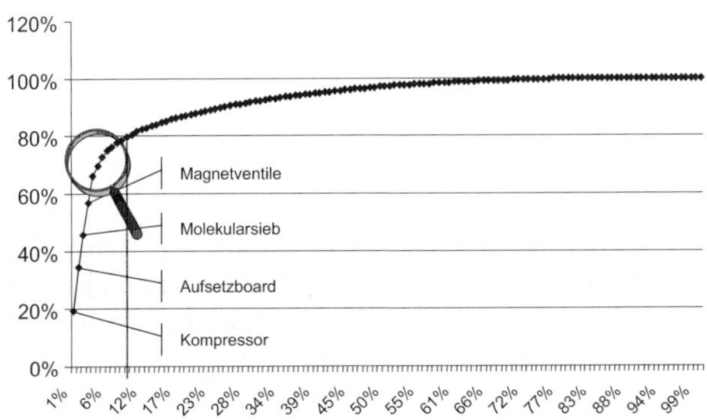


**Fig. 6.** Lorenz graph ABC analysis of material costs

The cost allocation of the cost drivers is illustrated in figure 6 with the help of the Lorenz graph of the ABC analysis method. The total cost reduction's target was set at 25% and redesign as based on this cost reduction target. The energy analysis showed that the best energy balance is possible by reducing the dismantling time and the use of required powered tools. For the given oxygen concentrator this was achieved by reducing the component number and avoiding bolted and pressed connections [7]. Product redesign allowed reducing the component number from 293 to 125 and the number of required dismantling tools from 45 to 9. Compared to the initial oxygen concentrator more components could be reused, increasing the weight of the reused parts.

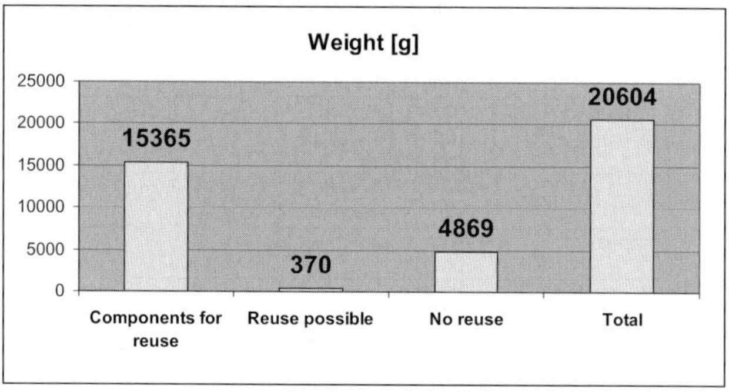


**Fig. 7.** Components for reuse

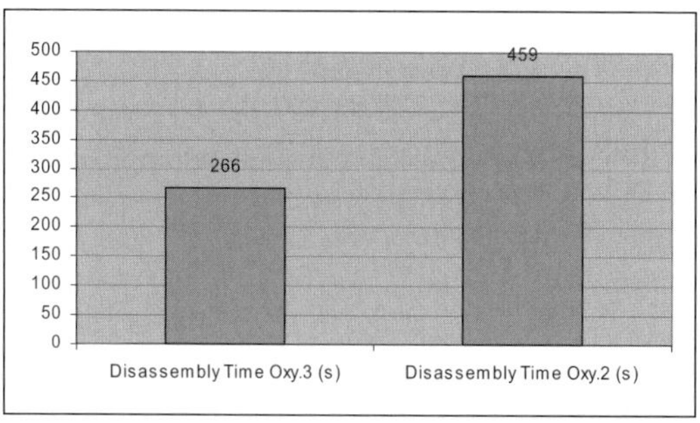


**Fig. 8.** Reduction of disassembly time

At the same time, the diversity of materials was reduced from initially 47 to 15 different materials. The redesign cut disassembly time by nearly 90% and assembly time by 50% [8]. Regarding the life cycle material costs of the oxygen concentrator, the redesign focused on the reduction of material costs, costs related to production processes (decreased component number and assembly time) as well as recycling costs. By reusing an increased number of parts the time for product design could also be reduced.

The new oxygen concentrator is an environmentally friendly product and achieved high customer satisfaction when put on the market. Better sales figures and increased benefits also reflect the effects of the successful product redesign.

## 4.2 Using TLCC on a washing machine

The use of the TLCC method for redesigning a washing machine followed the procedural steps mentioned above. The optimisation activities were mainly focused on recycling cost reduction while increasing the recycling quota. First, the functional structure of the washing machine was elaborated before breaking down the product into components.

| | Functional structure: Main functions |
|---|---|
| 1 | Disassemblability of the subassembly from the structure |
| 2 | Disassemblability of the parts from the subassembly |
| 3 | Toxic and harmful substances separability |
| 4 | Cleaning |
| 5 | Maintainability |
| 6 | Remanufacturing |
| 7 | Restoring |
| 8 | Material separability |
| 9 | Material recyclability |
| 10 | Quality control |

**Fig. 9.** Functional structure of the washing machine

The components were classified and weighted according to legal environmental requirements and guidelines as well as technical rules [9]. Figure 10 illustrates the weighting structure. To cover all environmental impacts during the life cycle phases, these impacts were divided as follows:

- Energy (energy consumption),
- Water (water consumption),
- Emissions into the air and water,
- Waste,
- Raw material.

In order to increase the recycling quota, we first determined which substances should be avoided and which must be separated by dismantling to avert their hazardous potential.

| 6. REMANUFACTURING | | | | | | |
|---|---|---|---|---|---|---|
| Functions | Grade of fullfillment (A) | A | Weighting factor | Points (P=I*A) | Points max (Pmax*I) | Assessment (V=P/Pmax) |
| Physical, chemical and mechanical stress of remanufactured parts | none | 5 | 5 | 25 | 125 | 0,20 |
| | acceptable | 4 | | 20 | | 0,16 |
| | low | 2 | | 10 | | 0,08 |
| | high | 1 | | 5 | | 0,04 |
| Wear of the reusable parts | none | 5 | 3 | 15 | 45 | 0,33 |
| | acceptable | 4 | | 12 | | 0,27 |
| | low | 2 | | 6 | | 0,13 |
| | high | 1 | | 3 | | 0,07 |
| Quality of the remanufactured part | high, like the original | 9 | 3 | 27 | 81 | 0,33 |
| | low , but acceptable | 3 | | 9 | | 0,11 |
| | not acceptable | 1 | | 3 | | 0,04 |
| Cost | not necessary | 9 | 2 | 18 | 36 | 0,50 |
| | low | 7 | | 14 | | 0,39 |
| | medium | 3 | | 6 | | 0,17 |
| | high | 1 | | 2 | | 0,06 |
| RESULT | | | | 175 | 287 | 3 |

**Fig. 10.** Weighting structure

| Economic Values | | | | | | |
|---|---|---|---|---|---|---|
| Significant Items | Values per Unit | Units | Minimum (Shredding) | Partial Disassembly | Significant Disassembly | Complete Disassembly |
| Steel | 0,10 | €/kg | 10,00 | 10,00 | 10,00 | 10,00 |
| Copper | 0,30 | €/kg | 1,50 | 1,50 | 1,50 | 1,50 |
| Aluminum | 0,20 | €/kg | 0,60 | 0,60 | 0,60 | 0,60 |
| PVC | 0,10 | €/kg | 0,00 | 0,00 | 0,50 | 0,75 |
| ABS | 0,12 | €/kg | 0,00 | 0,00 | 0,40 | 0,40 |
| HIPS | 0,05 | €/kg | 0,00 | 0,00 | 0,00 | 1,50 |
| Polycarbonate | 0,13 | €/kg | 0,00 | 1,50 | 2,00 | 2,00 |
| Polyurethane | 0,10 | €/kg | 0,00 | 0,00 | 0,00 | 1,00 |
| Nylon | 0,10 | €/kg | 0,00 | 0,15 | 0,15 | 0,15 |
| SAN | 0,03 | €/kg | 0,00 | 0,40 | 0,40 | 0,40 |
| Refrigerant | 0,30 | €/kg | 0,10 | 0,10 | 0,10 | 0,10 |
| Service Parts | Current | € | 0,00 | 0,00 | 0,00 | 0,00 |
| Landfill "Fluff" | -0,05 | €/kg | -4,60 | -3,28 | -2,67 | -0,55 |
| Disassembly | -10,00 | €/hr | -0,30 | -4,00 | -6,00 | -15,00 |
| Shredding | -5,00 | €/ref | -5,00 | -5,00 | -5,00 | -5,00 |
| Logistics | -13,00 | €/ref | -13,00 | -13,00 | -13,00 | -13,00 |
| **Profit/Loss** | | | **-10,70** | **-11,03** | **-11,02** | **-15,15** |
| **Total Weight** | | | **200,0** | **200,0** | **200,0** | **200,0** |
| **Pounds** | | | **108,0** | **134,4** | **146,6** | **189,1** |
| **% Recycled** | | | **54,00%** | **67,19%** | **73,28%** | **94,53%** |

**Fig. 11.** Recycling matrix washing machine

The hazardous potential is defined by legal frameworks such as the Directive on WEEE and RoHS, or by technical rules. For example, capacitors as well as PCBs count as components with hazardous potential. These substances should be separated by dismantling before disposal. The next analysis step included the breakdown of components into material fractions and the allocation of dismantling, recycling and disposal activities, as well as the assessment of necessary tools, time and incurred costs. As illustrated in figure 11 recycling and recovery costs (negative sign) and revenues (positive sign) as well as recycling quotas are mainly influenced by dismantling friendly design, the quantities of hazardous materials to be separated, the price of raw materials, the quality of the recycling fractions and the recycling infrastructure.

The formulated target for the cost reduction included a recycling cost reduction while an increased recycling quota per appliance. Concerning the product design it was possible to reduce the environmental costs and increase the recycling and recovery rates by

- Choosing materials appropriate for environment and recycling,
- Material standardisation,
- Dismantling and recycling friendly design,
- Conscious choice of auxiliary and operating substances,
- Using recycled materials and reuse of components and assemblies.

Production cost reduction could be achieved by using recycled polypropylene fractions for the production of the transport safety devices. The PP-fraction mentioned was produced by material recycling of used coffee machines and high-pressure cleaning machines cabinets. The quality of the secondary raw material fulfilled all quality requirements and facilitated a material cost reduction of 3%. The evaluation of the recycling cost achieved after redesign showed a cost reduction of nearly 14% for dismantling of hazardous substances and of 10% for the complete dismantling process. As shown is figure 11 by redesigning the washing machine, the complete disassembly into the above mentioned fractions was possible. The process time for the complete dismantling was no significant longer than before but the recycling quota could be increased by nearly 20%.

| | Complete Dismantling | | Dismantling of Hazardous Substances | | Disposal by another company | |
|---|---|---|---|---|---|---|
| | before [€] | after [€] | before [€] | after [€] | before [€] | after [€] |
| Disposal revenues | 1,89 | 7,35 | | | | |
| Disposal costs | | | 29,34 | 27,73 | 20,00 | 20,00 |
| Dismantling costs | 44,58 | 45,96 | 7,76 | 4,22 | 0,00 | 0,00 |
| Total costs of disposal | 42,69 | 38,61 | 37,10 | 31,95 | 20,00 | 20,00 |

**Fig. 12.** Dismantling cost reduction

## 5 Summary

The TLCC method is a combination of tools such as Target Costing and Life-Cycle Costing supporting cost optimisation, and tools for environmentally friendly product design such as Life Cycle Assessment.

It provides an assessment of the total cost of acquisition, ownership and recycling/disposal of a product and helps evaluate and optimise a product's life cycle cost, comparing it to the market-oriented target costs while satisfying specific environmental targets as well as performance, safety, reliability, maintainability and other requirements. The aim of TLCC is to equally fulfil market and environmental objectives. To this end, the business perspective comprises the complete value chain including the own company as well as suppliers, customers, recyclers and disposal companies. The analysis shows that implementing TLCC facilitates cost reduction by product redesign considering functional roles of product components from both an environmental and an economic point of view. This type of combination supports designers effectively, whereas a single tool could not achieve the same results.

## References

1. Coenenberg, A. G., 1999, Kostenrechnung und Kostenanalyse, Moderne Industrie, Landsberg/Lech
2. Seidenschwarz, W., 1993, Target Costing - Marktorientiertes Zielkostenmanagement, Franz Vahlen, München
3. Ewert, R., Wagenhofer, A., 2003 Interne Unternehmensrechnung, Springer, Berlin
4. Ehrlenspiel, K., Kiewert, A., Lindemann, U., 1998, Kostengünstig Entwickeln und Konstruieren – Kostenmanagement bei der integrierten Produktentwicklung, Springer
5. Janz, D., Kühne, F., Berger, S., Westkämper, E., Sihn, W., 2004 Customer-Oriented Life Cycle Costing, 5th International Continuous Innovation Network Conference, University of Western Sidney, Sidney
6. Friedel, A., 1999, Einfluss der Produktgestalt auf den Energieaufwand beim Recycling mechanischer Bauteile und Baugruppen, Springer, Berlin
7. Herrmann, C., 2003, Unterstützung der Entwicklung recyclinggerechter Produkte, Vulkan-Verlag GmbH, Essen Ruhr
8. Fraunhofer-Institut für Produktionstechnik und Automatisierung, 2002, Endbericht Medikreis, FhG-IPA, Stuttgart
9. Fraunhofer-Institut für Produktionstechnik und Automatisierung, 2003, Final Report: Environmental and Recycling Oriented Design of the Delta Washing Machine, FhG-IPA, Stuttgart.

# PLM to support hazard identification in chemical plant design

F. Giannini, M. Monti, S. Ansaldi and P. Bragatto
*Istituto di Matematica Applicata e Tecnologie Informatiche – CNR, Via De Marini 6, 16149 Genova – Italy, giannini@ge.imati.cnr.it, CAD/CAE/PDM Consultant, Via Oberdan 40, 00040 Monte Compatri (RM) – Italy, DIPIA - ISPESL, Via Fontana Candida 1, 00040 Monte Porzio (RM), Italy*

Abstract: A complete plant design is described through an huge amount of data organized in a big number of documents of different types logically linked each other by complex relationships; managing information about plants is a complex and expensive task. The techniques commonly adopted for hazard identification require high costs and time for their development; many specialists in different disciplines need to be involved for all the time necessary to complete the plant documentation analysis. The research project here presented is based on the idea of taking advantage of the organization of the information typically offered by Product Lifecycle Management (PLM) systems in order to allow designers to evaluate specific plant aspects, in particular those concerning possible hazards.

Keywords: Product Lifecycle Management, Plant Design, Hazard Identification

## 1 Introduction

The increased attention to safety risks in industrial sites, potentially dangerous for accident vulnerability, has to be put in relationships with various facts linked to the industrial development, such as the enlarged dimensions of the constructed plants, their aggregation in industrial area, the urbanization which extends the city agglomerations closer to industrial sites. For industrial hazard it is intended the possibility that at a certain moment of the activity, a susceptible situation evolves to an industrial accident with catastrophic consequences. In chemical plants, the accidents that can cause devastating dangers to humans and to the environment are linked to fires, explosions and toxic substances' delivery. Among the possible causes of such accidents within a plant, we have to take into account



*D. Brissaud et al. (eds.), Innovation in Life Cycle Engineering and Sustainable Development,* 349–362.

equipment malfunctioning, corrosion phenomena, maintenance faults and lines breakdowns. Anyhow design mistakes have not to be underestimated, such as, for instance, the inadequate dimensioning of the cooling for a reactor may activate important secondary exothermic reactions.

A common practice to evaluate plant hazard is to adopt a risk index, defined by the product of the frequency of the event and the level of produced damage [1]. Unfortunately it is not trivial to appraise these two quantities. Risk analysis can be performed according to different methodologies, applied in the different phases of the plant development process to find out various possible hazard sources. It is clear that taking into consideration all the safety aspects directly at the beginning of the plant design phase is crucial; it may avoid the need of successive corrections and the introduction of additional safety barriers, thus making the plant more complex and requiring additional maintenance operations.

Plant design is a very long and complex activity involving various disciplines: consequently many experts with different competencies and background are working on the different aspects. It is composed by several phases: from conceptual design to process detailing, from the engineering detailed design to the equipment selection. Decisions and solutions are guided by various factors, which can be economical, technical or related to safety, health and environment preservation. Analogies with collaborative product development in the mechanical sector can be identified, where at each development phase the related experts have to be involved to guarantee that all the related product and process aspects are analysed and conveniently satisfied. Correspondingly to the various aspects to be defined and analysed, different types of documents and tools have to be adopted that better support and describe the related data.

Engineering companies more and more adopt CAD systems providing capabilities to automate the design process and to link together the data of the various project development phases, such as the process diagram drawing, equipment design, electrical distribution and layout specification. These activities are supported by different software tools, which are developed around a modelling system, that in the most advanced configurations may be enhanced with knowledge based capabilities guiding the design of new project according to the provided rules as well as their validation on already existing documents. To proficiently support the management and maintenance of the huge amount of documents related to whole plant, in all its life cycle phases, CAD systems are integrated in the so-called PLM (*Product Lifecycle Management*) systems. PLM systems are the evolution of Product Data Management (PDM) systems and offer a structured database of the various information while handling the access rights and use of documents originated by different application tools, making easier the creation, modification and retrieval of the diverse data and information. They can be configured to provide customized views of the stored data according to the specific user profile for browsing the available documentation.

In this paper, we present the functionality of a software tool aimed at supporting the expert in performing hazard analysis and in verifying safety criteria in a plant project stored in a CAD/PDM system. The system is not aimed at substituting the expert in the evaluation but in supporting him/her by browsing the various plant data related to possible critical configurations. The paper is organised as follows. Section 2 introduces the most adopted methodologies for risk assessment. Section 3 focuses on the two methodologies supported by the software prototype and describe the main functionalities. Finally conclusions are given in Section 3.

## 2 Risk analysis methodologies

Increasing the reliability and security means to reduce the risks related to the plant construction and process. The main motivation enhancing the accident risks may be grouped according to the following modalities [2]:

- As a direct consequence of faults in parts or equipment of the plant.
- Due to fail of protection systems and equipments that could be inactive or out of order when necessary.
- Indirectly when highly risky operations are protracted along the time, as for example shut down and start up of the plant.

The first two aspects put in evidence the reliability aspects of the composing components as mandatory issues. Whereas the latest is more related to the adopted process and working procedures. It is then clear that taking into consideration all the details of a plant design since the beginning may have big impact in the reliability and security. To this aim the Inherent Safety Design approach has been proposed [2-6].

Inherent safety is similar in concept to pollution prevention or cleaner production. Both attempt to prevent the possibility of harm, from accidents or pollution, by eliminating the problem at its source. Both typically involve fundamental changes in production technology: substitution of inputs, process redesign and re-engineering, or final product reformulation [1]. Secondary prevention and mitigation, meanwhile, are similar in concept to pollution control and remediation measures, respectively, in that each involves only minimal change to the core production system. In particular, secondary accident prevention focuses on improving the structural integrity of production vessels and piping, neutralising escaped gases and liquids, and shut-off devices rather than changing the basic production methods. When plants expand beyond the capacity they were initially designed for, secondary prevention capacities may be exceeded. Sometimes, overconfidence in these added-on safety measures may invite an expansion of production capacity.

Inherently Safety Design is better applicable at the initial stages of design since there is bigger freedom in the choice of materials, equipment, layout and so on. Anyhow even if Inherently Safe Design seems conceptually simple, its implementation is quite complex, since it may require radical changes and anyhow some risk analysis has to be performed to check if changes are needed.

In literature and in common practise, several methodologies are considered for the risk analysis and assessment. They are using different types of information and may be applied in the various plant life cycle phases. Some of them require detailed plant description and may only be applied at the later stages of the design activity; others are considering more general process and aspects thus are suitable at the conceptual and preliminary steps. Among them we can mention the following:

- **Safety Review (SR)**, it adopts a qualitative description of the potential problems related to safety.
- **Checklist**, provides a list of good criteria to be followed in order to avoid critical configuration, anyhow the fact that they are not satisfied does not imply that the plant is unsafe.
- **Preliminary Hazard Analysis (PHA)**, general methodology that evaluates the whole system.
- **What-if**, analyses the consequences of the unwished events, possibly external to plant itself.
- **Hazard and Operability Analysis (HAZOP)**, systematic method for evaluating the effects and causes of the deviation from the planned behaviour of the plant equipment and process.
- **Fault Modes and Effects Analysis (FMEA)**, it mainly concentrates of faults modalities.
- **Fault Tree Analysis (FTA)**, it is a deductive approach which analyses the tree of faults by starting form a fault and going back to the original one.
- **Event Tree Analysis (ETA)**, it is an inductive approach that analyses the tree of faults starting from the beginning to the final fault.
- **Cause Consequence Analysis (CCA)**, it combines the trees of faults with that of events.

Table 1 summarizes which methodology is more suitable for any specific project phase.

**Table 1.** Risk Identification methodology in the different project phases

| | Safety Re-view | Chceklist | PHA | What-if | HAZOP | FMEA | FT | ET | CCA |
|---|---|---|---|---|---|---|---|---|---|
| Conceptual design | | X | X | X | | | | | |
| Detailed Design | | X | X | X | X | X | X | X | X |
| Extension and changes on the design | X | X | X | X | X | X | X | X | X |
| Accident analysis | | | | X | X | X | X | X | X |

Among the various hazard investigation methodologies, the following two methodologies have been chosen to be supported by the system prototype: Checklist, HAZOP.

The choice mainly derives from the type of information necessary to apply this methodology, which is fully available in a plant project developed in PLM environment. The checklist methodology is based on a written list of criteria to verify the safety of a plant, mainly from the layout point of view rather than the process. These criteria are commonly derived from past experience of plant design and maintenance and are to be considered as conditions to be complied in order to avoid safety problems; if a plant satisfies the list of criteria, then the hazard incidence degree is very low, even if it doesn't imply that the plant is totally out of risk [7]. The checklist procedure is applicable for conceptual design, detailed design and operation; even if it cannot substitute a more rigorous approach, the list of criteria reminds the reviewer of possible critical areas and in this perspective may be a valid basis from which to start for a more meticulous analysis. Hazard and Operability Analysis [8-10] is one of the most adopted safety analysis methods in the process industry; it was developed to identify safety hazards in a process plant and to focus on operability problems, which could compromise a plant's ability to meet the process design intent. Typically HAZOP analysis is conducted by a team of different specialists that examine the plant documentation from different points of view on the basis of their expertise [11–14]. The information required to apply such analysis to an existing plant or to a new design includes basically accurate process information and knowledge of instrumentation and operation, while the team performing the analysis needs a good expertise in design, operation, and maintenance. Some attempts in developing intelligent systems for automating HAZOP analysis have been more or less successfully conducted, but all weakly integrated in the plant development environment thus insufficiently supporting the whole project data browsing for a comprehensive risk evaluation. A more careful review on the developed systems is given in [15].

## 3 The system Architecture

The software prototype has been conceived as an add-on of an integrated CAD/PLM system for the design of industrial plant and the management of the design data and process; it provides functionalities capable to support hazard investigation and risk identification, by supporting the navigation throughout the plant project data and analyzing them, on the basis of the chosen safety analysis techniques, in order to identify possible critical situation in the plant. Critical situations are identified and highlighted, automatically when possible, or throughout successive specific queries to the plant design data. In the developed prototype the chosen CAD/PDM environment has been CATIA because it is the system adopted in ISPESL, it also offers the capabilities for accessing the stored data and for adding user defined evaluation rules. Figure 1 schematizes the application modules, the required databases and illustrates the dependency relationships; the arrow direction corresponds to the information flow.

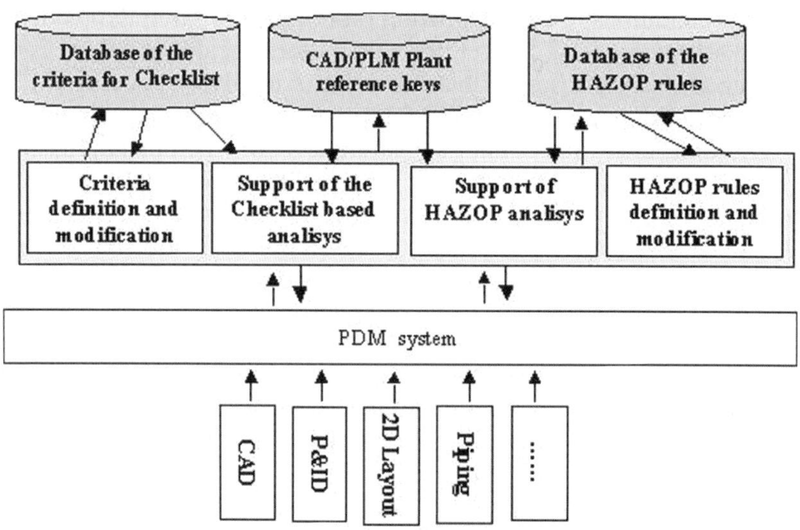


**Fig. 1.** The system architecture

The application includes the following main modules namely supporting:

- *Checklist based analysis:* it includes the functionalities necessary to apply the analysis based on the list of predefined criteria.
- *HAZOP analysis:* it includes the functionalities necessary to apply the HAZOP methodology.
- *Criteria editing:* it includes functionalities that allow the definition and modification of the checklist criteria.
- *HAZOP rules editing:* it includes the definition and modification of *guide words*, *parameters*, *causes* and *consequences* for the HAZOP methodology.

The links with the plant design data (i.e. P&ID, CAD, 2D layout,..), such as the identifiers and types of the elements, are created by reading the plant project master documents and stored in an ad hoc local data-base structure. Therefore the access to the CAD/PDM data is performed throughout these identifiers by using the related applications and the browsing capabilities provided by the hosting PDM system, In our prototype the master documents considered are the P&I and the 3Dlayout, assuming that the elements are semantically annotated and the links among the various representations of the same object and related elements are explicitly inserted by the user at the design stage.

The results of the performed hazard analysis become themselves part of the plant database and the PDM browser may manage the access to them; they illustrate possible critical situations in a specific design stage and consequently represent project data to be handled by the documentation manager (PDM). The database of the criteria for Checklist and database of the HAZOP rules contain all the data and relationships representing the knowledge, which they are based on; data have been structured in a way suitable to an easy modification and definition of new ones.

## 3.1 Checklist based analysis

The criteria considered are referred to problems that may arise in specific plant situations. Since not all the criteria may apply to all the situations, criteria have been classified taking in account different aspects. In particular the following issues have been considered [16]:

- single component that should comply to some specific norms,
- layout of components, building and walking ways,
- piping,
- existence of specific elements in specific process steps.

For each of the above listed items, criteria have been further classified and subdivided in categories sharing specific properties; finally for each of them it has been identified which data occur for a complete specification and evaluation. Often to complete the verification of a criterion it is necessary to navigate through different types of documents. For example, to verify if process units and service units are at a certain distance, first the user has to check in the process schema which kind of process is performed in the diverse units and then, once the proper units are identified, CAD data are necessary to effectively calculate the distances among them. The algorithms necessary for analysis and recognition of critical plant situations are developed on the basis of the defined criteria taxonomy. Table 2 shows some examples of categories of criteria. Once the user indicates criteria to be applied, the corresponding verification algorithms, if required, are activated; they operate on the PDM database, thus accessing directly, to the different design documents (P&ID, CAD, layout), depending from the criterion under verification.

**Table 2.** Taxonomy of criteria

| Type | Property | Criterion |
|---|---|---|
| Single Component | Parameters | Proportion between dimensions of tank and safety barrier |
| | | Proportion between dimensions of tank and safe storage container |
| | …… | |
| Layout | Distance | Between process units and other plant areas (e.g. services, offices, magazines, fire station and emergency services...) |
| | | Between process units and loading areas |
| | | height of pipes crossing roads |
| | | ……… |
| | Position | Distance between offices and main entrance |
| | | Sewers parallel to road system |
| | ……. | Accessibility to emergency hand valves |
| Pipes | Location | Adjacency of pipes depending on the material |
| | | ….. |

Not all the criteria imply quantitative results: many of them give rise to a qualitative evaluation. Since these kinds of criteria are not rules that have to be strictly respected but rather recommendations, it could be useful to provide more verbose results, rather than a simple yes/not. Results are presented at the end of the analysis process in a synthetic format by an electronic sheet that gives the user the possibility of easily accessing the related project documents.

According to the methodology supported by the developed prototype, at first the user has to indicate the context to which he intends to apply the analysis (for example a logic unit or a component or the whole project). Then for each element of the specified context the user has to indicate the criterion category he wishes to verify. In the example illustrated in Figure 2, the user selected to apply criteria for vessels and pipes and the system lists the criteria (on the right side of the application window in Figure 2) that are applicable to these kinds of elements.

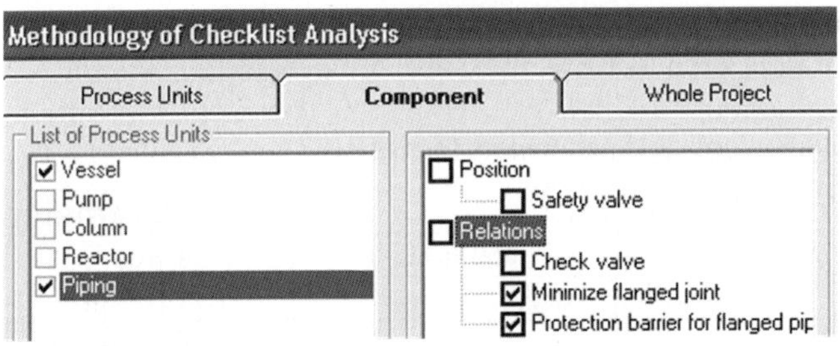


**Fig. 2.** Selection of the criteria to apply

Once the verification has been performed, a window is automatically displayed in which the list of the applied criteria appears with the results obtained for each of them (Yes/No/Percentage) (see Figure 3).

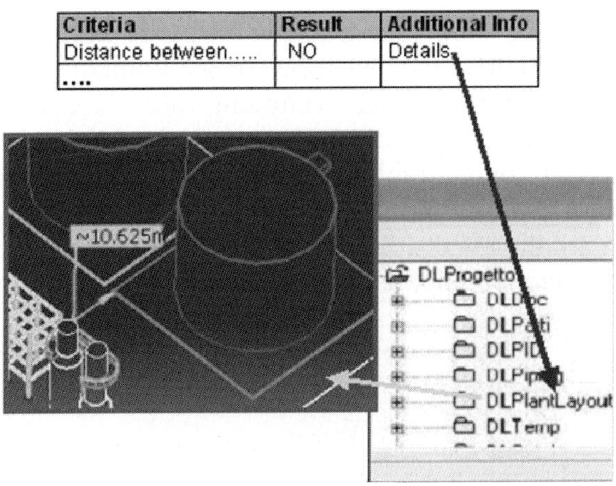


**Fig. 3.** Visualization of results of criteria verification

## 3.2 The HAZOP analysis

The HAZOP approach consists in carefully reviewing a process or operation in a systematic way in order to identify *deviations* from process design intent, which could lead to undesirable *consequences*. The results are identified hazards and operability problems, which can be the *causes* and/or *consequences* of other possible accidents. The methodology is applied to each component of the plant, or to each component of the selected logic unit. HAZOP is based on *guide words* that have to be combined with the component functional parameters to constitute the *deviation* to be analyzed. The *guide words* listed in Table 3 indicate the displacement from the foreseen values of the physical parameters that characterize a specific component [17] such as flow, pressure, temperature, level, etc.

**Table 3.** Guide words and parameters for HAZOP

| Guide Word | Description |
|---|---|
| **No or Not** | Negation of the intention |
| **Less** | Quantitative decrease |
| **More** | Quantitative increase |
| **Part of** | Qualitative decrease |
| **As well as** | Qualitative increase |
| **Reverse** | Logical opposite of the intention |
| **Other than** | Complete substitution |

(a)

| Parameter |
|---|
| Flow |
| Pressure |
| Temperature |
| Level |
| Corrosion |
| Substance |
| |

(b)

For example, the combination between the guide word "MORE" and the parameter "PRESSURE" indicates that in the component under analysis the pressure is greater than the maximum value foreseen for a regular working process. Since not all the parameters are meaningful for any component of the plant, only a limited set of deviation can be considered for a specific element. *Causes* are the events that may bring to a *deviation* and *consequences* are the effects of the *deviation*. More *causes* could be identified for one *deviation* and, similarly, more *consequences* may be linked to a *deviation*.

As illustrated in Figure 4, the system firstly requires the user to identify the plant's area, or process unit to be analyzed, and secondly to indicate the component from which to start.

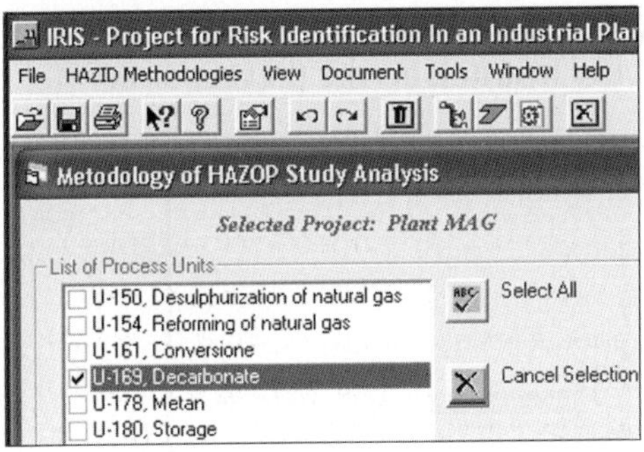


**Fig. 4.** Identification of the plant area to analyze

Analysis may be applied to specific components selected interactively by the user or sequentially to each component of the selected plant's area, following the process flow. The prototype drives the user throughout a sequence of steps, which are typical of an HAZOP study, i.e. to systematically analyze a component to identify possible *deviations* from the design intent. For each selected component the system provides the list of appropriate parameters and, once the parameter is selected, a list of guide words meaningful for it (see Figure 5).

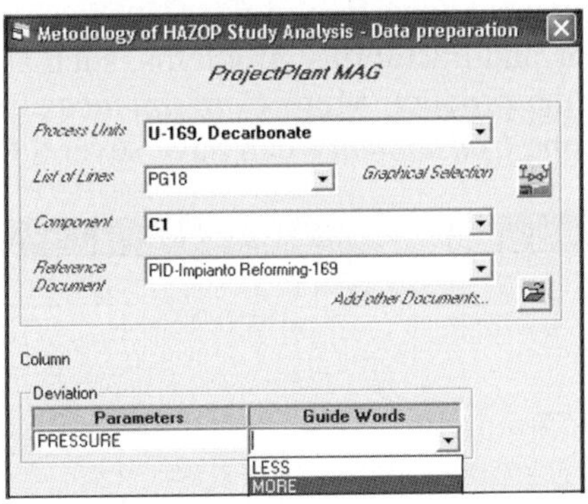


**Fig. 5.** Definition of the deviation to analyze

Once a component is selected, the system automatically presents the possible guide words and parameters to be combined to identify the interested *deviation*. Guide words and parameters appropriate for the type of the component selected are extracted from the system database, that the user may modify or extend at any time. For each *deviation* the system presents to the user a list of possible *causes* (see Figure 6), automatically extracted from the system database, taking into account the characteristics of the component itself and also those of the adjacent components; in addition it considers if specific instruments are present or not (for example valves).

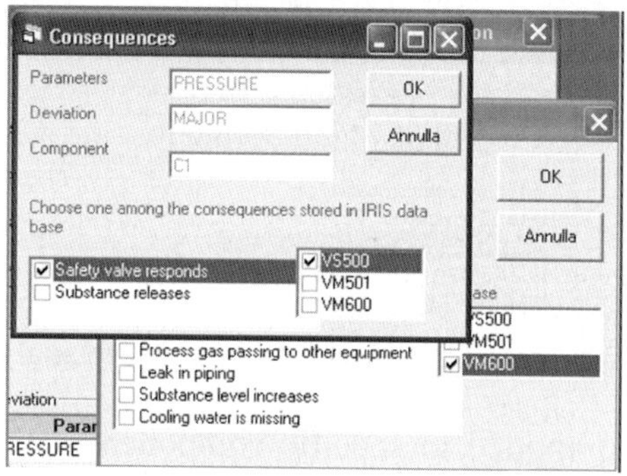


**Fig. 6.** Causes and Consequences identification

In this way, even if the explicit indication of the *cause* to be further analyzed is left to the expert, he is driven by the system in considering only the causes that are possible in the plant area under evaluation. Moreover he/she may take advantage of a fast and easy navigation throughout the design documentation. Analogously the system drives the user to individuate the possible *consequences* (see Figure 6). *Consequences* may both comprise process hazards and operability problems, such as plant shutdown or quality decrease of the product. More *Consequences* can follow from one *Cause* and, in turn, one *Consequence* can have several *Causes*.

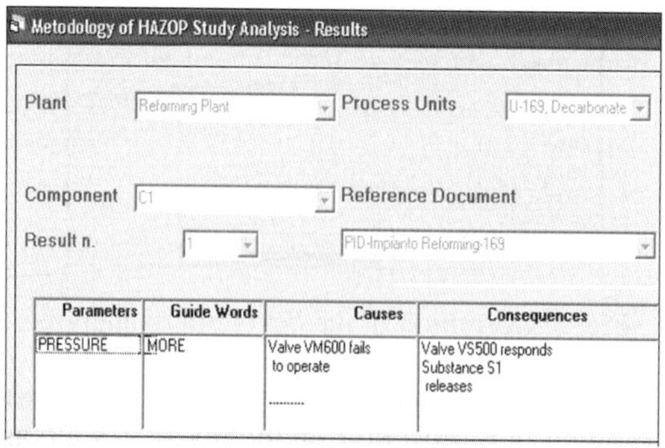


**Fig. 7.** HAZOP result

At the end of the HAZOP analysis, it will be possible to automatically produce an electronic sheet, similar to the one commonly produced in the HAZOP analysis, as showed Figure 7.

## 4 Conclusions

A 100% secure plant doesn't exist because unplanned external configurations can always happen that may cause accidents with possible high impact on the environment. In addition, several statistics have been shown that small accidents in chemical plants happen almost every day creating occupational injuries, illness and thus costing to the society billion of dollars every year [15]. Therefore, learning from experience and taking advantage of complementary technical expertise, accident hazards can be strongly reduced by analyzing all the various aspects and processes of the whole plant. To this aim, new computer-aided tools can help experts in easier examining the whole plant project and better exploiting all the available information and knowledge.

The software prototype developed in the framework of the research project presented in this paper, demonstrates that the digital technology currently available for plant design and for project documents and design process management may be exploited to develop software tools able to support expert in the hazard identification during the plant's design phases. The objective of the tool is not to automatically perform hazard identification but to provide the user with functionalities that drive him in the examination of a specific plant design, making available step by step the necessary data and automatically performing, where possible, verification on these data. In addition, since a considerable part of the knowledge necessary in the hazard identification process derives from past experiences of previous accidents, the database that contains the information on which the analysis methodologies are based on, may be augmented by the expert at any time, throughout functionality that easily prompts the user. In the current version of the system, the rules have been derived by literature, as future step experts will be invited to test the system on more complicated examples and to extend the rules' database. The prototype has been developed as application of CATIA V5 © Dassault-Systèmes, in MS Windows NT, by using the programming language Visual Basic 6. MS access is used for the databases. Future work will also deal with the problem of semi-automatic annotation of the various plant elements by taking advantage of feature-based descriptions, possibly obtained through recognition processes, and knowledge representation techniques, such as ontology.

## Acknowledgement

This work has been supported by ISPESL- Dipartimento Insediamenti Produttivi ed Interazione con l'Ambiente (contract ISPESL B64/DIPIA/02).

## References

1. Heikkila, A.M., 1999, Inherently Safety in Process Plant Design – An Index-based Approach, VTT Publications 384, PhD1999, ISBN 951-38-5372-1 http://www.inf.vtt.fi/pdf/publications/1999/P384.pdf
2. Hendershot, D.C., Post, R.L., 2000, Inherent Safety and Reliability in Plant Design. Proc. of Mary Kay O'Connor Process Safety Center, 2000 Annual Symposium: Beyond Regulatory Compliance, Making Safety Second Nature, Texas, October 2000
3. Hendershot, D.C., 1995. Some Thoughts on the Difference Between Inherent Safety and Safety. Process Safety Progress 14, October 1995
4. Kletz, T.A., 1992b, Inherently safer design – a review. 7th International Symposium on Loss Prevention and Safety Promotion in the Process Industries, SRP Partners, Rome, Vol. I, pp 1-13
5. Inherent Safety Approach – http://process-safety.tamu.edu
6. INSIDE Project, "The INSET toolkit" http://www.aeat-safety-and-rsk,com/html/inset.html
7. Hendershot, D.C., 2002. A Check List for Inherently Safer Chemical Reaction Process Design Workshop on Risk and Reliability. Proc. of International Conference and Workshop on Risk and Reliability, Center for Chemical process Safety, October 2002
8. Kletz, T.A., 1992. HAZOP and HAZAN- Identifying and Assessing Process Industry Hazards. 3rd ed. Rugby: IChemE.
9. Zwetsloot, G.I.J.M., Ashford, N.A., Perspectives for combining cleaner production with inherently safer production in industrial firms, UNEP IE, Nov. 2001
10. Howat, C. S., - Process Hazard Identification Using Hazard and Operability Studies – University of Kansas, 2002,lecture notes, http://www.engr.ukans.edu/~ktl/
11. Chung, P.W.H., Yang, S., HAZARD analysis and support tool for computer controlled processes, Journal of Loss Prevention in the Process Industries, November 1998
12. Chung, P.W.H., McCoy, S.A., 2001. Trial of the "HAZID" tool for computer-based HAZOP emulation on the medium-sized industrial plant, HAZARDS XVI, IChemE Symposium Series No. 148 , Manchester, UK, 2001, pp. 391-404, ISBN 0-85295-441.
13. Palmer, C., Chung, P.W.H., 1999. Verifying Signed Directed Graph Models for Process Plants, Computers and Chemical Engineering Supplement, Elsevier Science Ltd.
14. Palmer, C., Chung, P.W.H., 2001. Verification of process plant models – Proceedings of the 2001 AAAI Symposium on Model-Based Validation of Intelligence, Stanford, 2001, AAAI Press
15. V. Venkatasubramanian, J. Zhao, S. Viswanathan. Intelligent system for HAZOP analysis of complex plants. Computer and Chemical Engineering 24 (2000) pp. 2291-2302, Elsevier Science Ltd.
16. Ansaldi, S., Bragatto, P., Giannini, F., Ludovisi, G., Monti, M., 2003, Criteri di sicurezza intrinseca per la progettazione di impianti di processo attraverso strumenti CAD, CAE e PDM, Convegno Scientifico Nazionale "Sicurezza nei sistemi complessi", Atti del Convegno, Bari, October 2003
17. Lees, F.P., 1980. Loss prevention in the Process Industries – volume 1, Butterworths

# Smart machining systems: issues and research trends

L. Deshayes, L. Welsch, A. Donmez, R. Ivester, D. Gilsinn, R. Rhorer, E. Whitenton and F. Potra

*National Institute of Standards and Technology (NIST), 100 Bureau Drive, Gaithersburg, MD 20899*

Abstract: Smart Machining Systems (SMS) are an important part of Life Cycle Engineering (LCE) since its capabilities include: producing the first and every product correct; improving the response of the production system to changes in demand (just in time); realizing rapid manufacturing; and, providing data on an as needed basis. Thereby, SMS improve the performance of production systems and reduce production costs. In addition, an SMS not only has to improve a particular machining process, but it also has to determine the best optimized solution to produce the part faster, better, at lower cost, and with a minimum impact on the environment. In addition, new software tools are required to facilitate the improvement of a machining system, characterized by a high level of expertise or heuristic methods. A global approach requires integrating knowledge/information about the product design, production equipment, and machining process. This paper first discusses the main characteristics and components that are envisioned to be part of SMS. Then, uncertainties associated with models and data and the optimization tasks in SMS are discussed. Robust Optimization is an approach for coping with such uncertainties in SMS. Current use of machining models by production engineers and associated problems are discussed. Finally, the paper discusses interoperability needs for integrating SMS into the product life cycle, as well as the need for knowledge-based systems. The paper ends with a description of future research trends and work plans.

Keywords: Smart Machining Systems, Life Cycle Engineering, Robust Optimization, Knowledge bases, Ontologies

## 1 Introduction

Many manufactured products involve machining. For such products, machining systems play an important role in the product life cycle as part of the connection between design and the finished product. The time and



*D. Brissaud et al. (eds.), Innovation in Life Cycle Engineering and Sustainable Development,* 363–380.

cost of transition from specification/inception to commercial birth may significantly affect the remaining phases of the life cycle. Furthermore, the productivity and the responsiveness (agility) of production systems as well as the product quality are important factors affecting product life cycle. All these characteristics are critical outcomes of machining systems used in production. There has been a continuous improvement in machine tools and machining systems to respond to the needs for better quality products at lower costs. Evolution from manual machine tools to numerical control (NC) and computer numerical control (CNC) machine tools and introduction of various sensing and control improvements have enabled machine tools to be more capable, effective, and productive over the last several decades. Even after these improvements, machining systems still require long periods of trial and error to optimally produce a given new product design or component. They still require cryptic NC language to operate with limited knowledge of what they are producing or how well they are producing. Furthermore, they rely on inefficient vendor-specific interfaces to receive partial information about design intent and function of a product to be machined. They either break down unexpectedly or require costly periodic maintenance to avoid these breakdowns. These deficiencies cause significant delays in time-to-market, increase cost, and reduce productivity. Smart Machining Systems (SMS) are envisioned to have the capabilities of: self recognition and communication of their capabilities to other parts of the manufacturing enterprise; self monitoring and optimizing their operations; self assessing the quality of their own work; and self learning and performance improvement over time. These attributes can be realized by seamless integration of various hardware and software components into new or existing machining systems. Some of these components have already been incorporated in existing machining systems in a limited fashion. The current direction of SMS research at NIST [1] (National Institute of Standards and Technology) is to identify the barriers for complete integration and functioning of SMS with product life cycle and develop necessary tools to overcome these barriers. Within this context, a Smart Machining System (SMS) provides the following capabilities: 1) producing the first and every product correct; 2) improving the response of the production system to changes in demand (just in time); 3) realizing rapid manufacturing; 4) providing data on an as needed basis. These characteristics make SMS appealing for Life Cycle Engineering (LCE). The purpose of the SMS program at NIST is to lead the development of an infrastructural capability for realizing these SMS capabilities for a broad range of products and processes.

LCE [2] involves complex and timely communications of critical data on an as-needed basis. It also involves the large number of design and manufacturing tools, as well as architecture for product life cycle management [3]. Since the SMS is such a central part of the production system, it requires a high level of interoperability and communications infrastructure. This requirement is not trivial to fulfill because a broad range of information sources must be considered such as design specifications, process planning, machine specifications, cutting tool specifications, and cutting parameters as well as heuristic knowledge. Furthermore, SMS relies on a broad range of expertise from various disciplines, both internal and external to the company, to constantly improve its performance and produce innovative products, technologies, and methods. This drives the need for effective management of SMS-generated information for the product life cycle. Unfortunately, current machining systems are not capable of providing appropriate information for LCE. Given a means to share information appropriately, LCE tools should be able to capitalize on production capability more effectively. In this paper, a general view of SMS, their characteristics and functional components, along with the associated issues related to their development and integration into the PLC, are discussed.

## 2 Characteristics and components of SMS

SMS must address the communication of all information needed to fabricate a product that satisfies customer and market needs. A simple component can be produced easily through a conversation between a customer and a machinist, with the machinist operating the machine. As the complexity of the product, design and production process increase, the necessary scope of communications encompasses more people and more sources of information. For complex products, it is virtually impossible to fully encapsulate all information needed for an SMS using tools and technology readily available today. The machining strategy defines the collection of issues related to fabricating a part such as the machining process plan or NC tool path. A machining strategy can vary in complexity depending on the machining feature, costs, part geometry, technology, etc. The SMS must optimize machining process plans before and during their realization, i.e., during its planning, as well as during their execution. A machining process plan indicates the immediate objectives (i.e., the tactical choices), their priorities, executing times, and necessary resources. Machining optimization uses models and data that are incomplete or approximations, therefore any results will involve uncertainty. When these uncertainties lead to unexpected performance, the process monitoring and control

(PMC) will use adaptive control to return to optimal conditions. So for an SMS, design and development of optimization tools using robust methods coupled with on-line systems are a key issue. They would help enable an SMS to produce the first and every subsequent part on time and to specification through a science-based understanding and monitoring of the available machining processes and equipment without significant time spent on process development or setup. The Smart Machining Systems program at NIST aims to develop, validate, and demonstrate the metrology, standards, and other infrastructural tools that enable the manufacturing industry to characterize, monitor, and improve the accuracy, reliability, and productivity of machining operations. Figure 1 provides a view of the SMS components. From a machining system analyst point of view [4], a Conceptual Process Plan (CPP) is the main input to SMS. The role of the CPP is to determine which and when general resources will be used. It represents the company's strategy for manufacturing and adds important global constraints to the future optimization tasks. Based on this a Detailed Process Plan (DPP) is built whose goal is to determine optimal machining parameters, tooling systems, and fixturing elements in order to satisfy design specifications. Traditionally, DPP adjusts these parameters using different approaches, including: 1) physics based models; 2) numerical simulations; 3) use of heuristic models; and 4) trial and error. Generally, conceptual and detailed process planning are more effective when the company's experience about its machining capabilities is used. The proper use of company's experience is facilitated by LCE (see figure 1). In SMS, a Dynamic Process Optimization (DPO) will optimize a DPP. To maintain an optimized system, the SMS has to assess the quality of its work and outputs as well as improve itself over time. The DPO builds and then satisfies objective functions using Machining Models (MM in figure 1), including process, control, and machine tool. The DPO includes constraints from design such as: dimensional and geometrical tolerances, surface integrity, and surface quality. PMC modules execute these optimized solutions, i.e. the optimized DPP, and improve them over time. In the face of such optimization complexity, which can easily lead to ill-defined problems, we believe that a DPP will become more robustly optimized using an optimization umbrella that can incorporate information from different types of models, such as numerical, theoretical, experimental, or heuristic models, and represent it in such a format that it would be unambiguously understood by LCE. In addition, a knowledge base for the DPO itself is used to properly construct the set of objective functions and constraints. Consequently, development of optimization tools and associated models are key issues for SMS research.

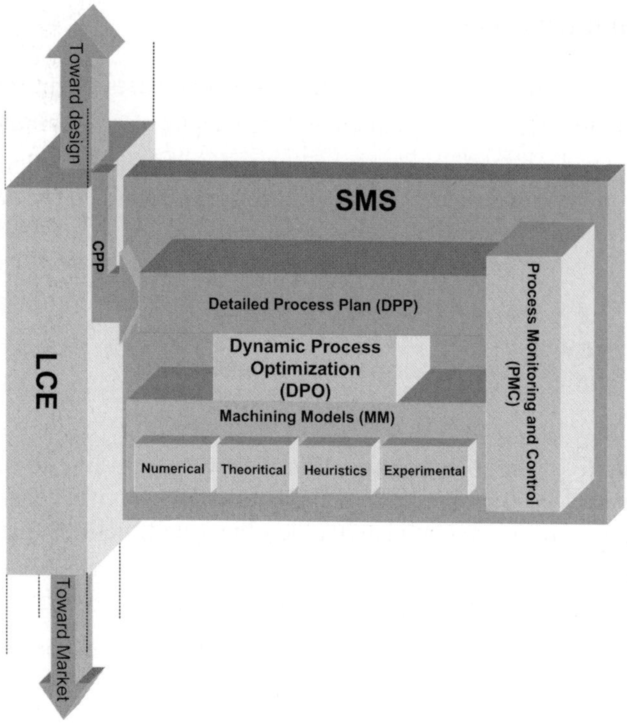


**Fig. 1.** Components of SMS

## 3 Optimization issues

Computer-based simulation and modeling will be an increasingly important part of the future in machining. There have been many studies looking at future manufacturing. One example from the 1990s states: "The Next Generation Manufacturing company will be characterized by its use of modeling and simulation, often coupled with agile and flexible manufacturing processes and equipment. ...*Modeling and Simulation* will be pervasive throughout the enterprise as a new way of doing business. The modern manufacturing enterprise is the sum of the large and small decisions made by people and so-called intelligent machines."[5] The Smart Machining Systems program at NIST will address some of the important decisions involved in improving and optimizing machining processes.

## 3.1 Optimization trends

A general machining optimization problem consists of determining decision variables $x_1, x_2, \ldots, x_n$, such as feed, depth of cut, spindle speed, in such a way that a set of given constraints are satisfied and a desired objective function is optimized. The constraints are determined by both empirical, heuristic, and theoretical considerations, and they can usually be expressed as a system of inequalities. If we denote by $x$ the vector of decision variables and by $f_0(x)$ the objective function, then the optimization problem can be written as

$$\text{Minimize } f_0(x) \tag{1}$$

$$\text{subject to } f_i(x) \leq 0, i = 1,2,\ldots,m \tag{2}$$

An example of an objective function that we would like to minimize in machining may be the cutting tool deflection. The above general form of an optimization problem can also handle objective functions that we would like to maximize, like the material removal rate. This is accomplished by replacing $f_0(x)$ with - $f_0(x)$ in (1). If the objective function $f_0(x)$, as well as the functions $f_1(x), f_2(x), \ldots, f_m(x)$ defining the constraints (such as cutting force, machine tool power and torque, tool life, surface roughness, and spindle speed) are linear in the decision variables, then the optimization problem (1)-(2) becomes a linear programming problem (LP) that has been extensively studied, and for which efficient algorithms are known [6]. However, in most applications both the objective function and the functions defining the constraints are nonlinear. By introducing an additional variable $x_0$, we can always consider that the objective function is linear. Indeed it is easily seen that the optimization problem (1)-(2) is equivalent to

$$\text{Minimize } x_0 \tag{3}$$

subject to

$$f_0(x) - x_0 \leq 0, \tag{4}$$

$$f_i(x) \leq 0, \; i = 1, 2, \ldots, m. \tag{5}$$

While in a traditional deterministic setting, where $f_0(x), f_1(x), \ldots, f_m(x)$ are considered determined precisely, the form (3)-(5) can be conveniently extended to deal with uncertainty in the data defining the optimization problem. Indeed, in real applications the functions $f_0(x), f_1(x), \ldots, f_m(x)$ depend on some parameters $\zeta_1, \zeta_2, \ldots, \zeta_p$ that are only approximately known. In some cases we can define an "uncertainty set", or set of possible parameter values, $U \subset R^p$ that contain all possible values of the parameter vector $\zeta$. If U contains a single vector then we are in the traditional deterministic setting. Otherwise, we consider the robust optimization problem

Minimize $x_0$ (6)

subject to

$$f_0(x, \varsigma) - x_0 \leq 0, \quad \forall \varsigma \in U \quad (7)$$

$$f_i(x, \varsigma) \leq 0, \quad i = 1, 2, \ldots, m, \quad \forall \varsigma \in U \quad (8)$$

The robust optimization problem above aims at determining the vector of decision variables x such that the objective function is minimized and the constraints are satisfied for all possible values of the parameter vector $\zeta$. Although this solution seems hopeless, recent progress in optimization theory and practice shows that for many engineering problems we can formulate robust optimization problems that can be efficiently solved by modern optimization algorithms [7]. An alternative way to deal with uncertainty is to consider that the parameters $\zeta_1, \zeta_2, \ldots, \zeta_p$ are random variables with given probability distributions. Then the optimization problem becomes a stochastic optimization problem (see the recent monograph [8] and the references therein). While for some problems of this type, like the stochastic linear programming problems, good solution methods are known, though they are in general more difficult to solve than their robust optimization counterparts. Another optimization paradigm in SMS is provided by multi-criteria optimization. In this approach one aims at determining decision variables that satisfy the given constraints and simultaneously minimize several objective functions. For example we would like to simultaneously minimize the cutting tool deflection and to maximize the material removal rate [9]. This can be accomplished by constructing a "master" objective function as a weighted combination of the given objective functions, or to formulate the problem as a Pareto optimization problem [10–11]. A recent application of Pareto optimization in machining is presented in [9]. In the following sub-sections we introduce the types of models we will use in our optimization in the future.

## 3.2 Machining models

The variety of criteria used in machining optimization involves material removal rates, chatter avoidance, dimensional and form accuracy, surface integrity of the machined part, and tool wear. Generally two main categories of machining models are used to represent such criteria; models describing the cutting process and models describing the machine tool and its components. These two categories are represented in figure 2 where the hatched zone represents the cutting process, i.e., the tool work piece interaction. Of course other models are used to describe the interaction occurring between those two categories. The cutting process is the interaction between the cutting tool and the work piece and is the excitation source for the machining system. Extensive research has been done in these two domains. Here we provide a brief overview of the approaches that will be used in SMS. The first set of models generally involves decisions in specifying cutting parameters by predicting cutting forces and controlling cutting process quality criteria. Thereby, one of the keys to making good decisions is to accurately predict cutting forces in a particular machining process. By accurately predicting cutting forces, the power and torque needed for a specific machining operation can be calculated. Knowing the cutting forces is also important for fixturing and tooling decisions. From cutting force predictions, the total energy input by the machining process can be calculated, which is essential for predicting cutting temperatures and tool wear. Approaches to predicting cutting forces include (a) traditional experience-based models; (b) machining experiment methods using mechanistic models; and, (c) models that use standard material properties rather than specific machining experiments. The experienced-based models, including the use of "physical models", are the oldest approach. Anytime there is an existing machining process similar to the one being studied, this may be the best approach to accurately predict cutting forces. Recently, a set of standards for describing a method to realize such experiments in industry has been developed [12]. This ideal scenario is rarely the case in product development. Models that use generally accepted material properties such as hardness or ultimate strength, rather than machining generated properties, have an advantage for predicting cutting forces in situations where the process or material are new to the company. Starting in the 1940s, approaches of this type have been developed that are based on simple geometry and rather simple material models, such as the well known Merchant model [13-14] or Lee and Shaffer's study [15], Since the 1980s, Finite Element Methods (FEM) have been developed that can handle both the complicated geometrical aspects of metal cutting as well as sophisticated material relationships. These FEM models [16] can be extremely accurate at predicting cutting forces, but require a great deal of knowledge about the cutting tool geometry and other machining parameters plus highly accurate constitutive models for the material. The FEM approach can also be very effective at predicting temperatures in the cutting zones and residual stresses in the work piece after machining. The use of FEM machining models may be limited in SMS applications because of the time required to solve a problem - several hours to days of computer time - and current limitations of adequate constitutive material models. Current work at NIST [17] and several others laboratories, is aimed at determining reasonable values of the flow stress and constitutive model parameters for machining modeling. The NIST work includes using a high strain-rate testing device, called a split Hopkinson Pressure bar or Kolsky bar, with electric pulse heating to determine flow stress values useful for predicting cutting forces [18]. This work is aimed at providing flow stress values to aid the simple model force predictions, as well as providing detailed constitutive models expressing the stress as a function of strain, strain rate, temperature, and heating rate that will aid the use of finite element approaches. The dimensional and form accuracy of a machined part is affected by the quasi-static performance of the machine tool, as well as environmental effects on both machine tool and the machined part [19–22]. Quasi-static performance of machine tools includes positioning accuracy and repeatability, geometric errors of linear and rotary motions, as well as alignment and locations of moving axes with respect to each other. The environmental effects are primarily in the form of temperature changes and gradients resulting in deformation of machine structure. In addition, errors associated with the coordination of multiple axes during the creation of complex tool paths are also contributors to machine performance. Furthermore, static and dynamic stiffness of the machine tool/cutting tool/work piece structural loop contributes to the accuracy of the machined part. Similar to modeling the cutting process, there are various approaches to modeling machine tool performance. Most models are based on a combination of a kinematic model and experimental data [23, 24]. Such a modeling approach uses sample performance data along the main axes of the machine, and then uses kinematic models to estimate performance in the whole work zone of the machine (in 2D or 3D). Other models rely more on correlating the representative measurable inputs to the machine performance [25]. These models are used for improving the accuracy of machine tools and processes [26]. Research at NIST aims to incorporate these diverse models and the associated measurement data into the available set of objective functions and contraints used by SMS and its dynamic optimizer. Ideally, a sophisticated Smart Machining System will utilize all of these types of approaches from time to time. It will also need to recognize the limitations and range of uncertainty based on the different methods. Finally it will have to identify what information is needed by each of the modeling approaches and how to exchange it effectively with the other parts of LCE.

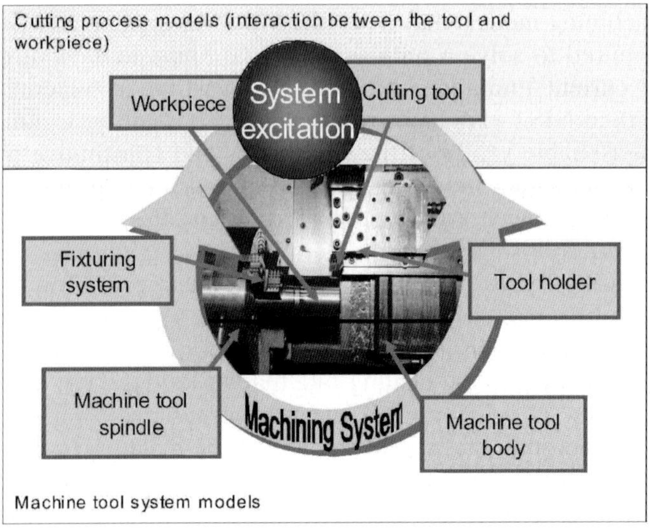


**Fig. 2.** Chain components machining system and its model categories

## 3.3 Improving robustness of machining process models

In order to provide meaningful improvement to the robustness of predictions from machining process models, we must first define and distinguish the various sources and forms of uncertainty and variability involved in the models, the predictions, and the machining process. The performance of any machining process has some level of inherent *variability*. This performance can be characterized through various measurements. There is *uncertainty* associated with how well the measurement represents the process performance. If the process is continued or repeated, additional measurements will demonstrate some degree of variability with some statistical distribution. If process parameters such as feed or depth of cut are adjusted, the distribution may shift or change shape. Empirical machining models are based on experimentally derived relationships between process parameters and process measurements. Given values for process parameters, this type of model typically produces an exact value. In order to relate this exact value to an expected distribution of future measurements, we must combine our uncertainty associated with the model and our uncertainty associated with the variability associated with the process itself. Balancing these factors against the various priorities of the overall strategy will be handled through robust optimization, as described in section 3.1. We are changing the form of mechanistic and empirical constraints to incorporate the inherent uncertainty of machining process models and the inherent variability of the machining process. For example, a typical force constraint has the following form:

$$F_c = K_f f^r a_p^s \tag{9}$$

where values for the constants $K_f$ (4500 N/mm$^2$), $r$ (0.7), and $s$ (1) have been determined experimentally. As a way of coping with the uncertainty in mechanistic and empirical models, we propose to redefine the constants as random variables with unknown distributions. Estimations of the distributions are then formed on the combined basis of the inherent variability, of the process and our uncertainty associated with the model's prediction of process behavior. With subsequent observations of actual process behavior, these estimations must be updated and anytime adequate information has been obtained to change our estimations of the expected distributions, the optimization problem can be revisited. Our initial formulation of the machining optimization problem will involve a weighted sum of production time, cost, and product quality, where the weighting of each factor depends on the strategy and priorities indicated by production and design. A collection of constraints of the same type as the force constraint in Equation (9) will be used to fully define the problem, including allowable cutting power and torque, a surface roughness constraint, and a tool life constraint. Additionally, upper and lower limits will constrain the cutting speed $V_f$ and the feed $f$. More details of this formulation are presented in [27].

## 4 Integration issues in SMS

This section gives a general description of the implementation to be accomplished with the intention to automate and integrate the SMS information within the product life cycle management tools. The most important issues concern dynamic modification and maintenance of the system according to heuristic knowledge, which is generally changing in production environment.

### 4.1 Software development needs

Each model previously described in section 3.2 and 3.3 presents a small part of an optimized solution. In order for an SMS to be robustly optimized over time, and in addition to the use of robust optimization algorithms, these models must be integrated such that the product life cycle engineer views one coherent optimization. This coherence must be performed automaticly selecting the particular models used into the optimization. In addition, the software that implements the previously described models must be able to work seamlessly as a unified package with other product life cycle software. To accomplish this goal of seamless integration, the software must be interoperable and there must be additional software that enables each company's strategy. SMS is envisioned to facilitate the information integration to the rest of the product life management tools. Interoperability is achieved through a common understanding of the semantics and the syntax of the data passed between software components. The easier problem is syntax, and today there are a variety of representations such as XML to handle syntax. The more difficult problem is semantics. Because people frequently communicate with each other[1], many have difficulty understanding why getting the semantics correct for software is so difficult. The reason is that software is precise. Any small differences in meaning may become exaggerated and cause software to cease working, or behave in an unexpected way. The current state of semantics in manufacturing and engineering is described in the next section. Enabling a company's strategy in machining to be followed means that there is a way for the product life cycle engineer to communicate with the SMS such that a coherent and global optimization will be formed by the machining decisions that are made at various points along the product life cycle. To bridge the gap between what is needed for each company's product life cycle, we believe knowledge-based software will require that can both communicate the strengths and weaknesses of the different models to a life cycle engineer, as well as capture what is important to the engineer, so that the best machining decisions can be proposed. Knowledge-based software is described in the last subsections.

## 4.2 Towards a semantic world in machining

Currently the most developed level of exchange for product information during its life cycle concerns data models and notably through ISO 10303, informally known as the STandard for the Exchange of Product model data (STEP) [28]. The objective of this standard is to allow the development of new application protocols, on the basis of integrated resources and by applying the STEP description and implementation methods. This standard must allow representation of product data from its conception, through its realization and ultimate recycling. Although work has been done to

[1] Getting the semantics right for people is also difficult. People regularly have semantic arguments and use dictionaries to make certain that they understand. Many interactions between people are simple and either do not require precise communication or are so simple that precision is easy to achieve.

develop STEP models for manufacturing and machining processes [29-31], it must be recognized that this approach is at present limited to modeling the product structure information such as geometry, dimensions and tolerances [32–34]. For this type of information, risks of nonsense (or miscomunication) are relatively limited. Information relative to different professions with high degrees of specific knowledge are difficult to represent in a uniform way. It is often very complex to represent a consensual schema of data which does not cause misunderstanding. The case of machining is concerned with such context [35] and to be competitive, companies need to represent their knowledge using different approaches [36]. This is mainly the reason why some researchers work on defining machining process concepts. In the USA, standards for data specifications, such as the ANSI B5.59-1[37] for machine tool performance tests and the ANSI B5.59-2 [38] for properties of machining and turning centers, represent data using the XML syntax which gives more definition to the concepts. For cutting resources the ISO 13399 [39] effort provides a glossary of terms for tooling and recently, data models have been proposed using some STEP parts and application protocols. The last example concerns STEP NC [31] whose role is to clearly define Numerical Control information in order to integrate it with STEP models. This approach seems promising for SMS to capture on-line information and make it available during the product life cycle. To exemplify the problem that these researchers are facing, the description of a cutting tool can be taken as an example. Some experts will see only the insert in the end of the tool, while others will be more concerned with the combined insert and tool holder, or still others will focus on their semantics for the active part of the tool constituted only of two faces and a cutting edge. This difficulty is more prevelant today due to the globalization of partners, which means that a company does not deal only with local partners but with worldwide practices and knowledge. Ontologies allow capturing both the semantic and the syntax description of the information. The purpose of ontologies engineering is to make explicit, for a given domain, the knowledge contained in engineering software and in business or companies' procedures [40]. An ontology expresses, for a particular domain; a set of terms, entities, objects, classes, and relations among them. It supplies formal definitions but also axioms whose role is to constrain the term interpretations. An ontology allows one to represent a very rich variety of structural and non-structural relations such as generalization, inheritance, aggregation, and instantiation. It can supply a precise model for software applications. Finally, an ontology is able to represent relations defined in taxonomic or data models by adding to it axioms which constrain the interpretation or implicit relations of terms. Ontologies are generally represented by using a wide variety of legible and logical languages which are understandable both by humans and machines. Such as shown in [41] Propositional Logic (PL) is one way to model ontologies, but PL lacks the expressive power to model concisely an environment with many objects and facts. First Order Logic (FOL) has much more expressivity and can represent much more complex relations between objects. The Ontology Web Language (OWL) is the language widely used by the semantic web community. In comparison to FOL, OWL is weightier and based on a taxonomic model. Some attemps concerning process ontologies [42] and machining [35, 43] have been done and are intended to be used as base models for the SMS implementation.

## 4.3 Knowledge-based models

In the previous section the concept of an ontology was introduced to help an SMS deal with interoperability problems. An ontology helps us deal with issues surrounding the semantics of terms and their precise usage. As such an ontology represents one kind of knowledge. Earlier in section 3, mathematical optimization trends and how an SMS might take advantage of those trends was discussed. This represents a different type of knowledge, that of solving problems that fall into the category of mathematical optimization. This is a critical component of machining process improvement. There is a third type of knowledge that helps with both the tactics and the strategy of improving the machining process. Tactics from the the point of view of helping a mathematical optimizer compute one of the functions the optimizer is evaluating. As described in section 3.1 there are different ways of computing cutting forces: (a) traditional experience-based models; (b) machining experiment methods using mechanistic models; and, (c) models that use standard material properties rather than specific machining experiments. When the mathematical optimization program needs a cutting force function, a tactical decision must be made about what is the best way to compute that function. The decision depends on: (a) what data is available; (b) how difficult is the data to obtain; (c) what stage of life cycle the product is in; and (d) various other questions that may depend on the organization's capitalized experience. This type of decision-making is based on heuristic knowledge. Frequently this knowledge is hard-coded into a system based upon interactions between domain experts and system analysts. The difficulty with this approach is that one ends up with a system that represents one view point, and for which there is difficulty representing multiple and changing viewpoints. The approach is to move away from capturing this knowledge as one unified program, but rather to use a mechanism that allows smaller fragments of knowledge to be captured at about the level of single sentences in a natural language. The formal languages used to capture this knowledge are the same as those used in ontologies described earlier. This has the advantage of allowing the domain expert and the system analyst to focus on single components of the knowledge base. When the tactical and strategic knowledge are captured, then generic programs (theorem provers, production rule / blackboard systems) can be used to make the tactical or strategic decisions. In conclusion, knowledge based systems will play two roles in SMS. First the role of helping to decide what to opitimize in the detailed process plan, and second what tactics to apply when optimizing the detailed process plan during its realization, i.e., using on-line monitoring and adaptive controls.

## 5 Conclusion and discussion of future work

The research presented in this paper is in its initial stages. Although significant information related to performance of machine tools, machining processes, cutting tools, and materials already exist, there is no unified methodology to combine this information to generate optimum machining conditions with expected outcomes. Our research aims to address this need by developing necessary tools and architectures. During this development, we plan to assess the robustness of available models and provide further improvements in models and data. For example, recent improvements in both FEM modeling techniques and sensors such as high speed visible light and thermal cameras are rapidly improving our ability to verify and improve machining models. Improvements in chip formation modeling are also expected as a result of this effort. Our immediate effort will be focused on demonstrating a simple version of the dynamic process optimizer with a specific example in turning including a knowledge base and mechanistic models. Loaded with high fidelity process and performance models and optimization tools, SMS will behave in a predictable and controllable manner well integrated with the rest of the manufacturing enterprise and life cycle engineering.

## Disclaimer

Any commercial equipment and materials are identified in order to adequately specify certain procedures. In no case does such identification imply recommendation or endorsement by the National Institute of Standards and Technology, nor does it imply that the materials or equipment identified are necessarily the best available for the purpose.

## References


1. Jurrens, K., Soons, J., Ivester, R., Smart Machining Research at the National Institute of Standards and Technology, DOE NNSA Small Lot Intelligent Manufacturing Workshop, 2003, proceedings number LA-14093, Los Alamos National Laboratory.
2. Wanyama, W., Ertas, A., Zhang, H.-C., Ekwaro-osire, S., 2003, Life-cycle Engineering: issues, tools and research, Int. J. Computer Integrated Manufacturing, 16/4-5: 307-316.
3. Fenves, S., Sriram, R.D. , Sudarsan, R., Wang, F., 2003, A Product Information Modeling Framework For Product Lifecycle Management, Proceedings of the International Symposium on Product Lifecycle Management.
4. Feng, S.C., 2003, Machining Process Planning Activity Model for Systems Integration, Journal of Intelligent Manufacturing, 14/6:527-539.
5. Jordan, J.A., Michel, F.J., 2000, The Next Generation Manufacturing: Methods and Techniques, John Wiley and Sons, New York: 50-51.
6. Robert, J. Vanderbei, 2001, Linear programming, International Series in Operations Research \& Management Science,37. Kluwer Academic Publishers, second edition.
7. Aharon Ben-Tal and Arkadi Nemirovski, 2000, Robust optimization---methodology and applications, Math. Program, ISMP 2000, Part 2, 92(3, Ser. B):453--480.
8. Fwu-Ranq Chang, 2004, Stochastic optimization in continuous time, Cambridge University Press, Cambridge.
9. Kurdi, M.H., Haftka, R.T., Schmitz, T.L., and Mann, B.P., 2004, Simultaneous optimization of removal rate and part accuracy in high-speed milling. Proceedings of, 2004 ASME IMECE, Anheim, November 17-23 2004.
10. Yann Collette and Patrick Siarry, 2003, Multiobjective optimization, Decision Engineering, Springer-Verlag, Principles and case studies.
11. Giorgi, G., Jimenez, B., and V. Novo, 2004, Minimum principle-type optimality conditions for Pareto problems, Int. J. Pure Appl. Math., 10(1):51-68.
12. NF E 66-520-1, Sept. 1997, Working zones of cutting tools – Couple Tool-material – part1 : general presentation, ISSN 0335-3931.
13. Merchant, M.E., 1945, Mechanics of the metal cutting process. I. Orthogonal cutting and a type 2 chip, Journal of Applied Physics, vol. 16: 267-275.
14. Merchant, M.E., 1945, Mechanics of the metal cutting process. II. Plasticity conditions in orthogonal cutting, Journal of Applied Physics, vol. 16 : pp. 318-324.
15. Lee E.H. and Shaffer B.W., 1951, The theory of plasticity applied to the problem of machining, J. Applied Mechanics, ASME, vol.18: 405-413.
16. Ng, E., El Wardany, T.L., Dumitrescu, M., Elbestawi, M.A., 2002, Physics-based simulation of high speed machining, Machining Science & Technology, 6/3, 301-329.
17. Burns, T.J., Davies, M.A., Rhorer, R.L., D. & al., 2004, Influence of Heating Rate on Flow Stress in High-Speed Machining Processes, Proceedings of the 7th CIRP Int. Workshop on Modelling of Machining Operations, Cluny, France, 2004, pp. 171-177.
18. Rhorer, R.L., 2003, Dynamic Material Properties for Machining Simulation using the NIST Pulse Heated Kolsky Bar, Proceedings of the American Society for Precision Engineering (ASPE).

19. Chatterjee, S., 1997, An assessment of quasi-static and operational errors in NC machine tools, Journal of Manufacturing Systems,16/1.
20. Donmez, M.A., K. Yee, B. Damazo, 1993, Some Guidelines for Implementing Error Compensation on Machine Tools. NIST. (U.S.) NISTIR 5236.
21. Hocken, R.J., 1980, Machine tool accuracy, Vol 5 of Technology of machine tools, A survey of the state of the art by the Machine Tool Task Force, Lawrence Livermore Laboratory, UCRL-52960-5.
22. James, B. Bryan, 1990, International status of thermal error research (1990), Annals of CIRP, vol 39/2.
23. Donmez, A., Blomquist, D.S., Hocken, R., Liu, C., Barash, M., Oct. 1986, A General Methodology for Machine Tool Accuracy Enhancement by Error Compensation._Precision Engineering, Vol. 8/4.
24. Soons, J.A., Theuws, F.C., Schellekens, P.H., 1994, Modeling of errors of multi-axis machines: a general methodology, Precision Engineering, Vol 14.
25. Gilsinn, D.E.; Donmez A., 1994, A Prediction of Geometric-Thermal Machine Tool Errors by Artificial Neural Networks. NIST. (U.S.) NISTIR 5367.
26. Donmez, M.A., 1992, Development of a New Quality Control Strategy for Automated Manufacturing, Proceedings of ASME Manufacturing International Conference.
27. Deshayes, L., Welsch, L., Donmez, A., Ivester, R., 2005, Robust Optimization for Smart Machining Systems: An Enabler for Agile Manufacturing, Proceedings of the 2005 ASME Internantional Mechanical Engineering Congress & Exposition (IMECE), Nov. 5-11 2005, Orlando (Florida), USA, Paper number IMECE2005-81537.
28. ISO 10303-1:1994, Industrial automation systems and integration Product data representation and exchange Part 1: Overview and fundamental principles.
29. Deshayes, L., Dartigues C., Ghodous P. & Rigal J.F. 2003, Distributed and standard system for cutting data management. CERA journal, 11/1:27-36.
30. Wolff, V.; Rigal, J.-F; Ghodous, P.; Martinez, M., 2001, Design and Machining data integration. Int. conf. emerging technologies and factory automation, IEEE:483-491.
31. ISO/FDIS 14649, 2001, Industrial automation systems and integration – physical device control - Data model for computerized Numerical Controllers. ISO.
32. ISO 10303-203, 1994, Industrial automation systems and integration - Product Data Representation and exchange – AP 203 : Configuration controlled design, International Organization for Standardization.
33. ISO 10303-214, 2001, Industrial automation systems and integration – Product data representation and exchange – Part 214: Application protocol: Automotive mechanical design process, ISO, Geneva.
34. ISO 10303-224, 2001, Industrial automation systems and integration – Product data representation and exchange – Part 224: Application protocol: Mechanical product definition for process planning using machining feature. ISO, Geneva.
35. Deshayes, L., 2003, Cutting Methodology Study, Link between Couple Workpiece Tool and Machine Tool Workpiece System (Méthodologie d'étude de la coupe, liaison entre Couple Outil Matière et Pièce Outil Machine)', Ph.D., Insa de Lyon, 271
36. Tichkiewitch, S., 2004, A Virtual Research Lab for a knowledge community in production (VRL-KCIP), proceedings of the 5$^{th}$ IDMME conference, Bath U.K.
37. ANSI/ASME B5.59-1 (draft), Data specification for machine tool performance tests
38. ANSI/ASME B5.59-2 (draft), Data specification for properties of machining and turning centers
39. ISO 13399, 2004, Cutting tool data representation and exchange – ISO, Geneva.
40. Uschold, M. & Gruninger, M., 1996, Ontologies: Principles, Methods, and Applications. Knowledge Engineering Review, 1996, Vol. 11:96-137.

41. Russel, S., Norvig, P., 2003, Artificial Intelligence A Modern Approach, second edition, ISBN 0-13-790395-2, 1080 p.
42. ISO/CD 18629-1,2001, Industrial Automation System and Integration – Process specification language: Part 1: Overview and basic principles, ISO.
43. Deshayes, L., El Beqqali, O., Bouras, A., 2004, the use of Process Specification Language for cutting processes, Int. Journal of Product Development (IJPD)

# Development of methods to support the implementation of a PDMS

J. Feldhusen, B. Gebhardt, N. Macke, E. Nurcahya and F. Bungert
*Chair and Institute for Engineering Design, RWTH Aachen University, gebhardt@ikt.rwth-aachen.de*

Abstract: The approach described in this paper comprises a set of methods that helps enterprises to decide whether a product data management system (PDMS) should be applied and how this could be done. On the one hand a precise and applicable procedure to implement the system is provided. On the other hand information whether an enterprise is prepared to the introduction of such a system is given already in preliminary stages. For both cases methods for planning and application will be presented to support a company's Product-Lifecycle-Management (PLM)-strategy.

Keywords: Implementation Set of Methods, Product Data Management System (PDMS), Product-Data-Model

## 1 Introduction

According to VDI 2243 every industrial product is subject to a Product Life Cycle including the phases "creation", "design & development" and "utilisation" [1]. Its Life Cycle Period is differentiated according to product type and industry sector. However, during the last years a constant truncation of this period has taken place. This significantly affects the company's organisational and business processes.

To master this challenge, many enterprises are currently planning to use a cost-intensive product data management system (PDMS) to make their business processes more efficient and to handle the diversity of their products [2]. This is not only motivated by the wish to retain the diversity of their products, but to lower the effort for administrating the diversity of product components [3]. A reduction of development and production cost as well as an enhancement of the efficiency of business processes within and between departments is anticipated.



*D. Brissaud et al. (eds.), Innovation in Life Cycle Engineering and Sustainable Development,* 381–395.

Although there are adequate proposals in the literature [4-7], they are only insufficiently realised [8-9]. The reason is on the one hand the lack of knowledge of persons who are involved in the introduction of the system. On the other hand the proposals are too abstract and fuzzy for the realisation. Therefore they are not supposed to be a useful guide for decision making and implementation.

The set of methods described below (Figure 1) is aimed to close that gap. It provides methods for analysis and implementation to support enterprises already in the run-up to an intended introduction of a PDMS. The emphasis is placed on preliminary inquiries to detect problems within departments of an enterprise and to balance them with the goals and requests of the enterprise. Misapprehensions and erroneous expectations of all involved persons, that will show up in general during the implementation phase can be minimised thereby. To reduce these risks caused by a constricted enterprise-specific view the project should be guided by an external consultant or a (neutral) university institute (in the following called "consultant").

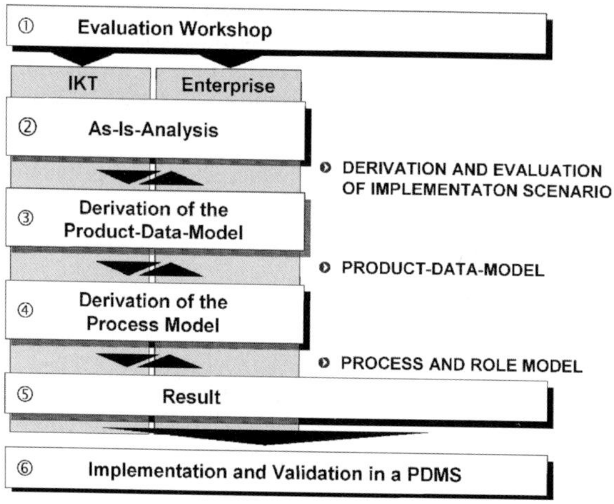


**Fig. 1.** Set of methods of PDMS implementation

Furthermore the emphasis is placed on the derivation and refinement of an enterprise specific product-data-model. This model not only results from development-specific procedures, but also accounts for department-spanning information, which matters to all people involved in the process. So the product-data-model amends the process analysis which is often considered to be paramount and will be addressed by the set of methods, too. The (system-independent) product-data-model as well as the (system-independent) process-analysis represents the content which will be implemented into a PDMS. The method addresses the management, the area managers and the project manager who plan, evaluate and realise the PDMS introduction. Although the set of methods addresses the department "development and design", it can be applied to other departments as well as to the whole enterprise to support the company's PLM-strategy in general.

# 2 Solution: Introducing set of methods

## 2.1 Methodological proceeding in the preliminary inquiry

The set of methods is roughly structured into two phases. These are the preliminary inquiry phase and the PDMS introduction and implementation phase. An important question during the preliminary phase is, how good an enterprise is prepared for the introduction of a PDMS. From that insight scenarios for the introduction are derived. They give a first overview about the effort and possible risks. After the evaluation of the scenarios the most appropriate ones will be selected and (PDMS- and product-neutral) criteria will be derived for the specification sheet. This is a basis for the decision, whether a system has to be purchased from an appropriate vendor or already existing software of the enterprise can be extended and adapted.

After the further proceeding of the project is specified this way, the phase of the PDMS-implementation follows. During this phase the selected processes together with the relevant part of the product-data-model (partial-model) are realised as a pilot implementation. The purpose is to get input for the assessment of the future effort for implementation of the system and of the possible (and reasonable) depth of system penetration as well as to prove the connection to existing systems (ERP, CAD etc.) of the future productive system. The conclusions drawn from the pilot implementation will be added to the requirement specification and serve as a basis for the acceptance criteria which the productive system has to fulfil after the pilot-phase. During the PDMS introduction project the presence of an intra-corporate project manager, who exclusively balances the project content and thus is freed from the day trade is of particular importance. He attends the project from the beginning on, coordinates and supervises the activities of all (internal and external) involved persons, works department spanning and reports directly to the management. Thereby the concerns of all departments are taken into account.

## 2.2 Evaluation of the enterprise goals, which are to be reached by the introduction of a PDMS

By the use of a PDMS enterprises expect distinct improvements concerning one of the following domains [10–11]:

- Improvement of the product quality
- Reduction of product and production costs
- Control of the version diversity
- Enhancement of the process transparency and efficiency
- Standardisation of processes, if possible and reasonable (for example for release or change-management processes)
- Control of the administration effort of current and future data (process and product data; in particular 3D CAD data) including the migration of legacy data
- Minimisation of the effort for data maintenance and simplified updates
- Creation of a role-model for project members
- Enhancement of the security through a corresponding administration of authorisation

In a preliminary talk the enterprise can present its requirements to the consultant. The initiators of the project as well as the management should attend the talk. Already existent conditions (for example a definition of development processes due to the ISO 9001 certification) can be identified by the consultant. This leads to a first estimation of the comprehensiveness of the necessary preliminary enquires. During the evaluation talk the requirements of the corresponding wishes have to be transferred into goals, which are realisable for the enterprise. It is important, that all possible groups and departments, which are affected by the PDMS introduction, attend. Thereby all requirements and views shall be registered to achieve an increased acceptance during the usage of the system. The registered requirements establish a base for the future specification sheet. Already at this time first problems and interfaces, which have to be taken into account during a later analysis can be identified. These goals have to be clearly documented for all people involved in the evaluation talk. To place an appropriate emphasis and to find antagonisms, the specified goals have to be reviewed in a subsequent workshop. Several and easily applicable assessment methods like a preference matrix [12] can be used for this review. Also in this case the attendance of all future key users is important, because now the different views are defined in terms of evaluation criteria. Often the most important criteria, which have to be taken into account during the next step, can be identified this way. An example for a result by dint of a preference matrix with adequate criteria is shown in Figure 2.

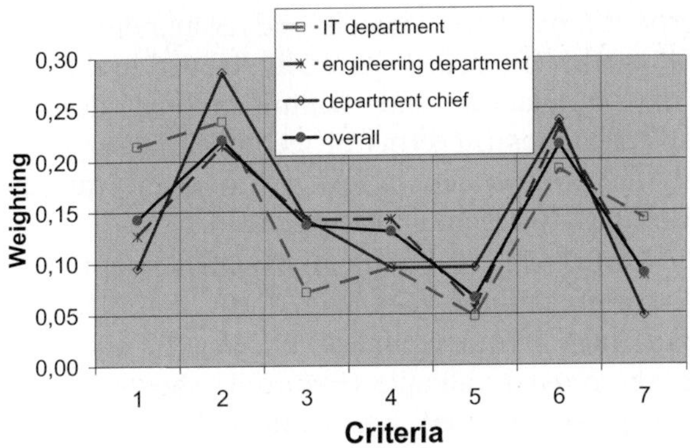


| | |
|---|---|
| 1 | possibility for self maintenance |
| 2 | short running time |
| 3 | product data model |
| 4 | data analysis possibility |
| 5 | number of user interface |
| 6 | process transparency |
| 7 | access control (user specific) |

**Fig. 2.** Identified criteria and its weighting from the fields IT and engineering

## 2.3 As-Is-Analysis as a conceptional base

Several surveys have to be performed to harmonise the enterprise goals with the requirements of a PDMS during the as-is-analysis. Thereby it has to be surveyed and documented, which fundamentals with what quality actually (!) exist in the enterprise. The following aspects are considered:

- Actual process analysis (among other things applying interview techniques): How precisely and detailed are particular processes defined (including their interfaces)? A feasible approach for the survey and analysis of processes provides ISO9001. From that individual survey tools in terms of process matrixes, which are adapted to the enterprise can be derived. These enable a computer based evaluation and comprise the input data, the actual process, the output data, the associated instances and the used tools.
- Actual product-data-model (among other things by examination of bills of material): How precisely and comprehensive is the product-data-model, which classifies and references all product data and contains appropriate attributes.

- Most enterprises have no explicit and comprehensive product-data-model. Product-specific data is generally located in the sets of drawings, in the bills of material and within product descriptions in the enterprises´ catalogues (in terms of technical and/or geometrical data). Accordingly the product data is scattered over the various departments. Often the user has divergent access rights.
- As-is-analysis of the IT-environment: To what extent offers the existing environment preconditions challenged by the efficient usage of a PDMS? For the accomplishment of the IT-analysis a top-down proceeding has proved valuable (Figure 3) shows this approach. Each structural layer is analysed and evaluated considering applicability criteria from "large" to "small" IT-structures to gain an overall structural overview

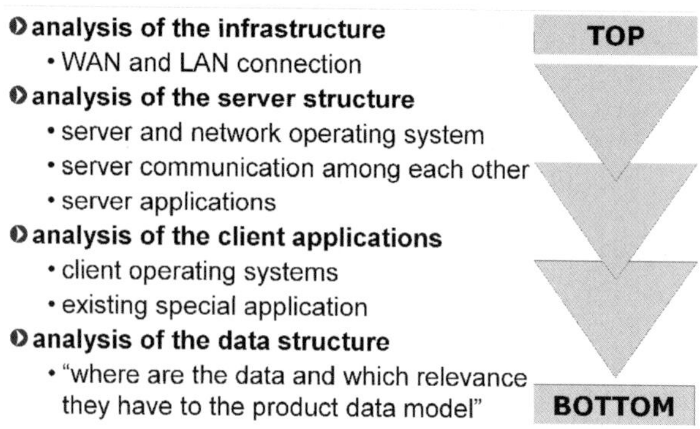


**Fig. 3.** Top-Down procedure for IT analysis

Concrete conclusions are derived for each single point. Together with identified criteria resulting from the evaluation workshop (see Figure 2) possible and, from the enterprises view, reasonable implementation scenarios are developed. They serve as a rough calculation basis for necessary resources, used time and emerging expenses. Because of that and further enterprise-individual criteria (such as the total number of users, depth of penetration etc.) the decision for further advancement is reached on the part of the enterprise in another workshop. The chosen scenario builds the foundation for the specification sheet constantly amended and rendered more precisely during the following project phases. The result of the as-is-analysis is a neutral documented present state of the three categories process, product and IT as well as the selection of a scenario for further orientation of advancement scheduling. This is put into writing in a first draft of a specification sheet.

## 2.4 Derivation of the product-data-model

Most enterprises manufacture physical products whose sale ensures market shares and therefore the enterprises livelihood. These products are described in entirety throughout the whole product lifecycle by product data. They come along with the product from its first draft to the final recycling. Thereby product data is not only used in each single product lifecycle phase but is also needed for preceding or subsequent phases and activities such as quality- or configuration-management [13]. Therefore the main focus for a PDMS-introduction has to be laid on a precise and extensive product-data-model. Such a model is missing in most enterprises and is perceived secondary in opposition to a process-model (although data associated to a given product should be identified and controlled as clearly desired by the enterprise [14]). Compilation of the product-data-model, describing the implementation content (including attributes, classes and references), is generated in three steps.

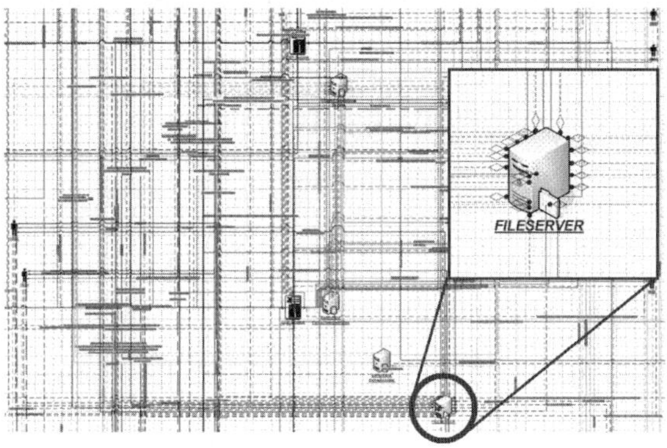


**Fig. 4.** Visualisation of information flow between IT systems and users (cut-out)

In the first step all product data has to be acquired, emerging from the product development process and accessed by the process parties (spanning departments) respectively. Identification and definition of the according interfaces is of special interest here and is taken into account in the process description later on. All manufacturing enterprises offer product descriptions serving as a first basis for the product-data-model. Further product data are acquired via interviews of employees. This is not only used for data gathering but also for visualization of data- and information-flux from persons involved in the product development process. By this means inconsistencies and problematic areas can be identified (Figure 4). Furthermore by distinguishing between analogue (paper-) and digital (document-) data redundancies can be discovered.

The second step comprehends consideration of the product itself on two levels. First the product is candled "theoretically" on the project-management level. This implies not only project-documentation but also the appropriate set of drawings and its derived bill of materials. On the other hand the physical product has to be stripped down to single parts. The main focus lies on the question if and in which representation each assembly and part appears in the set of drawings and the bill of material. In most cases it becomes evident that the traceability of the product development isn't unique or in the worst case impossible due to desiderative references.

By the means of this – very elaborate but worthwhile – product dissection product- and enterprise-typical "data-centroids" crystallise. By reason of these an enterprise-individual product-data-model can be laid down on the most basic structural layer. This model allows for conclusions concerning logical relations between the data-centroids. A further advantage in these data-centroids is a first rough classification on which basis a first attribution of the data generated in the first step can be found. Figure 5 and Figure 6 show the product-data-model as a STEP-model acc. to ISO 10303 [15] on the most basic structural layer (Figure 5) a detailed representation of the partial-model "assem(bly" Figure 6).

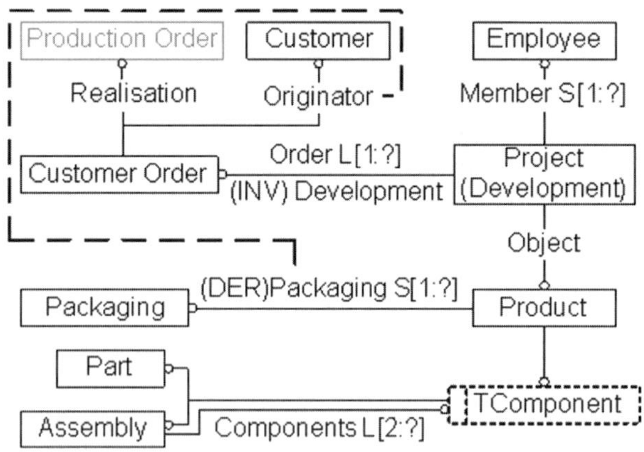


**Fig. 5.** STEP-product-data-model on the most basic structural layer

The result of this part of the set of methods is a common basis of a PDM-system-neutral product-data-model. This was either developed from an already existent product-data-model or compiled completely new. The enterprise is enabled to maintain this product-data-model and customise it for further needs. Corresponding experiences and criteria found during the compilation are registered in the specification sheet.

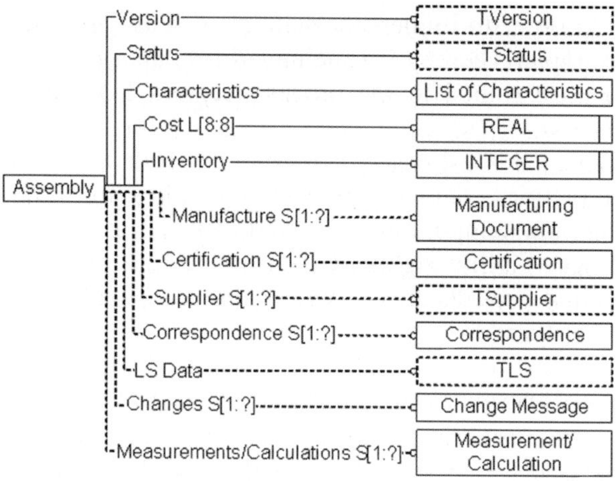


**Fig. 6.** STEP-partial-model of the data-centroid "assembly"

## 2.5 Derivation of the (whole) process-model and the rolemodel

For the implementation of a PDMS-capable process documentation is inevitable. The processes within the enterprise - if not already documented in the context of according certifications, in-house handbooks et al. - has to be newly recorded. Nonetheless these processes have to be conditioned for implementation purpose of a PDMS. Therefore an analysis and redefinition of the processes, incorporating intra-corporate standards, proceedings et al. and the compiled product-data-model, has to be made.

For the subsequent pilot-phase the redefinition of exemplary processes as basis for the following setup of a productive system is strongly recommended. In developing the (exemplary) process-model advancement in several steps is chosen, too. The fist step includes the identification of possible problem areas for the PDMS-implementation considering the conclusions of the previous pre-analysis schedule. Two or three processes which are particularly suitable for the following pilot-implementation result as recommendation. Two processes maximum are chosen after an evaluation workshop with the users and documented or worked off respectively as a base for workflow-implementation use within a flow-chart These workflows, which are a PDMS- compatible representation of the chosen processes, are implemented in the following pilot-phase with the appropriate process-relevant data (emanating from the product-data-model) By the process analysis (business-)processes can be standardised up to a designated degree in order to implement them as a basis for a workflow in the PDMS for example. Thereby it is achieved that the proper data, information and tasks are routed to the process-relevant persons at the correct point of time.

These activities and tasks of process-participating persons are described by the means of roles. Thus unique responsibilities and tasks are defined and redundant and temporally delayed tasks are avoided. Furthermore roles govern accessibility rights of data and assign process-participating persons explicit to projects and/or organisational structures. Table 1 points this up exemplarily.

**Table 1.** Roles and according activities (execution instructions)

| Activity/ Role | Description |
|---|---|
| Submit/ Proposer | The proposer defines, if and when an object (a drawing for example) is ready for status change and submits it. |
| Check/ Auditor | The auditor checks the object for correctness, gives annotations and votes pro or contra status advance. |
| Status advance/ Approver | The approver decides if the new status of an object will be advanced (accepted) or not (declined). He has the single authority to advance or degrade status, decline an object submission or totally remove the object from the process. |
| Observe/ Observer | The observer is notified as soon as an object is submitted for status change. He has access to the object itself, according input of the approver and the criteria for status change but is not able to make any decisions. |

Roles are generally defined within a role-model and will be assigned to specific persons by the administrator or the project or area manager.

Generally speaking a role-model incorporates the sum of and the interaction of all participants of a process (-step), which carry out defined and assigned tasks (execution instructions). Figure 7 illustrates the described correlations.

The role-model arises as a result of the product-data-model, the process analysis, the (general and specific) functionalities of a given PDMS and a considered project. All four factors interrelate with each other and therefore result in additional interdependencies. At this point in time it is only possible to show a general approach for compilation of different roles because of the strong interrelation of the compilation of a role-model with the according product-data-model (part of this project phase) and a consequential process analysis (part of the following project phase).

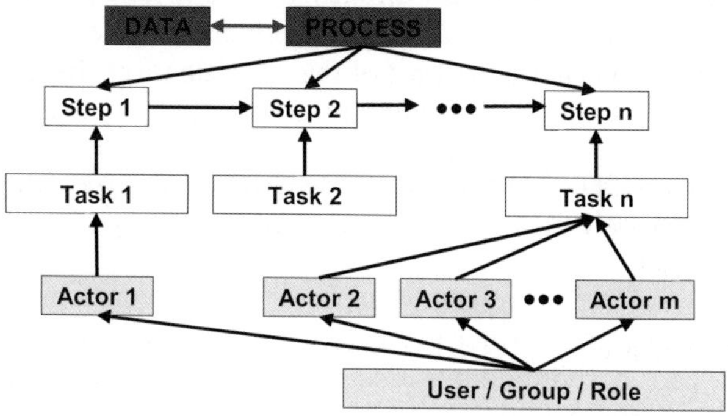


**Fig. 7.** Correlation between data, process and associated role

Depending on the functionalities of a given PDMS, certain standard roles can be derived. Concrete project-specific and organisational-structure bound roles can be assigned to these standard roles in form of responsible persons. Figure 8 illustrates some PDMS-functionalities, deriving standard roles and possible resulting roles for an enterprise.

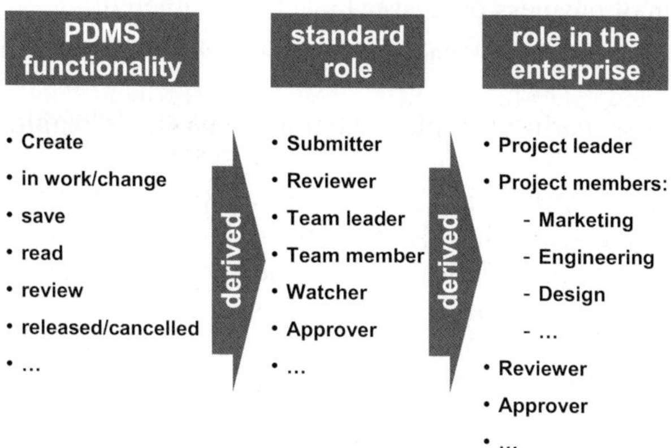


**Fig. 8.** basic PDMS functionality and possible deriving roles

Process-model and role-model are the result of this part of the set of methods. Most enterprises already have documented processes (ISO 9001, internal QM etc.) which have to be redefined and compiled more precisely in order to build a basis for later implementation.

Special attention in reference of the pilot-phase, has to be turned on the exemplary choice of two processes maximum which have to be up to the mark for the pilot-phase (see section 2.7). This choice has to be taken in agreement between the enterprise, the consultant and the implementation party (if a certain contractor has already been chosen) in order to acquire the best expressiveness for scaling the productive system. These requirements and selection criteria are part of the specification sheet.

## 2.6 Conclusion of the pilot survey phase: specification and system selection

Up to this point in the project all necessary preliminary inquiries making a statement concerning further project advancement have been concluded. Each result and deriving enterprise-individual requirements have been documented in form of criteria in the specification sheet.

For completeness and unambiguousness regarding the specification sheet it has to be verified if:

- necessary wishes if the enterprise have been considered;
- necessary criteria made by the enterprise on the PDMS have been listed (in the analysed departments, esp. IT-prerequisites; comp. criteria-determination and –evaluation [16];
- all functionalities supplied by the system which are necessary for succession of business processes have been gathered;
- all direct and indirect costs have been accounted for (or if hidden costs exist);
- the chosen scenario of implementation is realisable within the planned schedule.

The elaborate specification sheet serves as written basis of requirements for potential system vendors. These will check the listed requirements with regard to their product. Following that the enterprise is introduced to the possibilities of a given PDMS by the corresponding vendor in a first meeting (product-demonstrator). The prospects if one or more selected PDMS will be reviewed to it's adequacy, constituted in the specification sheet, by the means of a field test within operational practice will be argued in a second meeting (pilot-phase). Important in this context are discussions with reference customers of the given vendor already using the selected PDMS and therefore possession some experience according to this issue. Contact persons for reference customers can be found, asking the vendor itself or neutral institutions (such as universities). The result of this phase is the elaborate specification sheet which is used as a basis for the following pilot-phase of one or more system vendor(s) according the requirements of the specification sheet.

## 2.7 Implementation of a pilot- or starting-system

Within this phase the chosen PDMS will be evaluated on site considering preliminary defined criteria (chosen processes, required functionalities…) being used practically in the enterprise. This phase is very sensible according to two aspects: Firstly most vendors sense an opportunity in this phase to show an advanced product-demonstrator. Thus results the risk that enterprise-individual requirements of corporate day trade are not sufficiently considered. Secondly many enterprises cherish the illusion that the pilot can convert the bulk of their initial uttered wishes. Due to the temporal limitation of the pilot-phase (normally a timeframe in between three to six months) this is not feasible. Both aspects deliver no significant evaluation-bases for the productive system after completion of the pilot-phase.

Therefore the questions arises what the pilot-phase is good for? The pilot-phase allows for statements which are to be allied for later sizing of the productive system. The following points are of main interest here:

- How voluminous will the approximated effort be (mustered time; required cost, used resources…)?
- How deep should the penetration of the PDMS be within the enterprise (which processes can be implied to what extend in terms of corporate philosophy)?
- Is it possible to prove that the PDMS can be connected to the existing hard- and software-systems (ERP, CAD…)?
- Can the surplus value through implementation of a PDMS according to the enterprises wishes be truly achieved?

It is advisable to preposition contents, the amount of time and relevant evaluation criteria unique and in written form between all participants of the project. Part of this documentation is the constitution of a project core-team, the contact persons of all participants and the members of the project itself. This is mainly important for the affected users of the enterprise which are ideally freed from other duties and responsibilities for the duration of the pilot-phase. A supplement to the specification sheet proves a proper foundation for this written documentation.

The result of this phase is the deployment of the chosen PDMS within "pilot-conditions". All project participants gain initial enterprise-individual experience with application possibilities, exposure to the system by the employees as well as boundaries concerning the adoption. These results determine if further advance is reasonable of if (and which) modifications have to be made.

### 2.8 Implementation of the Working System

Due to the experiences of the preliminary implementation the relevant questions concerning effort, depth of implementation and connection possibilities can be answered and extrapolated to the extent of the whole implementation. Thereby the effort for the migration of legacy data and the training procedures for the staff have to be additionally taken into account. It is very important for the enterprise to know from the start of the evaluation talk, that the main effort has to be spent from now on. This not only comprises the whole implementation of the system, but also the administration and realisation of the process and data model. Furthermore both models have to be permanently updated and adapted to the actual condition of the enterprise. Dependent on the size of the enterprise this is a full time job for a PDM expert.

## 3 Summary

The set of methods, which is presented in this paper, provides a first scientifically developed and realisable planning tool for the introduction of a PDMS for a single department and/or a whole enterprise.

The planning tool emphasizes the importance of three dependant models influencing the success of a PDMS-introduction and therefore a company's PLM-strategy in general. These models concern

- the quality and the amount of identified product-data (including references);
- the accuracy of the identified and defined "core-processes" as a basis for PDMS-workflows;
- user's roles for the workflows handling the product-data (including appropriate right-management and actions).

Because of that, a high quality of each model is very important. Therfore, an "as-is-analysis" is also included in the set of methods to consider the individual situation of each company with respect to their product portfolio and their special organisational processes. Finally, this guarantees an excellent support of the company's general PLM-strategy.

## 4 Prospects

This planning tool will be supplemented in the future by a key data system to evaluate the enhancement of efficiency of the business processes due to the PDMS introduction (model approaches for monetary evaluations already exist [17]). Further on PDMSs will play a key role concerning the field of knowledge management [18] and Collaborative Product Commerce (CPC) [19], [20]. In terms of a holistic approach the presented set of methods considers even these fields, which will be more important in the future.

## References


1. VDI 2243, 2002, Recycling-oriented product development, Beuth
2. Huske, G., Volkel, B., Clayton, D., Thill, B., Chubrik, B., Hammond, M., 2002, PLM-Primer – The Next Big Thing in Enterprise Software?, Credit Suisse/First Boston Sector Review 4/2002; 6-10
3. Eigner, M., Stelzer, R., 2001, Produktdatenmanagement-Systeme, Springer; 29-80
4. Schöttner, J., Produktdatenmanagement in der Fertigungsindustrie, 1999, Hanser; 323-361
5. Eigner, M., Stelzer, R., 2001, Produktdatenmanagement-Systeme, Springer; 286-303
6. VDI 2219, 2000, Einführung und Wirtschaftlichkeit von EDM/PDM-Systemen, Beuth
7. Hartmann, G., Schmidt, U., 2004, mySAP Product Lifecycle Management, Galileo Press; 597-619
8. Abramovici, M., Ritter, H., Steyr, M., Naumann, U., 2004, Benefits of PLM, ITM Ruhr-Universität Bochum; IBM BCS
9. Feldhusen, J., Gebhardt, B., Nurcahya, E., Macke, N., 2004, Das unternehmensindividuelle Produktdatenmodell als Voraussetzung einer erfolgreichen PDMS-Einführung- ein Erfahrungsbericht. In: Stelzer, R., Bröckel, K., Engelmann, F., 2.Gemeinsames Kolloquium Konstruktionstechnik 2004, Saxoprint; 81-89
10. Schöttner, J., Produktdatenmanagement in der Fertigungsindustrie, 1999, Hanser; 26-51
11. Abramovici, M., Gerhard, D., 1997, Use of PDM in Improving Design Process: State of the Art, Potentials and User Perspectives. In: Riitahuhta, A. (Editor), Proceedings of the 11th int.l Conference on Engineering Design ICED 97, WDK 25/3; 317-322
12. Breiing, A., Knosala, R., 1997, Bewerten technischer Systeme, Springer: 238-241
13. Schichtel, M., 2002, Produktdatenmodellierung in der Praxis, Hanser; 17-35
14. Schichtel, M., 2002, Produktdatenmodellierung in der Praxis, Hanser; 5-17
15. ISO 10303, 1992ff., Industrial Automation Systems, Product Data Representation and Exchange, ISO
16. VDI 2212, 1981, Systematisches Suchen und Optimieren konstruktiver Lösungen, Beuth
17. Schabacker, M., Vajna, S., 2003, PLM: Bewertung und Optimierung von Prozessen, CAD/CAM-Report, 07/03; 40-43
18. Feldhusen, J., Gebhardt, B., Nurcahya, E., Macke, N., 2004, A Knowledge-Based-Engineering Design Process within Product Lifecycle Management – A Vision, Proceeding of the 5th Symposium on TMCE, Volume 2, 1109-1110
19. META-Group, 2003, Product Life-Cycle Management Tools, METAspectrum 45.0, Stamford
20. Feldhusen, J., Macke, N., Nurcahya, E., Gebhardt, B., 2004, Collaborative Product Commerce – Ready for the Global Cooperation using Systematic Engineering Design, Proceeding of the 5th Symposium on TMCE, Volume 2, 1179-1180

# The role of knowledge management in product lifecycle

G. Colombo and D. Pugliese
*Dipartimento di Meccanica, Politecnico di Milano, Italy, giorgio.colombo@polimi.it*

Abstract: The paper presents a discussion of some problems related to knowledge management in product lifecycle thought treating some applications developed in different aspects of lifecycle: conceptual design, detailed design, production, selling, maintenance, diagnostic and dismissing. In these applications, we show how Knowledge management is involved in all these phases: IT proposes several methodologies and tools to represent and manage knowledge to aid expert who operates in the domains related to product lifecycle. Methodologies such as Triz can be used to assist conceptual design, while KBE systems permit to automate design procedure of numerous products. Neural networks and fuzzy logic help to select right approaches to production. The paper ends presenting some applications performed by the authors, focusing aspects related to design.

Keywords: PLM, Knowledge Management, Knowledge Based Engineering

## 1 Introduction

This paper discusses the role of knowledge management in product lifecycle, focusing relations among phases, methods and tools. Discussion is based on examples documented in technical papers and on experiences that the authors have realized in the domain of Design Automation. In product lifecycle we underline following phases: conceptual design, detailed design, production, selling, maintenance, diagnostics and dismissing. For all of these, industrial and academic researchers have developed a lot of systems and applications to automate or assist the solution of engineering problems. They used different methods and tools but their target was the same: to build systems able "to think", to design or plan processes like a human being. To understand problems related to this domain, we can merge know-how and tools of specific engineering area (such as design, production, maintenance an so on) with Artificial Intelligence methods. Expert systems, rules, inferential engines, Object Oriented programming,



*D. Brissaud et al. (eds.), Innovation in Life Cycle Engineering and Sustainable Development,* 397–406.

graphs, reasoners, Knowledge Based Systems are in fact typical topics of AI. Presentation and discussion of case studies permit to understand how the previous methods and tool have been applied to solve problems in a specific phase of lifecycle. Our intention is also to highlight the fact that not only data are important in product lifecycle but also that knowledge plays a fundamental role. At the same time we investigate on the state of the art of the tools used to acquire, represent, manage and utilize knowledge because this is an important aspect that determines their dissemination in design context. An example of knowledge importance for product lifecycle is shown in the approach adopted by Boeing for its factories. Johns [1], in his paper, describes how a strategy named Lean is important in order to increase the quality of its products, decreasing costs at the same time. Knowledge and information deriving from finance, marketing, product development, suppliers and partners, aviation research, logistic management and security, research and development, are gained and shared among all departments in order to adapt the production in real time in accordance with costumers' requirements; consequently, cost, time and defects diminish, but quality remains high. In the following sections, examples of techniques and methodologies to represent and manage knowledge are shown in specific context of the entire product lifecycle.

## 2 Knowledge in conceptual design

In this section we start to discuss how Knowledge representation and management can affect the product lifecycle trough the presentation of some applications performed during the first phase of product development process, i.e. conceptual design. Conceptual Design concerns the identification in the space of possible solutions of that one corresponding to design targets. This activity requires a high level of synthesis capabilities and much information about existent products, parts and components, technical solutions and so on. It is very complicated to define a designer assistant for this activity because it is highly "intelligence driven". In spite of this, a lot of efforts have been spent in this direction and important results have been reached. A very interesting methodology for conceptual design is known as TRIZ (Russian acronym for "Theory for Inventive Problem Solving") and has been developed by Altshuller et al. [2]. TRIZ approach is particularly used when designers have to investigate an innovative design solution; an example of this use is proposed by Tsai et Alt. in [3]. A seat is a part of a valve; it is usually made from plastic or rubber, and can be damaged by high pressure or high temperature environments. TRIZ approach suggested designers to adopt a magnetic closing system to preserve the seat from wear and high torque, increasing its life. A lot of other methods and techniques such as Brainstorming, Mind Maps, Fermi's approach, Synectic Thinking and Axiomatic Design are presented and evaluated by Goel and Singh [4]. A very interesting and promising method is presented in [5]; the so-called 'graph representation' can be used for systematic design of engineering systems and also to find relations between different engineering fields. The basic idea of this method is very simple; let an original problem, we can transform this in another problem with well known solutions and then we transform that to original problem. The author presents also some examples related to static (trusses), kinematic (linkages) and electric circuits. Techniques to assist conceptual design are not used only in industrial engineering domain; also in civil engineering researchers work in this direction. An example of KBE applied to building design is presented in [6]; a software assistant permits to i) develop a specification of the design problem, ii) generate a schematic layout of the building and, finally, iii) generate a schematic 3D model of the building. In all these phases, the designer has a direct control of the process and can interact to him. Another example in building design is described in [7]; in this case, a framework allows engineers to explore design state space. The framework is based on an interactive hierarchical planning algorithm integrated with a description logic reasoner.

## 3 Knowledge in detailed design

In detailed design, selected solution for a product must be well defined. Probably, this phase of product lifecycle is the simpler to assist because it is based on activities that can be easily automated. Examples of such activities are: selection and sizing of the components, assembling of the parts by specifying constraints, analyses of the structural/kinematic/dynamic behaviour, and iterations of previous steps and so on. To aid designers in their activities, some techniques and tools can be used; first, parametric CAD systems. Parametric models and procedures written in programming or spreadsheet languages permit to automate design of a lot of parts and systems; this approach was developed during 80s of last century and is also today very popular, especially in SME. Colombo et al. presented an example of this approach in [8]; fig. 1 shows a more recent application concerning a storage tank. The previous approach is limited in managing complex products and different product architectures. Knowledge Based Engineering (KBE) systems were developed from the 80s of last century to overcome these bounds. First members of commercial systems of this type were *ICAD*® and *The Concept Modeller*®; they were based on object-oriented (O-O) language to define product architecture, parts parameters and methods to select, size and assembly them. A geometric modeling kernel, tools to define customized graphical user interface, interfaces to external data-base and programs complete the characteristics of these systems. Colombo and Cugini in first years of 90s presented an application focused on O-O description of the product architecture [9]. We think that this aspect, i.e. O-O representation of a product tree, is a very important contribution to represent knowledge to automate or aid design activities. Many industrial fields adopted this approach, like automotive or aerospace. Jaguar, from 1988, uses *ICAD*® system by KTI to design the headlight of his cars. In fact, every time car style designers change the form of the car body, it's necessary to modify the structure and the light projection angle in order to keep the right visibility of the road. Headlight data are made available for suppliers, in order to build and assembly the right components [10]. In aeronautic, Airbus uses *ICAD*® to generate a simple wing model for its airplanes, to be sequentially optimized in a FEM code [11]. The wing shape depends from data such as weight, dimensions, speed and similar information deriving from customer requirements. In the last years, the use of KBE systems has increased and several companies working in different fields have adopted this technology.

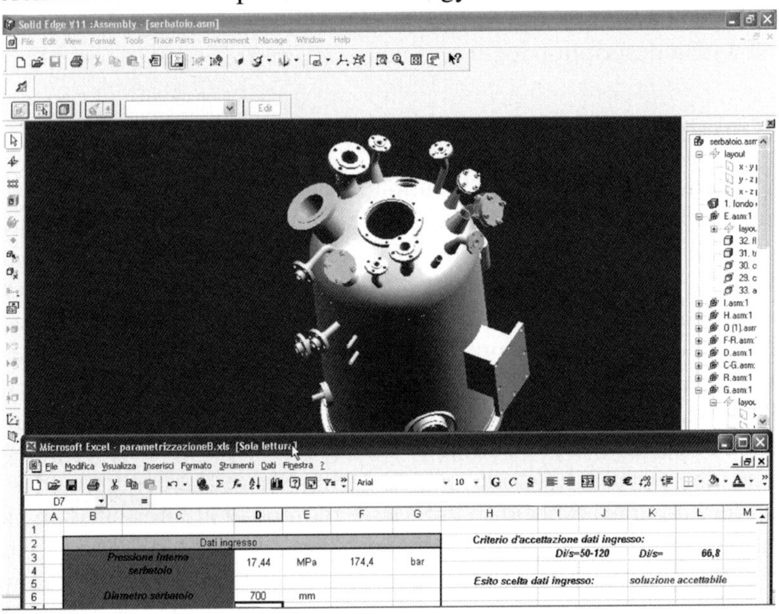


**Fig. 1.** Parametric model of a storage tank configured and dimensioned

KBE can be used also to plan and manage analysis activities; Craig et al. developed a KBE system named DART for automotive industry [12]. In this application, geometrical, materials and sizing information deriving from car components, as motor or gear box, are available to improve the shape of the chassis to be analyzed in a FEM code. An important aspect is related to control of activities performed by KBE applications; in most of them, the designer has not the control of the process; only at the end of the design process he/she can see the results of a lot of rules and procedures processed. This is a problem because professional engineers have responsibility of the design and "also because they like to do design" [13].

## 4 Knowledge in production

Knowledge Management in Production permits the planning of resources, tasks and overhauls, respecting time and restraining errors. Information deriving from customers' requirements, design specifications and law restrictions, has to be taken into consideration during the planning of production. Several methods and tools have been developed for this purpose. An example of interaction between production and customer was presented by Adenso Diaz et al. [14], who developed a fuzzy logic application to manage complex production planning. In their paper, they present a hierarchy of models for roll shop departments in the steel industry, focusing on the calculation of the priority of the rolls to be produced. A fuzzy-based model was developed and implemented in a real environment, allowing the simulation of expert behaviour, considering the characteristics of an environment with imprecise information. The main guidelines in this application were: i) to semi-automate the process of taking priority decisions; ii) this automation should be designed considering expert knowledge and master experience; iii) historical data and estimation should be used. On the basis of these premises, they developed an expert system that uses external database to store information. The role of inventory is not limited to conceptual design; an interesting application of TRIZ separation principles and theory of constraints (TOC) was developed by Stratton and Warburton, who studied a method to regulate flow fluctuations of inventories and to reduce waste [15]. Knowledge management in production field can be also reached using multi agent systems, a technology widely rife. Karageorgos et al. [16] used this approach to simulate behaviours to get values for better scheduling resources in production processes. In their paper, they present an application in which holonic agent systems are used to support non-trivial integration of manufacturing and logistics service planning. Archimede and Coudert present a work in [17], where they developed a system to solve the problem of scheduling flexible manufacturing systems. It starts from customers' expectations and it also takes into account the troubles at different planning levels. Their model guarantees the correct stability of the transmitted orders to the operators. The system architecture permits to consider several perspectives such as multi-sites scheduling, co-operation with other functions linked to the production management. Also neural network can be used to optimize processes. Sette et al. proposed a method to optimize generic product process using propagation neural network and optimising the resulting production process function with genetic algorithm. They applied their method [18] in a real test-case: the fibre-yarn production process. Using their method, choosing among several available fibre qualities, it's possible to predict and then optimize the resulting fibre in terms of tenacity and elongation. The resulting yarn is of superior quality against common fibres.

## 5 Knowledge in selling, maintenance and diagnostic

Also Sale and Support Divisions can derive profits from a KM approach. They are the final sectors managing the lifecycle of products and information deriving from them is very important for the entire product lifecycle management because it is directly linked to customers' and markets requirements. Changchien and Lin [19] have developed a Knowledge Based System to store, use and reuse marketing plans in which operations consist of four steps: information, analysis, decision, and action. Information refers to financial parameters, marketing audit, and omnibus research facilities; analysis includes brand feasibility studies, market analysis, and product mix; decision concerns marketing targets, marketing mix, and budgets; and action indicates the execution of the plan. Every plan is stored to be used in following market cases. An interesting experience of PLM implementation was developed by Kiritsis et al. (PROMISE project). This system enables feedback of data, information and knowledge, from service and maintenance and recycling experts back to the designers and producers [20]. Machine's dismissing is nowadays a felt problem and several applications were developed to well manage the disassembly operations. Apley and his colleagues have studied a system [21] able to diagnose what is happens when someone attempts to unscrew and remove a screw. Their paper presents an algorithm for diagnosing the unscrewing process and deciding what to do if the screw is not coming out. Information is got by measuring the unscrewing torque and shaft rotation and it's possible accurately diagnose which condition arises during unscrewing.

## 6 Our activities in design automation

In the domain of knowledge representation with IT techniques, our activities at Mechanical Department of Politecnico di Milano, Italy, concern principally the detailed design. In particular, from the last years of '80 we studied the problem and developed applications: first, using parametric and, then, KBE tools. First work is described in [8] and concerned a parametric design system integrated with a spreadsheet-like tool to represent design rules and procedures. Then we realised some application using *The Concept Modeller®* of Wisdom Systems; one is described in [9]. In the last years we developed KBE applications to configure a changing tools system [22], a tube heat exchanger [23], and a family of press brake. All these applications have been developed with RuleStream by homonymous Corporation (www.rulestream.com). In these applications, product architecture is represented with a tree structure; the design process is also defined and the user-designer is guided step by step during the configuration of the product thanks to the possibility of viewing and interacting with the system. Sellers, suppliers and designers can interactively modify the final configuration in accordance with their access right to the system. In Figure 2, the configuration environment for the Changing Tool System for a machine tool is shown with the process steps. In this application, data get from machine tool dimensions, customer requirements and suppliers products information, affect the design process in order to well resize and configure the changing tool system. Another application concerns a KBE prototype for a press brake producer Company. In this application, information about the design process, products data, costumers' requirements, production constraints and availability information of suppliers' products are usable by designer who is gradually guided into product configuration process. Suppliers can update databases where their product data are stored, customers can define their products configurations via web and designers can configure in real time the final product. In figure 3 the configuration environment is displayed. In the left side it is possible to select the steps of the process in order to guide the design process choosing the most suitable components, sizing them and producing constructive drawings and Bills Of Material. Information to correctly configure the final product is always available, correct and rationalized. We also developed a prototype for configuring industrial tube heat exchanger. Thanks to an iterative and automatic process and a structured representation of knowledge in design process, designer can arrive to final configuration also using other calculation tools like MS Excel. In Figure 4 a step of the configuration and the final result of the exchanger and the electronic sheet for tubes sizing are displayed. In all these applications, KM plays a fundamental role. In fact, the capability of having information and a structured know – how is important, in order to well configure the final products. To reach this aim, involved actors, as designers, producers, suppliers, sellers and also customers, have to play their contribution enduing right and update information.

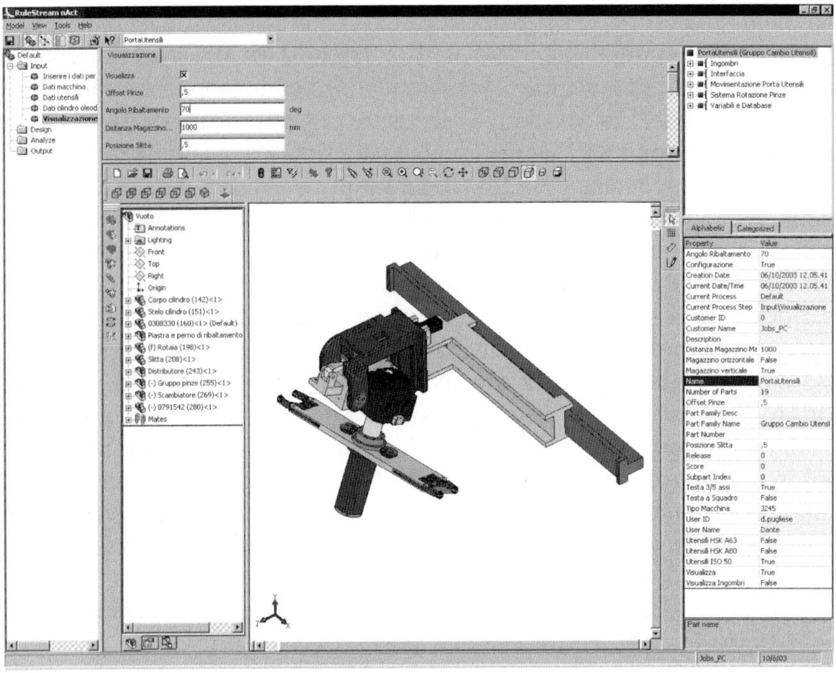


**Fig. 2.** Configuration Environment for a Changing Tools System for machine tool. On the left, the configuration steps are organized in a tree structure

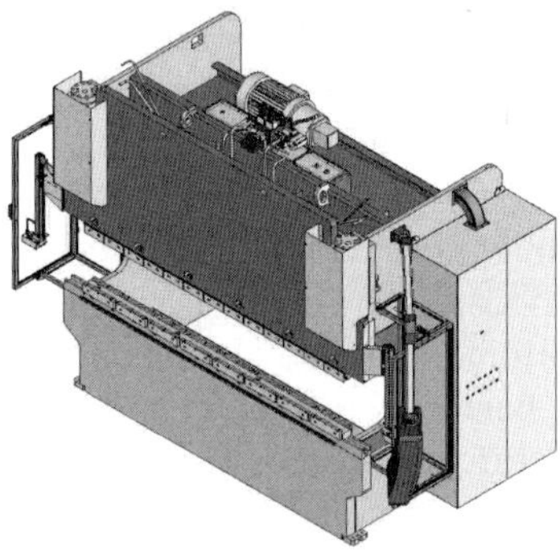

**Fig. 3.** A possible configuration of the press brake

## 7 Summary and next developments

In this paper we discuss which role KM plays in industrial product lifecycle. Several applications are presented to show how knowledge management can affect the various phases of design, production, selling, maintenance, diagnostic and dismissing. First consideration concerns the strategic role played by knowledge in all the aspects of product lifecycle;

after the problems related to digital representation of data, researchers in IT domain must face the problem how represent knowledge. Referred examples propose a great number of methods and tools to reach this objective; TRIZ, object-oriented programming, expert systems and so on have been used principally in a specific domain. For example, TRIZ is used in conceptual design while object-oriented tools are widely used in detailed design or expert systems in process planning and maintenance. Probably, a detailed analysis of tools, methods and problems can permit the identification of better way to solve specific aspect. Finally we present our activities in design automation, a specific domain of knowledge management. KBE systems have been realised to capture and use knowledge and know-how to design automatically products with well-known architecture. These applications are based in product architecture and design process representations; direct control by the users has been considered. Guided step by step during the configuration of the product, designer can view and interact with the subsequent tasks of the design process. Sellers, suppliers and designer can interactively modify the final products configurations in accordance with their access right to the system.

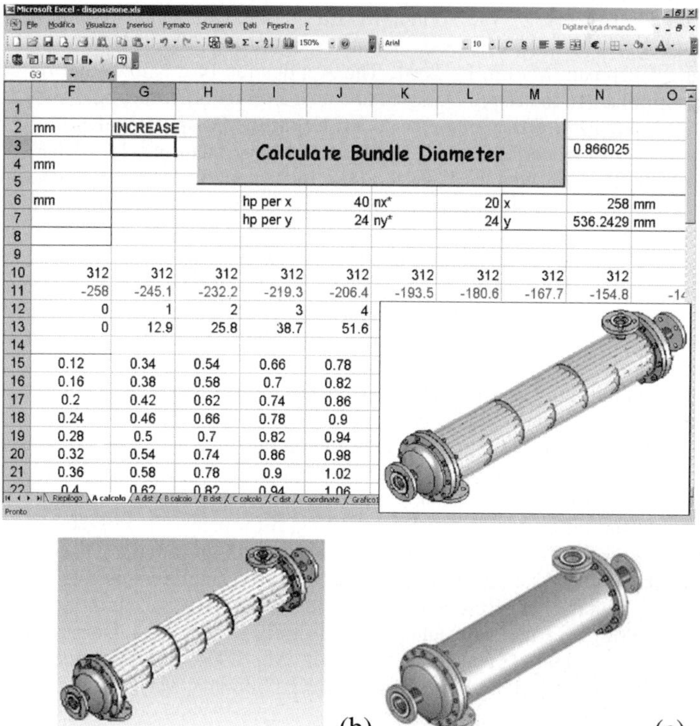


(a) (b) (c)

**Fig. 4.** (a) The electronic sheet for tubes sizing, (b) a step of the configuration of the tube heat exchanger and (c) the final configured product

## References

1. Johns, J., 2003, Factory of the future, AIAA/ICAS International Air and Space Symposium and Exposition: The Next 100 Y, Dayton, Ohio, USA
2. Altshuller, G., Zlotin, B., Zusman, A., Philatov, V., 1999, Tools of classical TRIZ. Ideation International, Southfield, MI
3. Tsai, C.-C., Chang, C.-Y. Tseng, C.-H., 2004, Optimal design of metal seated ball valve mechanism, Structural and Multidisciplinary Optimization, 26/3-4, 249-255
4. Goel, P.-S., Singh, N., 1998, Creativity and Innovation in Durable Product Development, Computer ind. Engineering, Vol. 35, Nos 1 – 2, 5 – 8
5. Shai, O., 2003, Transforming Engineering Problems through Graph Representation, Advanced Engineering Informatics, n. 17, 77-93
6. Flemming, U., Woodbury, R., 1995, Software Environment to Support Early Phases in Building Design (SEED): Overview, Journal Architec. Eng., 1, 147-152
7. Eisfeld, M., Scherer, R., 2003, Assisting Conceptual Design of Building Structures by an Interactive Description Logic Based Planner, Adv. Eng. Informatics, 17, pp. 41-57.
8. Colombo, G., Cugini, U., Ferretti, D. 1989, How to Represent Design Rules in a Parametric CAD System, Int. Symp. on Adv. Geom. Modelling for Eng. App., 271-282
9. Colombo, G., Cugini, U., 1992, An Example of Knowledge Representation in Mechanical Design Using Object-Oriented Methodology, EES 1992, Dresden, 208-214
10. Calkins, D., Learning All About Knowledge-Based Engineering, Product design and development CAD/CAE report
11. La Rocca, G., Krakers, L., van Tooren, M.-J.-L., 2002, Development of an ICAD generative model for blended wing body aircraft design, AIAA2002-5447
12. Craig, B., Chapman, Martin, P., 2001, The Application of a Knowledge Based Engineering Approach to the Rapid Design, Advances in Eng. Software, 32 903-912
13. Garrett, J.H.Jr., 1998, The Computer-Aided Engineer: Prospects and Risks, Artificial Intelligence for Engineering Design, Analysis and Manufacturing, 12, 61-63
14. Adenso-Diaz, B., Gonzales, I., Tuya, J, 2004, Incorporating fuzzy approaches for production planning in complex industrial environments, Eng. App. of AI, 17, 73–81
15. Stratton, R., Warburton, R.-D.-H., 2003, The strategic integration of agile and lean supply, Int. J. Production Economics n. 85, 183–198
16. Karageorgos, A., Mehandjiev, N., Weichhart, G., Hammerle A., 2003, Agent-based optimisation of logistics and production planning, Eng. App. of AI, 16, 335–348
17. Archimede, B., Coudert, T., 2001, Reactive scheduling using a multi-agent model: the SCEP framework, Eng. App. of AI, 14, 667–683
18. Sette, S., Boullart, L., Van Langenhove, L., 1996, Optimising a Production Process by a Neural Network/Genetic Algorithm Approach, Eng. App. of AI, 9/6, 681 – 689
19. Changchien, S.-W., Lin, M.-C., 2004, Design and implementation of a case-based reasoning system for marketing plans, Expert Systems with Applications, 28, 43–53
20. Kiritsis, D., Bufardi, A., Xirouchakis, P., 2003, Research issues on product lifecycle management and information tracking, Adv. Eng. Informatics 17, 189–202
21. Apley, D.W., G., Seliger, G., Voit, L., Shi, J., 1998, Diagnostics in Disassembly Unscrewing Operations, The Int.l J.l of Flexible Manufacturing Systems, 10, 111–128
22. Pugliese, D., Benassi, M., Bordegoni, M., Pulli, M., 2004, Application and evaluation of a methodology for candidate technology selection, Design 2004, Dubrovnik
23. Colombo, D., Colombo, G., Cosmai, G., Pulli, M., 2004, A KBE application to automateheat exchanger design in SME, CIRP Design Seminar, May, Cairo

# A product-process-organisation integrative model for collaborative design

F. Noël
*Laboratory 3S, University of Grenoble, France, Frederic.noel@inpg.fr*

Abstract: This paper deals with the interest of an integrative Product-Process-Organisation model. This model creates connections between a representation of product, a representation of design process and a representation of company capabilities. It also links various representations of a same product due to every life cycle step to various activities that are involved during design. This model is to be shared on a collaborative fashion by various design contributors who have very different knowledge, know-how and goals. This technology provides a virtual space where experts from various skills negotiate decisions about product design and manufacturing.

Keywords: Collaborative design, negotiation, integrative model.

## 1 Introduction

Various integrated models were proposed in last decades to efficiently [1] share information related to product definition during the design process. The first approach was reduced to a geometric model often called the CAD model. Then models [2–4] integrated more technology information about the product and its environment. Other models were defined to support remaining informal discussions [5].

A PPO model (Product Process Organisation) integrates a Product definition with a design Process definition [6] and an Organization definition of the enterprise. Here, process stands for a sequence of activities involved during design and does not refer to the manufacturing steps. Integration in the PPO model manages relationships between the three previous dimensions to keep an efficient trace of the design activity. This trace can be used to get handles about how the product is defined but also why it was defined in such a way. These handles give opportunities to efficiently solve



*D. Brissaud et al. (eds.), Innovation in Life Cycle Engineering and Sustainable Development,* 407–418.

conflicts occurring between experts from various skills. To complete this model an integrative approach [7] is proposed. An integrative model is a model adaptive enough to integrate new concepts and know-how whenever they appear. This property is fundamental whenever design activity is based on negotiation between experts from different know-how and knowledge. Thus models are adapted to current collaboration and negotiation process integrating on the fly the points of view of every partner.

In this paper, the author discusses the interest of an integrative PPO model respect to other various existing models for the purpose of collaborative design. It highlights the main benefits of this new information structure and how it can be used to support negotiation about product definition. Next Section presents the new integrated PPO model dedicated to share information for collaboration. Section 3 deals with a technology to make this model reactive enough for integration of new expert services. Before the conclusion, Section 4 describes a case study where a mechanical analyst is working in collaboration with a process manager and a manufacturing expert.

## 2 An integrated model

### 2.1 A product model

Industrial PDM systems provide a representation of product, which is based on a hierarchical decomposition of its structure. Every component of the hierarchy can be associated with various numerical documents. The structure is supposed to be agreed by every designer. Each designer attaches his own documents to components of the hierarchy as a specific representation of the component. Hence the part is the smallest atom of information that is shared through PDM.

In the IPPOP product model, four trees of information are provided:

1. Component tree defines a partition of the product into sub-components. A usual decomposition is based on assemblies, sub-assemblies and parts definition. However, in the PPO model, this partition may be deeper than the part definition to support any useful partition of the part.
2. Interface tree defines the interfaces of components at various levels of details. The geometric boundaries of a component are usual interfaces. Nevertheless, interfaces refer to various types of boundaries: as an example, a component may also have electrical interfaces that do not match the partition of the boundary geometric representation.

3. Function tree defines the role of components by associating almost two interfaces and a sub-component corresponding to the technical solution chosen to achieve the function.
4. Behaviour tree defines the states of the product during its life cycle and which are considered during the design step. It refers to other modelled entities that are active during this state. With this behaviour object, product may be represented in various states and common components, interfaces and functions are associated together.

Figure 1 shows a simplified UML representation of the whole integrated model. Part surrounded by the dotted rectangle describes the classes corresponding to the product model.

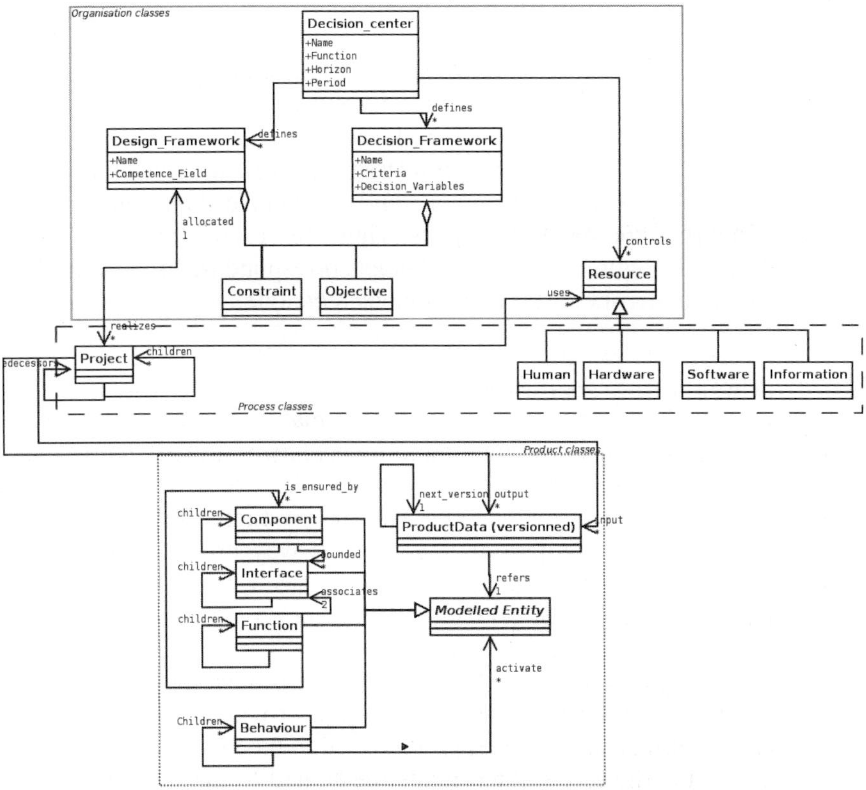


**Fig. 1.** A new integrated model; the PPO model

## 2.2 A process model

The process may be represented in various ways. In project management tools an IDEF fashion model is used [8]; a tree of projects describes the process. A sheet of this tree is a basic task. The project/task object is the main object of this model. The tasks are sequenced by links to predecessors. Obviously a project is achieved with a set of resources that may be human, hardware, software or information resource.

In Figure 1, the dashed rectangle, defines the main objects involved in process definition. Indeed, other objects complete the process definition (triggers, etc.) but are not represented here to simplify the discussion.

## 2.3 An organisation model

Unlike process and product models, the organisation model is rarely defined in commercial software. Here the idea is to define decision centers (based on some decision framework with objectives and constrains). These centers manage the design activity at a strategic level. A decision center is in charge to define design framework and to allocate their roles and allowed capabilities for specific projects. Thus the design framework has a technical role. In Figure 1, the main classes corresponding to the organisation model are underlined with a continuous rectangle.

## 2.4 A new integrated model

The previous sections described three independent models. These models are integrated as follow:

1. A product data, which is an handle over the product model, defines the input or output of a project or task: by this way process and product models are connected together. The process has a view about involved technical data while technical data are attached to the various tasks and projects working on them.
2. The organisation model is in charge to allocate project to various design frameworks and to ensure that resources are provided for projects. By this way the organisation model is connected to the process model.

This model is a new integrated model dedicated to the management of innovative design. Compared to existing PDM systems a deeper definition of component tree is provided. Other fundamental views as the function and interface tree complete product definition. The decomposition of these trees is deeper than in current PDM systems and deals with any detail that designers want to share. The organisation model is sharper than in PDMs, which is usually limited to the representation and allocations of resources. The process model connects organisation with product definitions. The STEP standard also provides an integrated model, which could have been used in place of the presented one. The STEP model takes advantage of its standardization that makes it more easily agreed by a large community. However a few works provide dynamic applications dealing with a complete STEP model [9]. At last the way in which a standard is revised implies STEP to be frozen with very slow evolutions. The model proposed in this paper is to be modified very dynamically through the concept of integrative technology [7].

# 3 An integrated technology

## 3.1 About integrative environment

The previous section deals with various integrated environment. Usually integration provides static environments: to add a new service into the integrated environment requires several months to specify and encode the necessary adaptations of both the integrative environment and the new integrated service. Moreover, the political decision before this process may take a while. Then integrated environments do not appear to be reactive enough to the evolution of services.

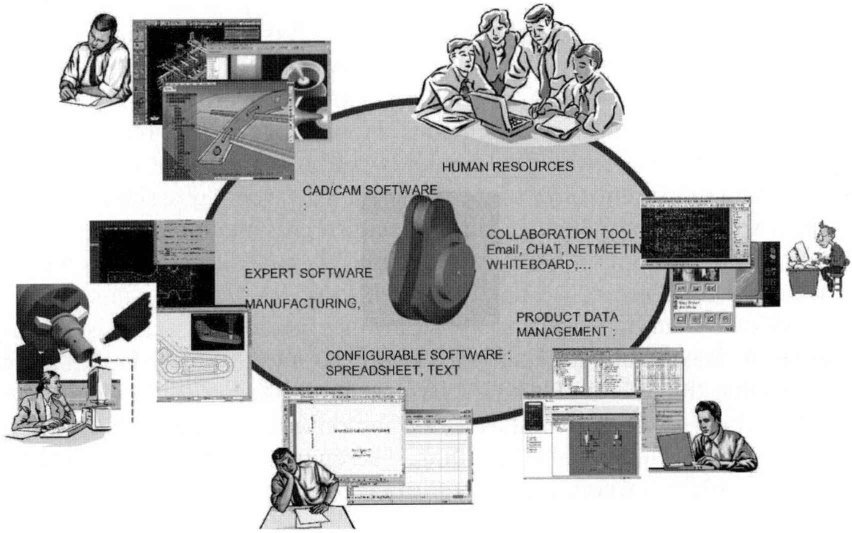


**Fig. 2.** Collaboration around the PPO model

In [10–11] the authors analyse the interest of dynamic technologies to provide integrative environment. On one hand, integrated environment are usually based on databases or object-oriented model. These models are defined by a set of object classes as defined in a UML class diagram [12]. Each class has a finite set of attributes and methods to access the attributes and define its behaviour. On another hand integrative environments must provide editable models. In the IPPOP specification the concept of fuzzy attribute is used to make the PPO model an integrative one.

## 3.2 Fuzzy attributes

A fuzzy attribute is defined in the same way as a usual attribute. A special tag is appended to set this property. When a class has almost one fuzzy attribute, it becomes fuzzy itself. A fuzzy object is an instance of a fuzzy class. This kind of object provides a usual access to its non-fuzzy attributes. These attributes are permanently defined. Nevertheless, fuzzy attributes may be or may not be defined. A fuzzy attribute has a value whenever the user - working on this object via a specific service - specifies a value for the attribute, but not before. Moreover the list of fuzzy attributes belonging to a fuzzy class may be completed at any time while services are already running and working with defined classes.

## 3.3 A self-learning environment

With fuzzy classes an application handles a basic model and completes this model by adding new fuzzy attributes. At the moment, the component or interface class in the PPO model have no specific attributes: just a few relations with interfaces and functions. However it is declared as a fuzzy class. Then, whenever the PPO model is working, the list of attributes defining the component object is completed. Any service assisting experts from a specific know-how handles special representations of the objects defined in the PPO model. An expert about manufacturing process will associate the interface of a component with the tolerance for this surface. The software service used to assist the manufacturing analysis usually handles a class defining the surfaces of objects and this class has an attribute to value the tolerance on this surface. In the fuzzy PPO model, tolerance attribute is added to interface definition at any time without a long process of specification and encoding. The environment learns new attributes for its objects whenever these attributes are required.

### 3.4 Sharing heterogeneous models

The PPO model is created for collaborative activities. Thus a server must share the PPO model over a network and provide multiple accesses to experts from various skills and knowledge. Figure 2 describes the corresponding organisation. In this figure experts continue to work with their usual services that may be very heterogeneous. They share a common representation at the centre of the ellipsis, supported by the PPO model. Each service (CAD-CAM modeller, Analysis software, Configurable Tools, PDM, Collaboration tool, but also organisation tools, etc.) may send the definition of attributes they use to define the various entities of the PPO model: component, interface, function, behaviour, project, product data, decision centre, etc. The PPO model is completed on the fly to take into account these new definitions through fuzzy attributes. Without any modification of the integrated PPO model, the concerned services are able to share information together. On the service side, just a plugging module must be provided to communicate with a predefined API of the PPO model. This API provides access to the management of the PPO model via a network connection and allows building three methods: 1) a method sends to the PPO model expected attributes for the current service – 2) a method checks out the PPO model: it retrieves information of the PPO model and updates the service model respect to the PPO evolution – 3) a method checks in the service model: it updates the PPO model respect to the service model evolution. These three methods are sufficient to manage an effective collaboration between experts using very heterogeneous models. Moreover this plugging module does not alter the model and functions of the service. It is not a complex specification for each service and it should be encoded as a simple add-on. Obviously some rules should be added to control some translations between attributes issued from several services. At this time rule definition was not specified but it should be a natural and necessary evolution of the current project.

## 4 Negotiation between several experts

### 4.1 Case study

This section proposes a case study to demonstrate the interest of the PPO model in collaborative design activities. Four experts are involved in the current case study: 1) project manager is in charge of the definition of the design process and resource allocation – 2) designer provides a CAD model of the product – 3) mechanical analyst is in charge to check the robustness of the product and – 4) manufacturing expert is in charge to define how to manufacture and to assemble the product.

Figure 3 presents the system under focus. This system concerns the assembly of a column and a rack for a mixing machine.

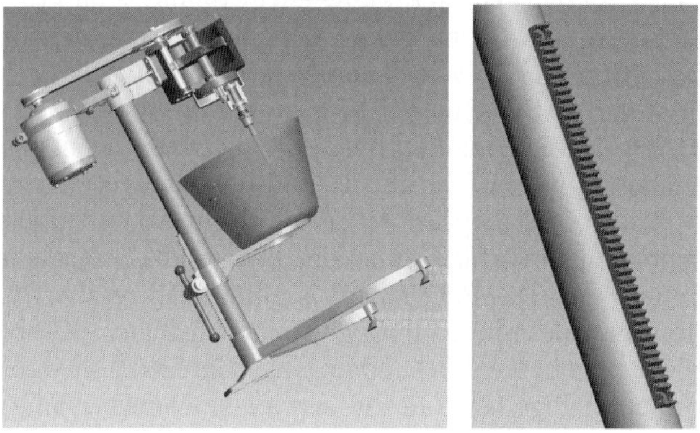

**Fig. 3.** The mixing machine and a zoom on the assembly of the column and rack

Several software services are involved during the study. Project manager uses a planner tool. For experimentation an Open-Source tool, named planner [13] is used. Figure 4 shows (top view) a snapshot of the main view of planner where sequence of task are defined and viewed as a GANT model. Allocation of resources and analysis of the resource charge are also available under planner. Figure 4 shows three other graphic user interfaces. Bottom left one is a tool dedicated to the management of mechanical analysis. This tool provides a framework to declare an analysis to be performed, the goal of the analysis and the main assumptions associated with it. Analysis task may be described in an IDEF fashion to record every steps of the analysis. At last the tool records the conclusion of the analysis that impacts the design activities. Internal steps are not shared in the PPO model since they only concern the mechanical analyst. However the analysis goal and its conclusion are collaborative information that must be shared. The bottom central graphic user interface of Figure 4 recalls the usual interface of the CAD modeller. A CAD modeller mainly provides a component tree and their interfaces as boundary surfaces. The bottom right graphic user interface is dedicated to manufacturing expert. This tool requires the definition of functional surfaces that must be either milled or drilled. Then it assists expert in the creation of milling process. Interface tree is the main element this service shares with others. It also provides some recommendations about design modifications to ease design process and make manufacturing cost lower. These recommendations are of great interest for other designers.

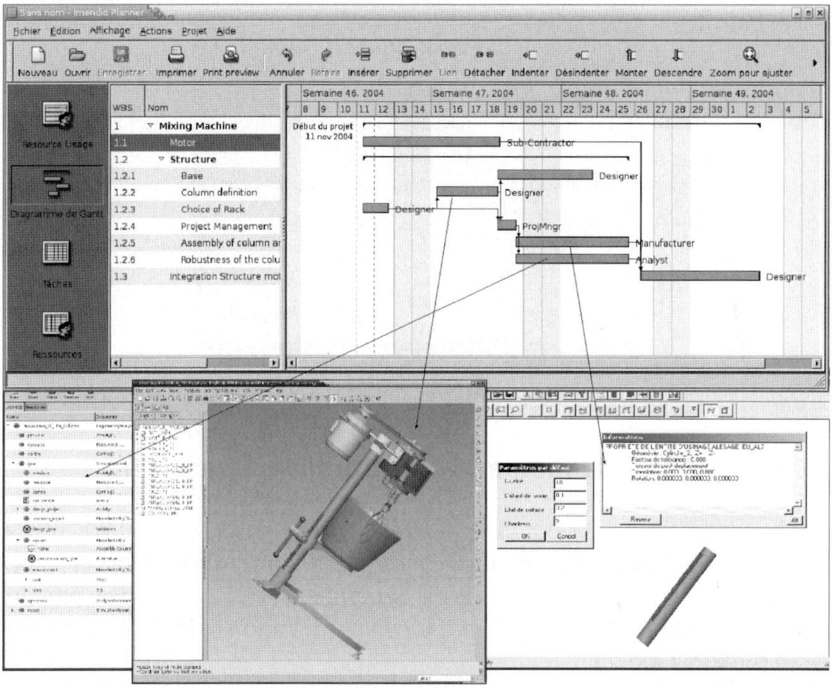


**Fig. 4.** A view of planner graphic user interface

## 4.2 Synchronisation of model services through the PPO model

Figure 5 is a snapshot of the plugging dialog that should be added to every services involved in the collaboration.

The current view is dedicated to the planner tool. Two buttons ensure the connection with the PPO model and the planner model. At connection, the first method sending the list of attributes required for each PPO entities is automatically integrated in the PPO model. Then, when models are connected only three functions are provided to the expert: 1) check-in and - 2) checkout methods, allowing integration of evolutions of models and 3) update button updating the state of models to aware the expert what was modified either the PPO model, or the expert model or both. Thus, services dealing with heterogeneous purpose are synchronized whenever the designer decides this synchronisation. The two next sections present a scenario where designers take advantage of this collaborative environment to detect a conflict and negotiate a common solution to this technical conflict.

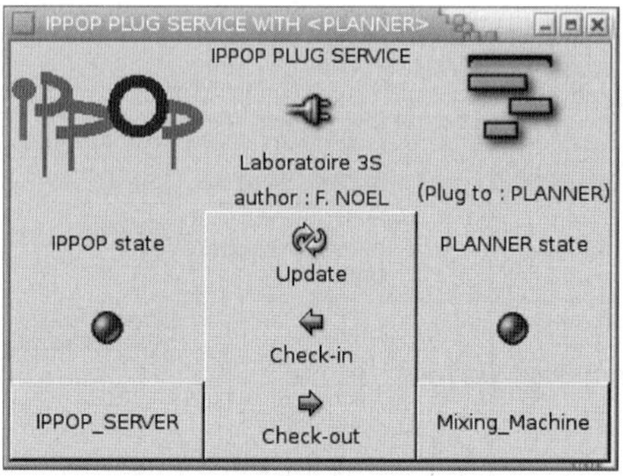


**Fig. 5.** plugging dialog to synchronize the PPO model and an expert service

## 4.3 Conflict detection and collaborative negotiation

Following sequence of actions leads to a conflict detection. We just assume that: 1) Two technical services share a same fuzzy attribute about material properties associated with a component – 2) A designer prepared a first definition of the product in a CAD model and shared the component tree in the PPO model.

Now, project manager requires analysis to be performed about this first design:

1. The project manager proposes the sequence of tasks described in the GANT of Figure 4. He models this sequence in his usual tool (planner) and checks-in this representation towards the shared PPO model.
2. Since manufacturer task and analysis task are concurrent the two following sub-activities are run in parallel:

- On one side, analyst checks-out the PPO model and thus loads on its own desktop a new analysis to be performed. The goal of this analysis is the task objective set in the planner model. Analyst completes this definition by requiring some information about the product definition from the PPO model as for example the component tree. If a function tree is already defined in the PPO model functions provide working conditions of the system. An analysis is then performed and its conclusion is that a surface Heat Treatment should be necessary to ensure toughness. Now he checks-in his new model into the PPO model and modifies the material properties of the rack component.
- On another side, manufacturer also checks-out the PPO model and starts his own study. Manufacturer proposes to drill the parts in order to screw them together. He creates a representation of the drilling process and

checks-in his new model into the PPO model adding new sub-components corresponding to holes.
3. Both, analyst and manufacturer, check-out the new PPO model to be aware of what was done by each other. At this step, manufacturer discovers the change in the material property that is incompatible with the drilling process at low cost. Thus a basic conflict is easily detected [14].
4. Through the PPO model the manufacturer knows who is responsible of the change of material properties, when he did it and for which goal but also who is responsible of the current project.
5. With all this information he will propose a review meeting to negotiate a new solution or a new sequence of tasks to find the solution. Every decisions coming out at this step are easily shared in the PPO model and each designer goes back to his office and checks-out the PPO model once again.

## 5 Conclusions and perspectives

This paper presents a new environment for collaborative design. This environment is integrative rather than integrated and thus is able to integrate easily any new expertise involved in the design process with a minimized effort of adaptation of tools assisting the expertise.

The defined PPO environment provides the opportunity to establish collaborations between Product and Process definitions in a concurrent engineering environment.

With a basic scenario, the paper demonstrates the potential to detect conflicts and to prepare technical negotiations in the project reviews. Few services, which were plugged to the PPO model, take a rapid benefit by increasing the participation of their experts in the design activity.

Nevertheless, the API of this environment needs to be provided as a standard to allow an increasing set of tools to be connected together. Moreover some functions remains to be implemented. For example the ability to add inference rules inside the PPO server to help translations between attributes associated with various and heterogeneous services are now under focus.

## Acknowledgment

This paper refers to the results of the IPPOP project where IPPOP means Integration of Product Process and Organization for the Performance in design. The author thanks every partners of the IPPOP project [15] for their active contribution to this work. The IPPOP project was founded by the French Network of Computer Technologies: RNTL: Réseau national des Technologies Logicielles.


## Acronym Table

API: Aplication Protocol Interface
PPO: Product-Process-Organisation Model
PDM: Product Data Management
CAD: Computer Aided Design

## References


1. Kimura, F., Lipson, H., Shpitalni, M., 1998, Engineering environments in the information age: Researches Challenges and Opportunities", Annals of the CIRP, 44/, 87-90.
2. Krause, F.-L., Kimura, F., Kjellberg, T., Lu, S.C.-Y., 1993, Product Modelling, Annals of the CIRP, 42/2:695-706.
3. Roucoules, L., Tichkiewitch, S., 2000, CODEMO: A cooperative design environment - a new generation of CAD systems, Concurrent engineering : research and applications, 8/4, 263-280.
4. Krause, F.-L., Kieswetter, Kramer S.,1994, Distributed product design, Annals of the CIRP,43/1, 149-152.
5. Kimura, F., Hiromasa, S., 1995, Representing background information for product description to support product development process, Annals of the CIRP, 44/1, 113-116.
6. Tichkiewitch, S., Brissaud, D., Co-Ordination Between Product and Process Definitions in a Concurrent Engineering Environment, 2000, Annals of the CIRP, 49/1, 75-78.
7. Noël, F., Brissaud, D., Tichkiewitch, S., 2003, Integrative Design Environment to Improve Collaboration between Various Experts, Annals of the CIRP, 52-1, 109-112.
8. IDEF, http://www.idef.com /Downloads, visited on 2005, January
9. Urban, S., Ayyaswamyn K., Fu, L., Shah, J., Liang, J., 1999, Integrated product data environment: data sharing across diverse engineering applications, International Journal of Computer Integrated Manufacturing, 12/16, 525-540.
10. Domazet, D.S., Choong, F.N., Sng, D., Sho, N.C., Lu, S.C.-Y, 1995, Active data driven design using dynamic product models, Annals of the CIRP, 44/1, 109-112.
11. Noël, F., Tichkiewitch, S., 2004, Shared dynamic entities technology to support distant coordination in design activity, Annals of the CIRP, 52-1, 109-112.
12. Object Modelling group,http://www.omg.org, visited on 2005, January.
13. Planner, http://www.imendio.com/projects/planner, visited on 2005, January.
14. Bettaieb, S., Noël, F., Synchronisation of heterogeneous geometric models, 2004, Proceedings of 2nd Int. Conference On Digital Entreprise Technology – Seattle.
15. IPPOP, http://www.opencascade.org/IPPOP, visited on 2005, January.

# Dynamic life cycle performance simulation of production systems

J. Niemann and E. Westkämper
*IFF Institute of Industrial Manufacturing and Management, Universität Stuttgart, Fraunhofer IPA, Nobelstrasse 12, 70569 Stuttgart, Germany*

Abstract: The paper presents a framework for controlling the life cycle of so called "integrated manufacturing systems". In the process, the economic efficiency of continuous improvement measures taken during its further life are evaluated in a simulation and monitored using real production data. To do so the paper identifies relevant data sources and describes the integration into a simulation model. To improve the performance of the manufacturing system the model describes general starting points. Before implementation the changes will be simulated and evaluated. After implementation the data from reality will be compared with the forecast to improve the model.

Keywords: Production Planning, Life Cycle, Simulation

## 1 Introduction

Today's manufacturing companies have the option of evaluating and optimizing the usage of investments from a life cycle point of view. In order to avoid outage losses, investment goods need to be continuously adapted during their life cycle to compensate for the technical, economical and organizational drivers of change which occur. To achieve this, influencing factors critical to success need to be identified and reproduced in order to benefit from the behavior of the overall system in a simulation for dynamically evaluating and optimizing a series production. To do this, segments need to be structured hierarchically in model cells and their effects in the network depicted. In this way, using an effect model, segment activities can be optimized pro-actively over their life cycle. In order to meet the requirements of modern manufacturing systems, the model is based on a system-related, structured and integrated production segment. By integrating a planning concept based on experience curves, it is possible to achieve



*D. Brissaud et al. (eds.), Innovation in Life Cycle Engineering and Sustainable Development,* 419–428.

continuous technical/economical controlling to maximize segment usage over the entire life cycle.

## 2 Integrated manufacturing systems

In the 90's, production-orientated topics such as the modularization or segmentation of manufacturing execution and planning processes were met with increasing resonance. The concepts of the "fractal company" [1] or of the "modular factory" [2] with "customer-orientated manufacturing by segmenting production" are management concepts with systematic integrated thinking in business processes. Technologically-orientated production segments are manufacturing systems or product- and market-orientated organizational units possessing extensive autonomy and integrated responsibility as far as costs, deadlines and quality are concerned. The term "integrated manufacturing system" describes the comprehensive integration and optimization of all company activities with the central aim of producing goods or services in a profitable and customer-orientated manner.

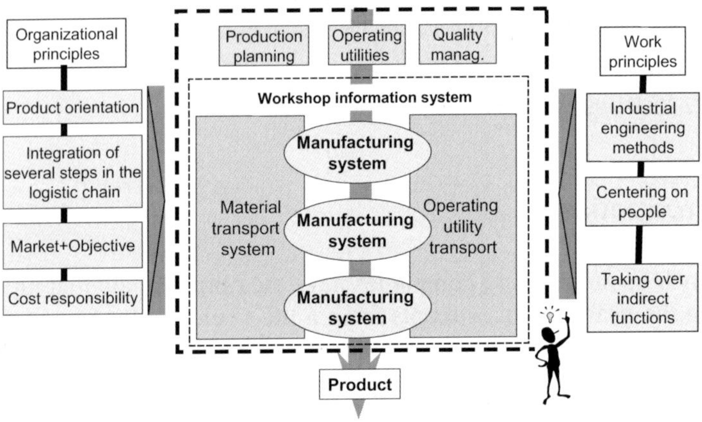


**Fig. 1.** Operating principles of an integrated manufacturing system

This is attained by focusing on methodical principles and strictly applying them in order to achieve the objective set and also by optimizing employee potentials when initiating and taking improvement measures. Through this form of understanding, the term not only addresses the actual production field but also includes and integrates the peripheral areas of technical and organizational support. [3-5] As a result, this concept does not only take into consideration the actual production and the areas providing direct support (maintenance, control, etc.) in optimization processes but also includes logistics, work preparation and quality assurance (Figure 1).

Therefore, system limits are no longer fixed for integrated manufacturing systems according to space- or time-related points of view. The main criterion when setting up the system is much more the necessity or function of a process in creating and supplying a product with regard to the criteria of costs, quality and time. In this way, an integrated manufacturing system can be viewed as being a networked unit of all the steps and elements of the production process involved in creating and supplying the product in a more far-reaching sense. The optimization of this extended overall system promises greater potentials in comparison with those which could be attained by improving individual elements singularly. The optimization of the overall system can be seen as being a continuously-running process which is essentially controlled by man and which functions according to well-known process-oriented and industrial engineering principles and methods. The system is controlled and (continuously) improved using robust processes, standardized work sequences and methodically-supported learning processes. [3–5]

## 3 Maximum exploitation of usage of manufacturing systems over their life cycle

Such types of manufacturing system invariably require a substantial capital investment. However, these machines and systems usually have a considerably longer life span than the products or generations of product they produce. Therefore, the life cycle of integrated manufacturing systems is not so much influenced by technical wear but rather by the danger of the system ageing technically as a result of rapid market developments. For this reason, permanent planning activities over the life span of the manufacturing system are required in order to guarantee investment security. Optimum economic decisions regarding system configurations need to be made constantly in order to achieve maximum system usage.

Figure 2 shows the planned life of an investment. The investment is calculated and made based on an assumed production program and an assumed lifespan. However, when the system is actually in use, numerous unexpected changes take place regarding order numbers and specifications. Due to limited forecasting, these factors cannot or can only marginally be calculated. It is for this reason that continuous adjustments and alterations which are life cycle-orientated need to be made. Their benefit (effect) must then always be evaluated with regard to the remaining lifespan of that particular manufacturing segment. Any measures or investments carried out beforehand represent "sunk costs" for the periods of time taken into consideration and are not relevant to later planning measures.

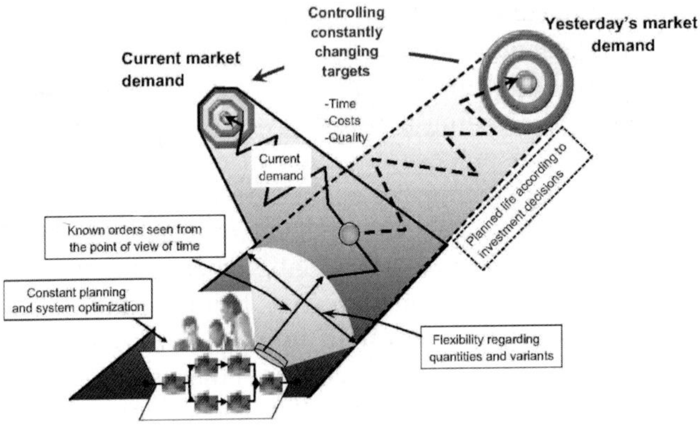


**Fig. 2.** Controlling volatile products and production targets (altered in accordance with [6])

## 4 Hierarchical structure of integrated manufacturing systems

Because system limits have been fixed very broadly, a deep multilayered interlocking of effect relationships now exists between the individual system elements. Due to the high level of complexity, it is wise to evaluate the integrated effect of individual optimization measures with the aid of simulation studies. However, for the simulation, the individual objects (system resources) need to be structured horizontally and vertically according to system-related criteria in order to describe the effect relationships according to type, intensity and effect. In this way, subsystems are formed which reduce the degree of complexity. As the target function is to optimize the overall life cycle usage of the integrated manufacturing system, the individual objects (resources) are described according to the functions of costs, quality and time. Figure 3 describes the horizontal and vertical structure of an integrated manufacturing system. Processes requiring resources with corresponding process times are placed at the bottom level. Planning activities are placed at the top hierarchical level as they are usually integrative, i.e. are active for one, several or all of the subsystems [7]. By using such a formal structure, it is possible to configure integrated manufacturing systems and to formally describe their composition.

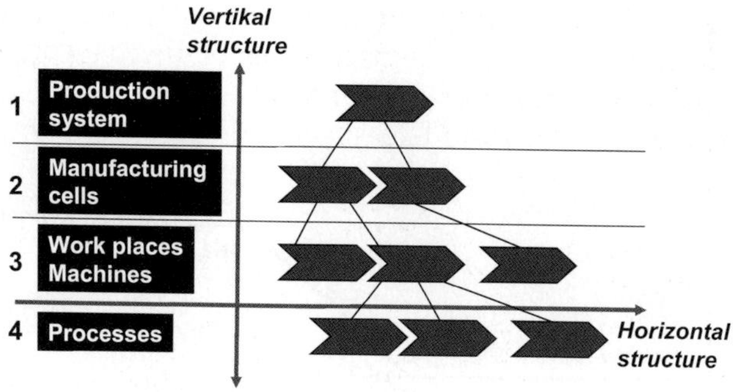


**Fig. 3.** Structural layout of an integrated manufacturing system

## 5 Experience curves for life cycle planning

Experience curves have the effect of reducing the costs, amount of time, etc., required to manufacture a product unit. Costs are reduced by a factor in percent each time the quantity manufactured is doubled. This effect was first discovered in aircraft manufacturing and has since been proven empirically. However, nowadays the knowledge obtained from the experience curve theory has been widely applied in practice and is used to forecast piece costs over time. It is essential that these effects do not occur automatically, but rather are always based on operational measures which result in cost reduction. [8] These measures are induced and implemented by production employees or planning departments to achieve an economizing effect in later production. In order to implement experience curves in planning the life cycle of manufacturing systems, effects which up till now have always been product-orientated need to be transferred to a plan for an integrated manufacturing system which is life cycle-orientated. Figure 4 lists the general points associated with the concept of such improvement measures. In order to be able to describe and model these effects, it must also be clarified which sub-systems (segment elements) within the system-related structuring of the integrated manufacturing system will be affected by the measures planned. Therefore, for each effect point, the consequence of the measure must be described according to the dimensions of costs, time and quality. (Cost) Savings may be the result of rationalization effects, scale effects, learn or practice effects or of technical advancements. [8] However, the description of the effects on the individual objects (partial models) only represents a statistical description of the situation; the main aim is to evaluate the consequences of the measures on the overall system and over its entire life-cycle.

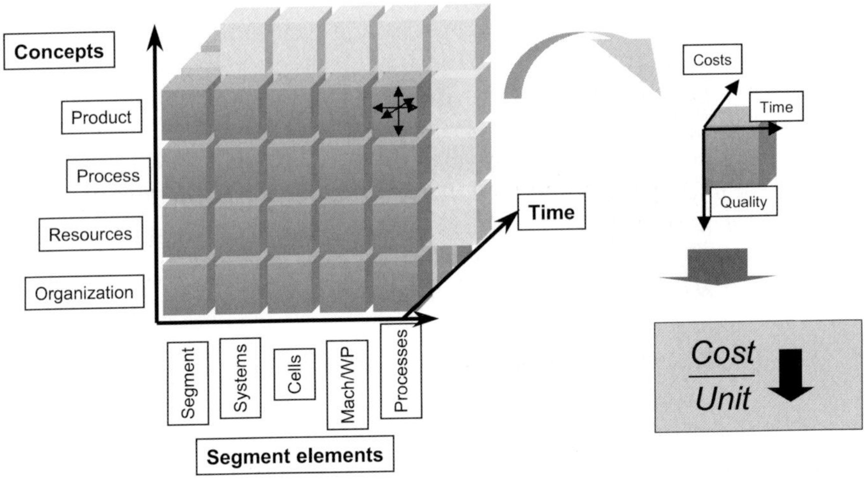


**Fig. 4.** Sources and reductions in experience curve effects over time

Each total process time is thus made up of the execution time and the preparation time [9]. If the sum of all total process times from all the orders is calculated and then multiplied by all of the resource costs involved, it is possible to depict the segment-related costs. Experience effects are then finally expressed as a reduction in production costs (cost-evaluated total process times). If this is documented continuously over time and also forecast for future orders using the work plans, life-cycle related segment activities can thus be optimized. Figure 5 illustrates this behavior in the form of a graph. In order to create effect models, the elements (resources) of the manufacturing system first need to be defined. The set-up can be made as detailed as desired. In order to derive core statements, it is sufficient to classify resources into the categories of personnel, operating utilities and auxiliary means (tools, large transportation aids, etc.) [10].

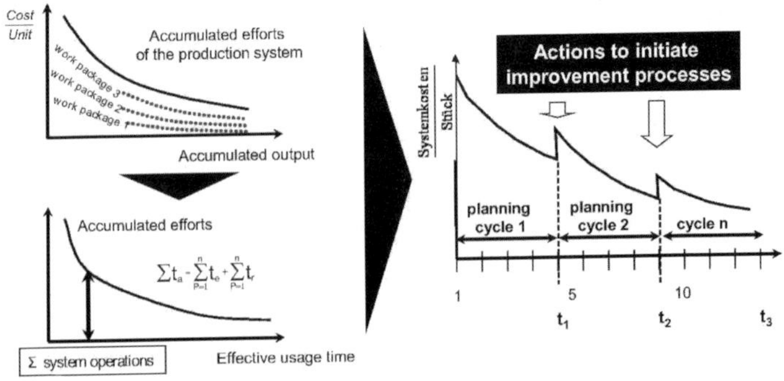


**Fig. 5.** Segment-related forecast of costs

These are then assigned to being time-related resource cost functions with the result that operating utilities and auxiliary means are made up of hourly machine rates and hourly wage rates for personnel. Quality defects, maintenance activities, etc., are integrated by norming activities to the "effective system usage time". Such disturbances represent unproductive periods of time and their elimination serves to increase productive usage time. An increase in productive time periods represents a reduction effect in relation to fixed costs, thus inducing a decrease in the experience curve.

## 6 Integration into simulation environment

By setting up a simulation for integrated manufacturing systems, it is possible to create a controlling model which enables usage to be planned, controlled and evaluated over the entire life-cycle. Planning is carried out based on the simulation of operating sequences and is constantly verified. The elementary model cells form the basis for modelling segment activities. The aim is to optimize the maximum usage of the overall system over the planned life cycle. Maximization of usage is defined as being the maximum economization in costs of the aggregated life cycle costs compared with the ex-ante life cycle cost calculation at T (Figure 6).

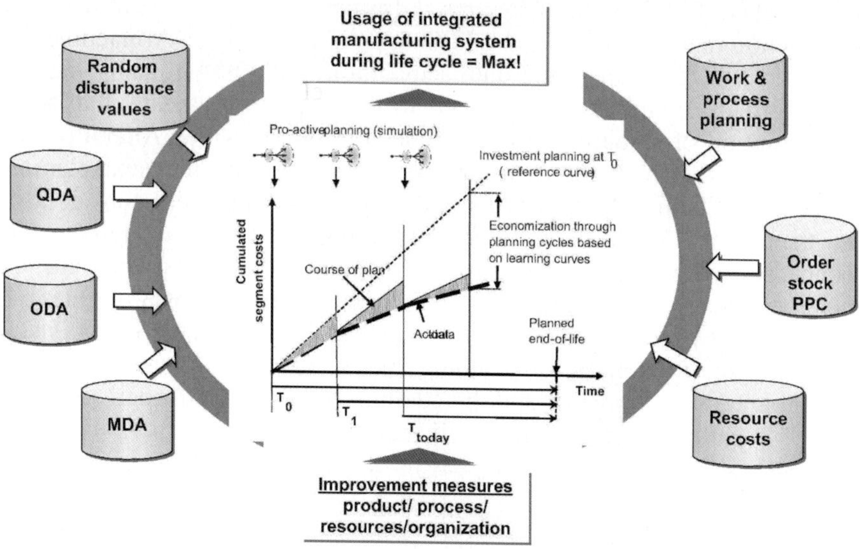


**Fig. 6.** Controlling the life cycle of integrated manufacturing systems

The reference curve created as part of the investment planning is based on the assumption of a future production program right up to the planned end of the manufacturing system's life. All current and future manufacturing orders are then continuously planned into the usage phase of the system, aggregated at segment level and then compared with the reference curve. Time- and cost-orientated assessments of rationalization measures to avoid opportunity costs make up simulation input values. In this way, cost effects due to order fluctuations, product variants, program changes and alterations in segment configuration (change drivers) can be evaluated over its life cycle and an increase in usage continuously induced by implementing appropriate measures. Thus by making events dynamic, it is possible to make a cost-benefit estimation of the measures over the planned remaining lifespan of the manufacturing system. The method to be developed starts in the phase where an order quote is made. As part of the pre-calculation, it provides a cost forecast based on experience curves for expected average piece costs per batch. Planning is carried out using previous production data filed in the PPS system from identical or similar orders ("starting point of the experience curve"). Planning data from work and process planning and also resource-related planning values or disturbance values (e.g. shift models, maintenance, set-up times, assumptions concerning unplanned production stoppages) is then entered into the simulation-aided plan of the experience-based curve expected for the production orders. The data basis for this is formed by real data recorded from production data acquisition. In addition, data from the MDA(machine data acquisition)/ODA (operating data acquisition) and QDA (quality data acquisition) is compressed in order to be able to evaluate the actual consumption of resources from a monetary point of view. Alterations are then entered as new forecasts or planned values into the next planning cycle whilst taking the new order and preparation times into consideration in the altered work plans. The controlling model thus supplies data concerned with events, planning and evaluation by integrating the experience curves into a time scale over the entire lifecycle of a segment. In this way it is possible to create a closed control loop which constantly supplies updated references for planning cycles. The continuous recording of time-related developments and the addition of all batches/orders produced by the manufacturing system over its entire life cycle thus permits the life cycle of the system to be evaluated in detail and to be dynamically controlled.

## 7 Simulation results for continuous life cycle planning

During the simulation several lists of different results are generated. According to the range of different simulation functions the simulator evaluates and visualises the previously determined and desired statistics. The results of the different examined scenario deliver a variety of reports about the states of the model elements. The analysis discovers bottlenecks and thus identifies optimisation potentials. The simulation delivers the following key figures:

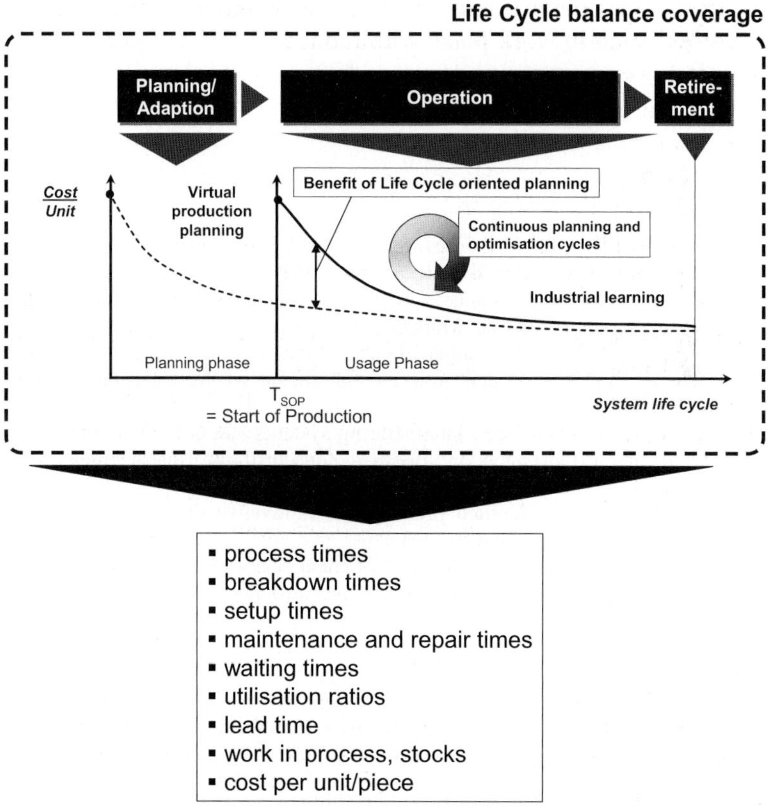


**Fig. 7.** Continuous Life Cycle Planning

These results pave the way for a continuous loop of ex-ante calculation, production and ex-post calculation. The permanent optimisation loops along the adaptive system's life span leads to an adaptive system monitoring to actively improve and optimise the system benefit.

## 8 Summary

The development of an experience curve-based method for controlling the life cycle of integrated manufacturing systems permits the consequences of technical, organizational and product-related drivers of change to be continuously simulated and evaluated over a system's complete life cycle. In order to achieve maximum usage of a production segment adapted to the situation at that moment in time, the effects and overall benefit of improvement measures taken can be assessed in advance. By coupling this with real production data, the planning model can be verified based on a closed control loop. The industrial engineering improvements made are reflected in work plans with reduced order and preparation times and also in lower resource costs.

## References


1. Warnecke, H-J.; Hüser, M.: Revolution der Unternehmenskultur : Das fraktale Unternehmen. 2. edition Berlin et al.: Springer, 1993
2. Wildemann, H.: Die modulare Fabrik: kundennahe Produktion durch Fertigungssegmentierung, 5.edition, München, 1998
3. Korge. A., Scholtz, O. Ganzheitliche Manufacturing systems, Produzierende Unternehmen innovativ organisieren und führen, in: wt Werkstattstechnik online vol 94 (2004) H.1/2, p. 2-6.
4. Hinrichsen, S.: Ganzheitliche Manufacturing systems - Begriff, Funktionen, Stand der Umsetzung und Erfahrungen. In: FB/IE - Zeitschrift für Unternehmensentwicklung und Industrial Engineering, Darmstadt, 51 (2002) 6, p. 251-255.
5. Spath, D. (Hrsg.): Ganzheitlich produzieren. Innovative Organisation und Führung. Stuttgart: LOG_X Verlag GmbH 2003
6. Schuh, G., Sesterhenn, M., König, R., Lebenszyklusgestaltung kollaborativer Produktionssysteme, in: zwf Zeitschrift für wirtschaftlichen Fabrikbetrieb, 98 (2003) 1-2, p. 17-21
7. Westkämper, E.: Das Stuttgarter Unter-nehmensmodell: Ansatzpunkte für eine Neuorientierung des Industrial Engineering. In: REFA Landesverband Baden-Württemberg: Ratiodesign: Wertschöpfung - gestalten, planen und steuern. Bodensee-Forum. 17. und 18. Juni 2004, Friedrichshafen. Mannheim, 2004, p. 6-18
8. Hieber, W.-L.: Lern- und Erfahrungs-effeke und ihre Bestimmung in der flexibel automatisierten Produktion. München: Vahlen (Controlling Praxis) 1991
9. REFA Verband für Arbeitsstudien und Betriebs-organisation: Methodenlehre des Arbeits-studiums - Teil 2: Datenermittlung. 6. volume München : Hanser, 1978
10. Lorenzen, J.: Simulationsgestützte Kosten-analyse in produktorientierten Fertigungsstrukturen. Berlin u.a.: Springer, 1997 (iwb Forschungsberichte 107). München, Techn. Univ., Fak. für Maschinenwesen, Diss. 1996.

# LC universal model for the enterprise information system structure

A. Bernard, M. Labrousse and N. Perry
*IRCCyN, Ecole Centrale Nantes, Nantes, France, alain.bernard@irccyn.ec-nantes.fr*

Abstract: The analysis of the product lifecycle processes is one of the crucial industrial concerns. It highlights the role of knowledge capitalization and valorization. In this work, a generic model is proposed: the FBS-PPRE model (Function, Behavior, Structure - Process, Product, Resource, external Effect). The main scientific contribution consists in a conceptual unification making it possible to increase the homogeneity of the represented objects as well as the completeness of the modeling, when addressing all the life-cycle situations of the product, the technologies and the enterprise. In order to allow a description of the various objects handled by the company according to the same formalism, the approach highlights the distinction between the concepts of nature (material, temporal, software, organizational or energy) and of role. Four different roles are distinguished: the role of product (object resulting from the process), of resource (means used in a process), of process (temporal, spatial and hierarchical organization of activities) and of external effect (constraints influencing the course of a process). The same object can play successively various roles during its lifecycle. In addition, an effective management of behaviors is proposed: a behavior of an object is always combined with a process element, which makes it possible to define the context. It then becomes easier to manage the evolution of the object and of its representations, namely their transformational dynamic. This approach was implemented within the framework of a demonstrator of information system of a company designing and producing vehicles.

Keywords: Life-cycle, knowledge, information system

## 1 Context and difficulties inherent to knowledge management

The general reference system (Figure 1) [1], shared between all the actors, is the base of the enterprise information system for facilitating life-cycle engineering [2] It also makes the concurrent engineering easier: each department can intervene at any time on the design process, before choices



*D. Brissaud et al. (eds.), Innovation in Life Cycle Engineering and Sustainable Development,* 429–446.

become very costly or irreversible. The aim is to find the best compromise between cost, quality, delays and risk when considering the different constraints related to the product life-cycle: production service, end-of-life [3]. This approach should be interesting for designers to improve products sustainability [4] by integrating the different life-cycle points of views.

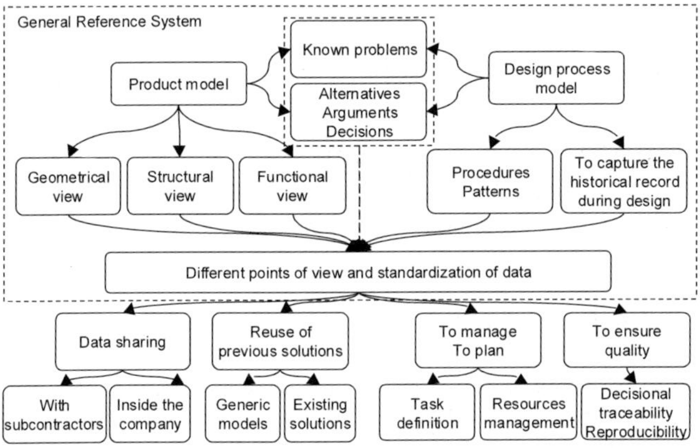


**Fig. 1.** Content and applications of the general reference system [1]

The general reference system is composed of two main models: the product model and the process model. The product data changes during time and highly depends on the context. A real linkage of the product and process data is desirable: in fact, the transformation of a product data can be considered as the result of a process applied to this product. The general reference system is made of different models to improve the management of different types of knowledge concerning the life-cycle of the products, technologies and enterprise processes. Managing life-cycle issues are linked to the deployment of a knowledge management approach to let the designers use models, data and constraints related to the different life-cycle stages. This kind of need implies a wide range of problematic areas:

- Transmission and acquisition of knowledge are complex processes;
- The influence of human resources is real and lot negligible: a knowledge management methodology is not exclusively the result of a particular conceptual model: human factors have to be taken into account. In order to achieve the goals, it is very important to explain the objectives of the routine changes. The role of the managers is also crucial;
- Modeled objects are of high complexity: the granularity of the systems and the interactions between its elements implies emergent properties difficult to grasp. The understanding of the whole did not enable the understanding of its parts. Thus, the model should be multilevel and homogeneous.

- The system is always evolving: the knowledge about this system can be uncertain or incomplete. It would be interesting to provide a dynamic management of the information. A coherent and homogeneous management of the "Process/Product/Resource/External effect" information may help.

## 2 Industrial context : the French company Voxan

The company VOXAN, created in 1995 by Jacques Gardette, is a French motorcycle manufacturer. The business plan of the company points out a range of seven motorcycles sharing a common base.

### 2.1 Industrial objectives

The objectives of the company are quite usual:

- To reduce design times and costs: to define reusable processes, to capitalize and to reuse past designs;
- To reduce manufacturing costs: to increase sales, to find new subcontractors and to renegotiate contracts, to improve the synergy between the departments and between the employees;
- To set up improved quality management processes: to define procedures and provide decisional traceability, to maintain the information coherence during the whole lifecycle, to provide the information timelessness.

### 2.2 Synthesis of the data exchanges inside the company

The data exchanges between the different departments of the company were checked of in order to define conceptual data models. These models were formalized using the Merise method.

- The conceptual product data model constitutes the basic structure used to represent the designed product. It includes the structural view, the functional view, the geometrical view, but other views may also be included.
- The aim of the product model is to favor knowledge exchange by a better interoperability and an easier reuse of previous designs.
- The conceptual process model ads the dynamic aspect to the information. It describes not the product, but how, why and by whom it was designed. In particular, this includes the lifecycle constraints that have to be considered during the design phase.

The process data model enables the recording of specific processes. The processes themselves have nevertheless to be defined. The processes, particularly the design processes remain undefined in the company. In order to overcome this problem, it was necessary [5] to:

- Define the different ways of designing the different types of parts,
- Specify the departments implied at the different stages,
- Define the different design and industrialization processes.

### 2.3 Industrial difficulties

The product models and the software tools were quite satisfying. The objective was then to model the processes of the product lifecycle and to link the product and the process views. Other processes of the company was also modeled: management processes (restructuring of a department, quality management, etc.) and complementary processes (employees training for example), because their influence on the costs, on the delays and on the quality is not negligible. However, the modeling of the product and of the processes was not sufficient: other objects affect the company such as the resources and the external effects. It was necessary to model in detail these different types of objects. For example, the structure and the behaviors of the resources are rarely defined. It was proposed to define models PPRE oriented (Process, Product, Resource, External effect) taking into account the different views.

## 3 Different approaches to model the enterprise objects

The objective was to model the information of the whole product lifecycle. To reach this goal, different objects had to be modeled. Before achieving the models it was necessary to propose clear and precise definitions for the objects [6].

The Process object can be defined as a sequential, spatial and hierarchical organization of activities using resources to make products (or outputs). The Resource object is considered as an object contributing to the process without being its purpose. The Product object is the result of the process, the object that the process intends to modify. The External effect object is acting as a constraint (positive or not) on the process/product/resource system. It is a part of the context, which is foreseeable or not and that can disturb the process progress.

The functional view refers to the different functions. The functions describe in an abstract way the aim of an object. The operational functions are formulated independently of any particular solution. On the opposite, the technical functions depend on the technical choices. The behavioral view describes the dynamic of the objects. It includes a set of rules (continuous models) and sequential states graph (discrete model) representing the transformation of an object stimulated during a process. The structural view defines the elements that are parts of the objects. It defines also the attributes of these parts.

## 4 Proposition of a generic model: the FBS PPRE model

The FBS (Function, Behavior and Structure) approach seems to be an interesting base for enterprise object modeling. However, the models proposed by the litterature remain incomplete [7–10].

### 4.1 FBS-PPRE model

In this work, an extension of the FBS approach is proposed, the FBS-PPRE model. It integrates the fundamental concepts needed to model the different enterprise objects. The FBS-PPRE (Function, Behavior, Structure – Process, Product, Resource, External effect) is the result of the use of FBS modeling for the four objects views: the process view, the product view, the resource view and the external effect view. The external effect view is interesting to define the context. These views are based on two main concepts, "Entreprise object" and "object" concepts.

« Enterprise object » is a generic concept dedicated to encapsulate the concepts of process, product and resource. This encapsulation enables a more homogeneous and efficient management of these three concepts. Enterprise object is defined as an enterprise entity or an entity controlled by the enterprise. It has a real influence on the company behavior. Nevertheless, the external effects can not be included in this generic concept (they're not always controlled by the company). The more generic concept of object is thus used. An object is an entity playing a role in the company. This concept aims at managing the different objects independently of their roles [6]. It is also necessary to add the following definitions concerning the roles of the enterprise objects. We postulate that the concepts of process, product, resource and external effect are abstract and circumstantial. This implies that an object cannot be considered during its whole lifecycle as a process, a product, a resource or an external effect. Actually, an object can have different roles during its lifecycle. For example, we can consider a designer creating the design of a part with a CAD tool. The CAD drawing will be the *product* of its study. Nevertheless, this object will be considered as a *resource* of the manufacturing design *process*. We consider now a sub-contractor acting usually as a company *resource*. It can also be an *external effect* if it is not able to deliver the order in time. Thus, an object is not, but plays a role of process, product, resource or external effect at a particular stage of its lifecycle. The object role concept defines the circumstantial use of an object. The roles can be process, product, resource or external effect. In addition, we propose to define the nature of the objects. Indeed, the objects cannot be classified by roles (they are circumstantial). They will be classifed by nature. We consider five object natures: Material (object that can have a concrete form); Organizational; Temporal (object that can play a process role); Software (including the generated documents); Energetic. The object nature concept is linked to intrinsic parameters of the object. The nature can be temporal, material, software, organisational or energetic. It is important to formulate the following remark: to say that an object is material will not implicate that this object physically exists. Only at the end of the processes (design, manufacturing) will the material product be created. Example: a car is material object (its main function "move people and objects" implies that this object is material). Different representations (often of software nature) will help to define the object before it becomes concrete. The object structures could be mainly considered as the decomposition of an object into sub-objects. The main classes are thus *object* and *assembly* (used to define decompositions). Nevertheless, these two classes are not sufficient. To take into account the sequence of temporal objects, the class *next* is used. The functions of an object are independent of the roles. The functions describe the behavior of the object when it is used (the object has a resource role). The functions will be linked directly to the object, not to their roles. The behavior management is an originality of the FBS-PPRE model. The Figure 2 presents a graphical representation aiming at making this management more explicit.

In order to simplify the script, an "object playing a role of process" will be replaced by "process element" and the script simplifications will be the same for the products, the resources and the external effects. The process element (in the center of the representation) plays a major role in the FBS-PPRE model:

- It links the different objects playing a role in the activity. It helps at defining the context;
- It also helps at managing the behavior of the product, of the resources and of the external effects.

The states are representative of the structure changes. They are the inputs and outputs of the process element. They represent the discontinuous behavior of the objects.

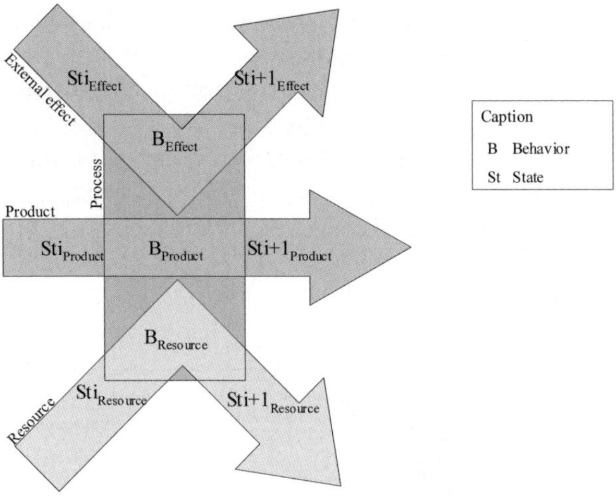


**Fig. 2.** FBS-PPRE behavior management

## 4.2 UML Class diagram of the FBS-PPRE model

The different objects share the same UML class diagram tough they can play different roles and they can have different natures (Figure 3). It avoids any objects conversion from one role to another. This is the same object, even if its role can change during its life-cycle. So, it is logic to use the same model, independent of the role.

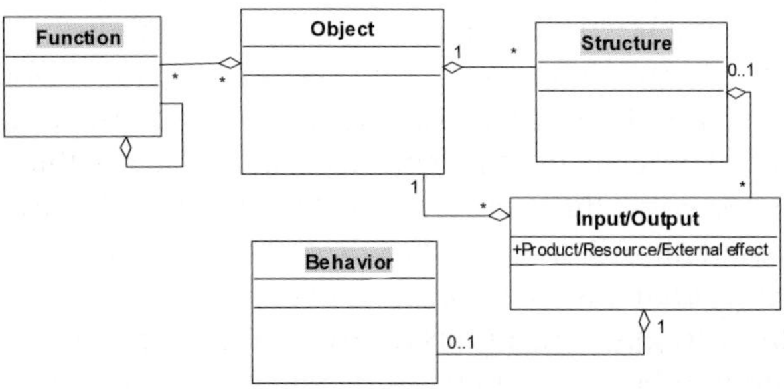


**Fig. 3.** Simplified UML class diagram of an object

Each object is represented using the usual views of the FBS approach: the *function*, the *behavior*, and the *structural* views. As previously mentioned, the *behavior* is not directly linked to the *object*: the link is made through the *input/output* class. This class is dedicated to associate a temporal object (which could have a process role) to other objects (having a role of product, resource or external effect). The behavior of an object is thus relative to a particular process object: the objects can have as many behaviors as links to temporal objects.

The Figure 4 presents the UML class diagram of the structural view.

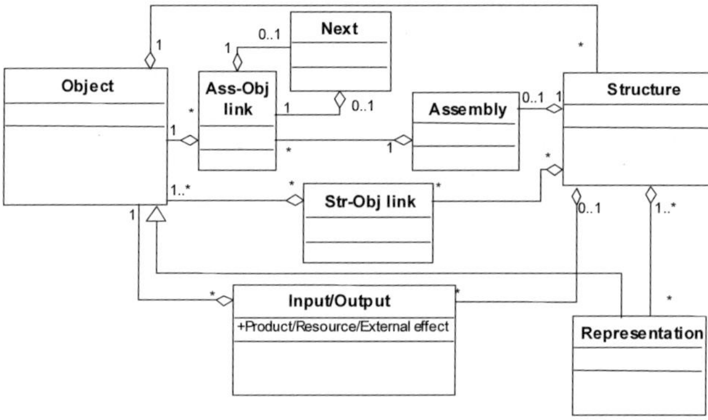


**Fig. 4.** UML class diagram of the structural view

The main classes of the structural model are *structure* and *assembly*: an object is composed of sub-objects, which could also be composed of sub-objects. The usual assembly types are *AND, XAND, OR, XOR*. For the temporal objects is made the distinction between the *synchronous AND* and the *asynchronous AND* (in the same way the *synchronous OR* and the *asynchronous OR*).

The *ass-obj class* is dedicated to the linkage of an *object* to an *assembly*. It includes the cardinality of the element in the *assembly*. The class called *next* enables the modeling of temporal objects sequences.

The *str-obj link* class is used to define transversal links between objects included or not in the considered structure (the structure would be just a hierarchical tree if only the *assembly* and *ass-obj link* classes were used): it enables the modeling of mechanical links, of electrical links, of interactions between activities of a temporal object, etc. It could also be useful to define variants of an object.

The *representation* class enables the definition of particular views of an object structure. A representation is a particular type of object.

When a representation is modified, it will influence the knowledge related to the parent object. For example, the CAD file of a part increases the knowledge related to the part. As a consequence, a process related to a representation is a sub-process of a process related to the parent part.

The Figure 5 presents the UML class diagram of the functional view of an object. The main class of the functional view is the *function* class: an object has functions that can be split in sub-functions. The *feature* class is linked to the *structure*. In fact, a part can be designed using predefined *features* that have the requested *functions*. A feature includes a set of *solution patterns* (geometric shapes, manufacturing process, etc.).

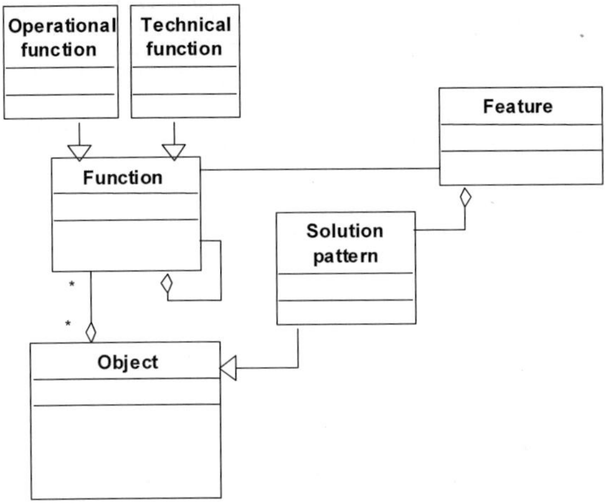


**Fig. 5.** UML class diagram of the functional view

The Figure 6 presents the class diagram of the behavioral view. The behavior is contextual: an object behavior is always linked to a process element, thus improving the context definition. The class called *Input/Output* is used to link the object to its *behavior* and to the *structure* of a temporal element. The *expected* or *real behaviors* include *input* and *output states,* a *status* (under process, completed, cancelled, etc.), a *shutter release* (describing the requested conditions to begin the activity) and *behavior laws* representing the continuous part of the behavior. The *states* are inputs and outputs of a process element. The states are instant views of the structure: a state can refer to a particular *representation* of the structure (for example the position of the arms of a robot), but it can also refer to different versions of the *structure* (for example an arm of the robot can be replaced by another one with different properties). The *performance indicators* are defined in a simple way as a comparison between the *real behavior* and the *expected behavior*. The model includes also the *external effects*. They are linked to the process element using the *Input/Output* class. They have a real influence on the behaviors. In fact a behavior is always the result of a particular context. The complete class diagram of the FBS-PPRE model presents the integration of the different partial view already described [11].

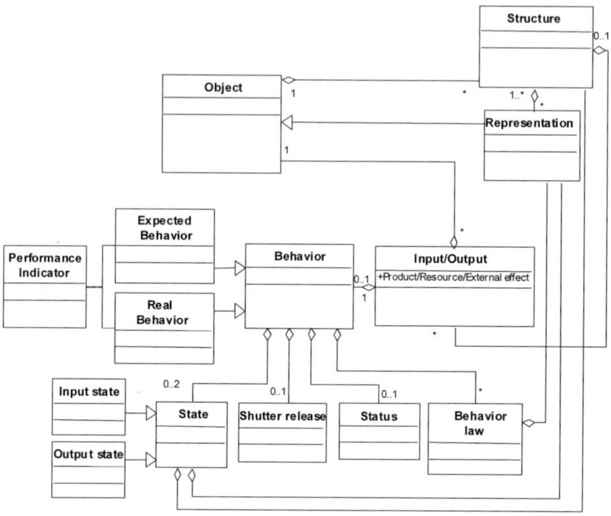


**Fig. 6.** UML class diagram of the behavior view

# 5 USE of the FBS-PPRE model

FBS-PPRE modeling should be considered as a knowledge-structuring tool. It is compatible with different knowledge management methodologies: the high completeness of FBS-PPRE offers a wide-open connectivity and adaptability to particular needs.

## 5.1 FBS-PPRE instantiation examples

The scriptural conventions has to be first defined. The representation of the five objects natures is proposed in Figure 7. Objects can be refined in sub-objects; the *decomposition* operator (Figure 8) realizes the link between levels. The *representation* operator (Figure 8) realizes the link between an object and its representations (CAD files, spreadsheets files, etc.). The roles are related to a temporal object (cf. Figure 9). The discontinuous behaviors are represented through their states, status and transitions (cf. Figure 10).

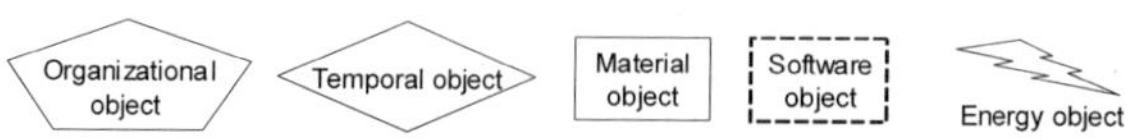


**Fig. 7.** Representation of the objects natures

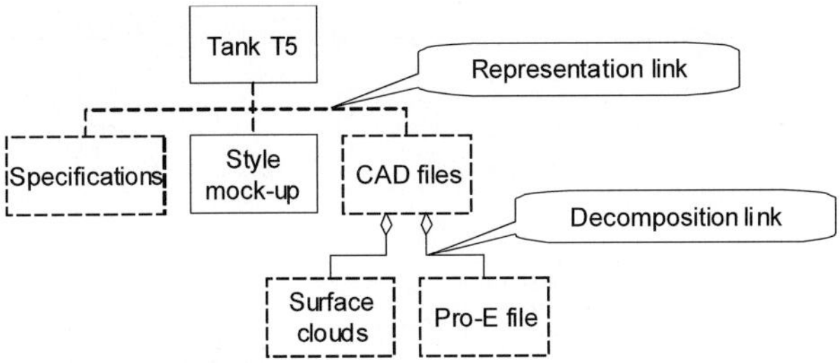


**Fig. 8.** Representation of the structural view

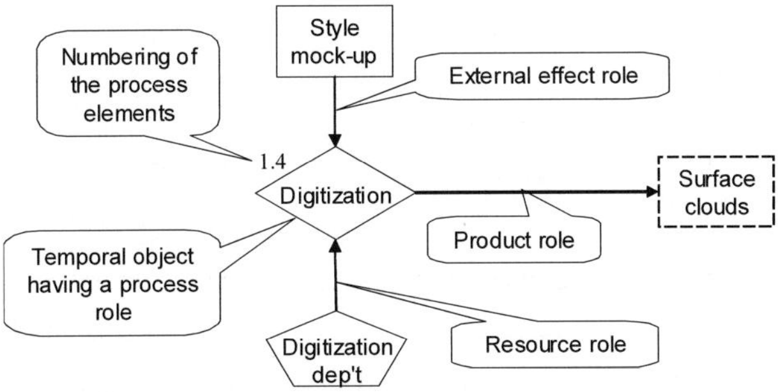


**Fig. 9.** Representation of the objects roles

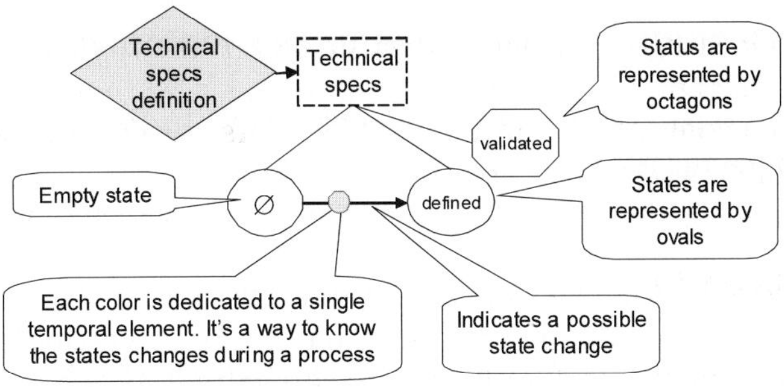


**Fig. 10.** Representation of the discontinuous behaviors

The representation of functions and behavior rules is proposed in Figure 11. As commented in Figure 11, expected and realized behavior rules and laws are represented by frames. The Figure 12 presents the representation of the links between temporal objects. The FBS-PPRE representation is now defined. In the following paragraphs, instantiations of the model are proposed through the study of industrial examples.

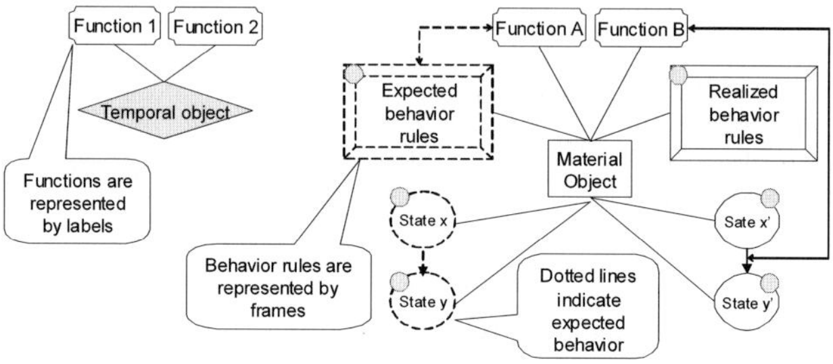


**Fig. 11.** Representation of functions and behaviors

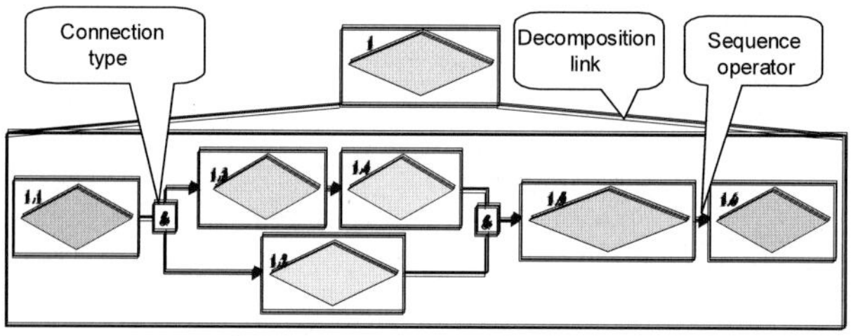


**Fig. 12.** Representation of the links between temporal objects

## 5.2 Case study number 1: design of a motorcycle tank

The tank is a part on the top of the motorcycle. Its design is constrained by different parameters:

- Its shape needs to be perfectly complementary to the other parts in order to maximize its capacity;
- The top part of the tank remains visible (the part is just painted). The shape needs to look really nice and to be compatible with the ergonomic constraints (it influences the driver position);
- The legislation affects also the design.

In order to validate the ergonomic and style constraints, the reverse engineering process is selected to realize the top part of the tank shape. Nevertheless, the bottom part of the shape can be designed directly using a CAD tool. The amount of production is small. For this reason, a rotomolding process is used. This manufacturing process will add specific constraints. A FBS-PPRE instantiation example of the tank design is proposed in Figure 13 and Figure 14. In order to improve the legibility of the representation, the process elements sequence (Figure 13) is not represented on the diagram including the product elements, the resources and the external effects (Figure 14).

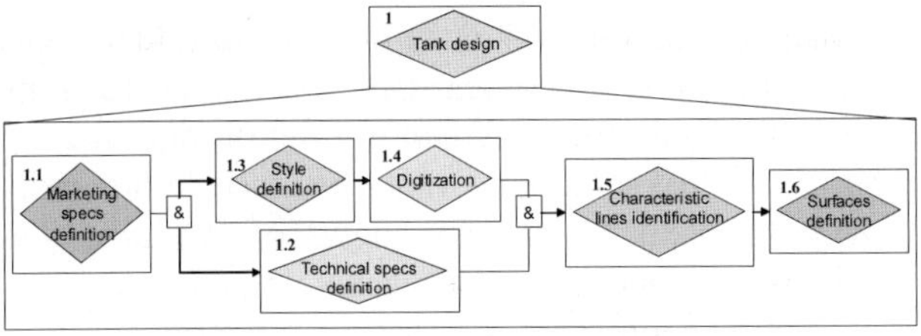


**Fig. 13.** Activities of the tank design

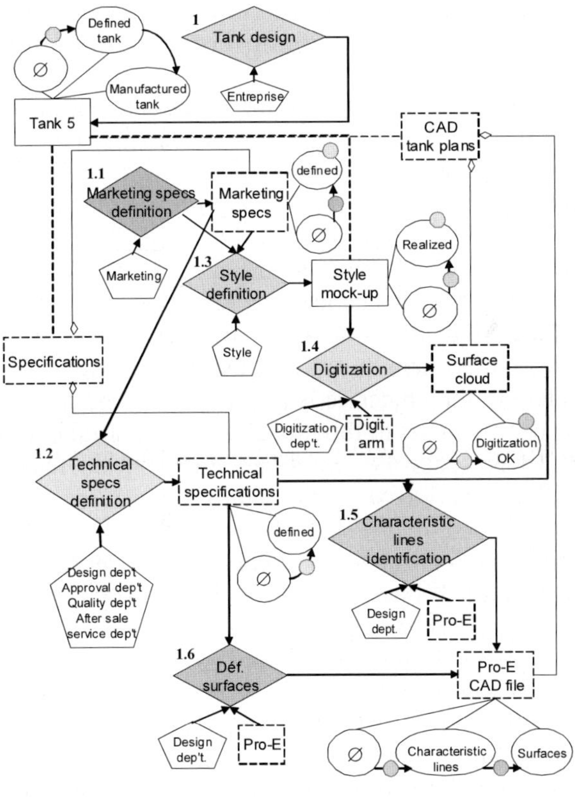


**Fig. 14.** FBS-PPRE of the tank design

In fact, the completeness of the model is very high and it is thus necessary to extract different partial representations to avoid too much links crosses. The Figure 16 highlights the interest of the representation class of the FBS-PPRE model:

- Top-level processes interact with the product (in this example, the tank) and its states;

- Sub-level processes enable the definition of product representations. The products of these processes (the outputs) are representations (physical mock-up, CAD file, etc.) but not the product itself.

The *representation* links indicate that the knowledge included in the various documents describes particular views of the tank. An interesting point is that a partial uncoupling between levels and activities is possible: the users of a particular representation (for example the activity of digitization) do not need to have in mind all the knowledge related to the tank.

## 5.3 Case study number 2: design of the tank specifications

This case study presents the use of the functional and behavioral views of the FBS-PPRE model. In the first case study, the specifications of the tank were just composed of two documents: the technical specifications and the marketing specifications. If this approach remains compatible with the FBS-PPRE model, it is not optimal. In this case study, the functional views and the expected behaviors of the model will be used. This other modeling option has the advantage of presenting the specifications elements in the object lifecycle i.e. in their context.

The example studied here is not exhaustive (very few functions of the tank are considered), the objective is just to present the transcription methodology from specifications to functional and behavioral entities.

The Figure 15 presents the simplified transcription result of the tank specifications. The function *to contain fuel* is considered. It can be split in *vehicle autonomy* and *standard conformity*:

An expected behavior linked to *standard conformity* is the *leak flow* when the tank is full and turned over: for this reason, it is quite logical to link the *standard conformity* function to the *leak flow* expected behavior. This behavior is defined relatively to the *homologation* process and constrained by the *NF standard.* This process also indicates the tank state during the *homologation* process: *full and turned over.*

The *vehicle autonomy* function results from a wide range of parameters. Nevertheless it can be considered as resulting from the *vehicle consumption* and from the *tank capacity*. The consumption will be evaluated in order to calculate the expected tank capacity:

The *consumption* is linked to the *motorcycle* object. At this stage of the design process, a material representation of the motorcycle already exists: the *rolling demonstrator*. It is a prototype using the majority of the mechanical parts of the final motorcycle. Nevertheless, the shapes of the aesthetic parts are not finalized at this stage.

The demonstrator consumption can be measured using the standards conditions. The real behavior of the demonstrator (linked to the *consumption measuring* process) is split in three representations elements: the *average consumption*, the *90 km/h consumption*, the *120 km/h consumption.*

The average consumption is used to calculate the tank capacity (it is a representation of one of the tank structural view). The *average consumption* and the expected *autonomy* are constraining the *capacity calculation* process.

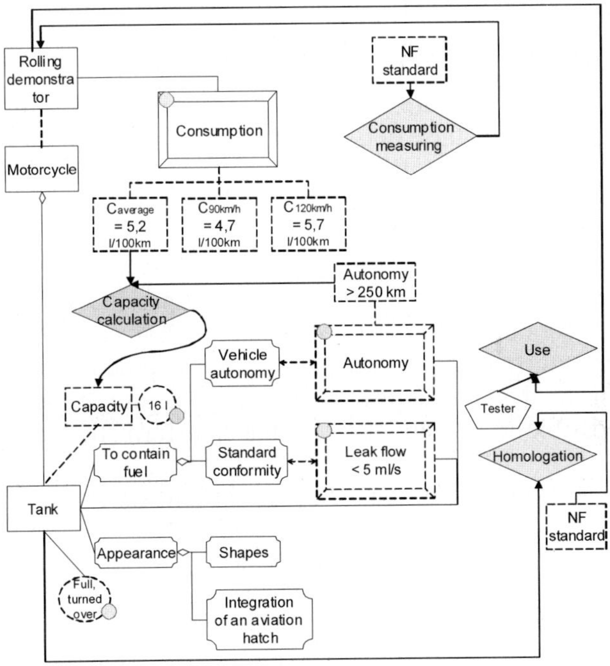


**Fig. 15.** FBS-PPRE transcription of the tank specifications

## 6 Summary

The study of the products lifecycles is one of the main industrial issues. It highlights the role of knowledge capitalization and management. The significance attached to the processes, to their aim and to their modeling enables the identification of the needs and of the knowledge flows.

The development of the content implies a real integration: the main difficulty remains the representation of the knowledge in order to insure its interpretation, its sharing and its keeping. The FBS-PPRE model allows advances principally in three domains: the completeness of the modeling, the management of the dynamic of the objects, the conceptual unification.

As regards completeness, the model offers a wider view than the usual approaches. Each object having an influence on the enterprise processes (defined as a temporal, spatial and hierarchical organization of activities) is taken into account trough its role of product (the object is then the result of the process), of resource (the object is then contributing to the process without being its purpose) or of external effect (the object is then acting as a constraint - positive or not - on the process). Furthermore, each object is modeled with the same views: the functional, the behavioral and the structural views.

The process elements are playing a structural role: they link the products, the resources and the external effects. Due to its definition, the behavior of an object is always linked to a process element: it enables a fine modeling of the dynamic transformation of the modeled elements. In fact, different behaviors can be taken into account: the model makes it possible to characterize the behavior of an object during its use process (the object has then a role of resource), but it also makes it possible to keep the evolution of an object during a process of design, of realization, of dismantling, of recycling, etc (the object has then a role of product). Moreover, this modeling of the dynamic applies whatever the nature of the object: it also makes it possible to manage natively the transformation of the temporal objects.

This strongly increased completeness, in particular with the management of the dynamic of all the objects, could have been reached by increasing the complexity of the model. That is not the case: due to the generic views of the FBS-PPRE modeling, the enterprise objects can be described according to the same formalism independently of their circumstantial roles of process, of product, of resource, or of external effect.

Furthermore, this conceptual unification is not challenged by the existence of objects of various natures (temporal, material, software, organizational or energy). The model is thus very compact and easier to apprehend: its implementation and its maintenance will also be easier.

These conceptual elements can thus constitute an essential support for the representation of knowledge for the life-cycle engineering information system definition.

The adoption and the deployment of the FBS-PPRE model can contribute to the analysis, the specification and the follow-up of the enterprise processes. They can lie with an ISO 9001 quality certification process: the objectives of the new standard are indeed in perfect adequacy with FBS-PPRE. To use this model in the company, a reliable and effective information system should be implemented. A demonstrator already allowed checking the interest of the handled concepts, but it remains inadequate in an industrial context.

The model needs to be validated in large companies in the future. In particular, the companies' profits have to be evaluated. The human and economic parameters as well as the deadlines are generally the limiting factors in this type of projects.

Because of its good completeness, the model can be used as a support for more specific methodologies. One can for example think of:

- The automated reorganization of process elements in order to optimize their execution;
- The extraction of reusable knowledge from the capitalized knowledge thanks to dedicated processes;
- The definition of particular views concerning the different phases of the product, the technologies and the enterprise life-cycles;
- The specification and the instantiation of specific objects;
- The definition of performance indicators based on the expected and realized behaviors of the FBS-PPRE model.

The avenues worth exploring to this work are thus diversified, and fit in the current scientific trends.

## References


1. Bernard, A., Labrousse, M., Véron, P., Numerical reference system for motorcycle development: proposition of a base conceptual model, Annals of 2002 Int'l CIRP Design Seminar, Hong Kong, May 16-18, 2002.
2. Hauschild, M., Alting, L., Poll, C., Life-cycle engineering - from methodology to enterprise culture, Keynote paper, 11th International CIRP life-cycle engineering seminar, Belgrade, Serbia, 20-22 June 2004.
3. Zwolinski, P., Lopez-Ontiveros, M.A., Brissaud, D., Products end of life characterisation for integrated design, 11th International CIRP life-cycle engineering seminar, Belgrade, Serbia, 20-22 June 2004.
4 Fargnoli, M., Di Latta, E., Notarnicola, S., The improvement of products sustainability in design stages, 11th International CIRP life-cycle engineering seminar, Belgrade, Serbia, 20-22 June 2004.
5 Labrousse, M., Bernard, A., Véron, P., Enterprise processes traceability based on a FBS approach, 5e Congrès International de Génie Industriel, Québec, Canada, October 26-29, 2003.
6 Labrousse, M., Bernard, A., Véron, P., Generic, FBS concept for Process /Product/Resource integration, in Tools and Methods of Competitive Engineering, Edited by Imre Horvath, Paul Xirouchakis, Volume 1, pp.384-394, Millpress Rotterdam Netherlands, ISBN 90-5966-018-8, 2004.
7 Qian, L., Gero, J.S., Function-Behavior-Structure Paths and Their Role in Analogy-Based Design, AIEDAM 10 (4), pp. 289-312, 1996.

8 Umeda, Y., Ishii, M., Yoshioka, M., Shimomura, Y., Tomiyama, T., Supporting conceptual design based on the function-behavior-state modeler, Artificial Intelligence for Engineering Design, Analysis and Manufacturing, Vol. 10, No. 4, pp. 275-288, 1996.
9 Umeda, Y., Tomiyama, T., Yoshikawa, H., Function, Behavior and Structure, in J.S. Gero (ed.): Applications of Artificial Intelligence in Engineering V, Vol 1: Design, Proceedings of the Fifth International Conference, Boston, USA, July 17-20, 1990, Computational Mechanics Publications, Southampton, Boston, Springer-Verlag, Berlin, pp. 177-193, 1990.
10 Magnet-Canet, S., Yannou, B., IfNot: a Function-based Approach for both Proactive and Reactive Integrations of New Productive Technologies, FLAIRS'98: Special Track on Reasoning about Functions, in The Eleventh Florida Artificial Intelligence Research Symposium, Sanibel Island, Florida (USA), 1998.
11 Labrousse, M., Proposition d'un modèle conceptuel unifié pour la gestion dynamique des connaissances d'entreprise, PhD report, University of Nantes and Ecole Centrale de Nantes, defended on 13th July 2004.

# Authors index